Discovering the Solar System

Barrie W. Jones
The Open University
Milton Keynes, UK

JOHN WILEY & SONS
Chichester · New York . Weinheim · Brisbane · Singapore · Toronto

Copyright © 1999 John Wiley & Sons Ltd,
Baffins Lane, Chichester,
West Sussex PO19 1UD, England

National 01243 779777
International (+44) 1243 779777
e-mail (for orders and customer service enquiries): cs-books@wiley.co.uk
Visit our Home Page on http://www.wiley.co.uk
or http://www.wiley.com

Other Wiley Editorial Offices

John Wiley & Sons, Inc., 605 Third Avenue,
New York, NY 10158-0012, USA

WILEY-VCH Verlag GmbH, Pappelallee 3,
D-69469 Weinheim, Germany

Jacaranda Wiley Ltd, 33 Park Road, Milton,
Queensland 4064, Australia

John Wiley & Sons (Asia) Pte Ltd, Clementi Loop #2-01,
Jin Xing Distripark, Singapore 129809

John Wiley & Sons (Canada) Ltd, 22 Worcester Road,
Rexdale, Ontario M9W 1L1, Canada

Library of Congress Cataloging-in Publication Data

Jones, Barrie William.
    Discovering the solar system / Barrie W. Jones.
        p. cm.
    Includes bibliographical references and index.
    ISBN 0-471-98243-1 (cloth : alk. paper).—ISBN 0-471-98648-8
    (pbk. : alk. paper)
    1. Solar system.   I. Title.
    QB501. J65   1999
    523.2—dc21   98-39573
                                                                    CIP

British Library Cataloguing in Publication Data

A catalogue record for this book is available from the British Library

ISBN 0 471 98243 1 (cloth)
ISBN 0 471 98648 8 (paper)

Typeset in Great Britain by Dobbie Typesetting Limited, Tavistock, Devon
Printed and bound in Great Britain by Bookcraft (Bath) Ltd
This book is printed on acid-free paper responsibly manufactured from sustainable forestry,
in which at least two trees are planted for each one used for paper production

Discovering the
Solar System

To my family

Contents

List of Tables   xiii
Preface and Study Guide   xv

1   The Sun and its Family   **1**
   1.1   The Sun   1
      1.1.1   The Solar Photosphere   1
      1.1.2   The Solar Atmosphere   3
      1.1.3   The Solar Interior   6
      1.1.4   The Solar Neutrino Problem   9
   1.2   The Sun's Family — a Brief Introduction   9
      1.2.1   The Terrestrial Planets   10
      1.2.2   The Giant Planets   12
      1.2.3   Pluto, and Beyond   14
   1.3   Chemical Elements in the Solar System   15
   1.4   Orbits of Solar System Bodies   15
      1.4.1   Kepler's Laws of Planetary Motion   15
      1.4.2   Orbital Elements   19
      1.4.3   Asteroids, and the Titius–Bode Rule   21
      1.4.4   A Theory of Orbits   22
      1.4.5   Orbital Complications   25
      1.4.6   The Orbit of Mercury   28
   1.5   Planetary Rotation   29
      1.5.1   Precession of the Rotation Axis   33
   1.6   The View from the Earth   34
      1.6.1   The Other Planets   34
      1.6.2   Solar and Lunar Eclipses   36
   1.7   Summary   40

2   The Origin of the Solar System   **48**
   2.1   The Observational Basis   48
      2.1.1   The Solar System   48
      2.1.2   Other Planetary Systems   49
      2.1.3   Star Formation   52
      2.1.4   Circumstellar Discs   54
   2.2   Solar Nebular Theories   55
      2.2.1   Angular Momentum in the Solar System   55

		2.2.2	The Evaporation and Condensation of Dust in the Solar Nebula	59
		2.2.3	From Dust to Planetesimals	62
		2.2.4	From Planetesimals to Planets in the Inner Solar System	64
		2.2.5	From Planetesimals to Planets in the Outer Solar System	69
	2.3		Formation of the Satellites and Rings of the Giant Planets	72
		2.3.1	Formation of the Satellites of the Giant Planets	72
		2.3.2	Formation and Evolution of the Rings of the Giant Planets	73
	2.4		Successes and Shortcomings of Solar Nebular Theories	77
	2.5		Summary	78
3	**Small Bodies in the Solar System**			**80**
	3.1		Asteroids	80
		3.1.1	Asteroid Orbits in the Asteroid Belt	81
		3.1.2	Asteroid Orbits Outside the Asteroid Belt	84
		3.1.3	Asteroid Sizes	86
		3.1.4	Other Properties of Asteroids	88
		3.1.5	Asteroid Classes	90
		3.1.6	Asteroid Classes Across the Asteroid Belt, and Asteroid Differentiation	92
	3.2		Comets	94
		3.2.1	The Orbits of Comets	95
		3.2.2	The Coma, Hydrogen Cloud, and Tails of a Comet	97
		3.2.3	The Cometary Nucleus	99
		3.2.4	The Death of Comets	101
		3.2.5	The Sources of Comets	102
	3.3		Meteorites	105
		3.3.1	Meteors, Meteorites, and Micrometeorites	105
		3.3.2	The Structure and Composition of Meteorites	106
		3.3.3	Dating Meteorites	109
	3.4		The Sources of Meteorites and Micrometeorites	110
		3.4.1	The Sources of Meteorites	110
		3.4.2	Martian and Lunar Meteorites	113
		3.4.3	The Sources of Micrometeorites	113
	3.5		Summary	115
4	**Interiors of Planets and Satellites: The Observational and Theoretical Basis**			**117**
	4.1		Gravitational Field Data	119
		4.1.1	Mean Density	119
		4.1.2	Radial Variations of Density: Gravitational Coefficients	120
		4.1.3	Radial Variations of Density: The Polar Moment of Inertia	124
		4.1.4	Local Mass Distribution, and Isostasy	126
	4.2		Magnetic Field Data	127
	4.3		Seismic Wave Data	129
		4.3.1	Seismic Waves	129
		4.3.2	Planetary Data	133
	4.4		Composition and Properties of Accessible Materials	135
		4.4.1	Surface Materials	135

4.4.2   Elements, Compounds, Affinities   135
4.4.3   Equations of State, and Phase Diagrams   137
4.5   Energy Sources, Energy Losses, and Interior Temperatures   141
4.5.1   Energy Sources   142
4.5.2   Energy Losses and Transfers   146
4.5.3   Observational Indicators of Interior Temperatures   151
4.5.4   Interior Temperatures   152
4.6   Summary   153

5   Interiors of Planets and Satellites: Models of Individual Bodies   155
5.1   The Terrestrial Planets   155
5.1.1   The Earth   158
5.1.2   Venus   161
5.1.3   Mercury   162
5.1.4   Mars   163
5.2   Planetary Satellites and Pluto   164
5.2.1   The Moon   164
5.2.2   Large Icy–Rocky Bodies   167
5.2.3   The Galilean Satellites of Jupiter   169
5.2.4   Small Bodies   173
5.3   The Giant Planets   173
5.3.1   Jupiter and Saturn   175
5.3.2   Uranus and Neptune   179
5.4   Magnetospheres   180
5.4.1   An Idealised Magnetosphere   180
5.4.2   Real Magnetospheres   182
5.5   Summary   184

6   Surfaces of Planets and Satellites: Methods and Processes   187
6.1   Some Methods of Investigating Surfaces   187
6.1.1   Surface Mapping in Two and Three Dimensions   187
6.1.2   Analysis of Electromagnetic Radiation Reflected or Emitted by
        a Surface   190
6.1.3   Sample Analysis   191
6.2   Processes that Mould the Surfaces of Planetary Bodies   191
6.2.1   Impact Cratering   191
6.2.2   Craters as Chronometers   196
6.2.3   Melting, Fractional Crystallisation, and Partial Melting   200
6.2.4   Volcanism and Magmatic Processes   203
6.2.5   Tectonic Processes   204
6.2.6   Gradation   206
6.2.7   Formation of Sedimentary and Metamorphic Rocks   209
6.3   Summary   209

7   Surfaces of Planets and Satellites: Inactive Surfaces   211
7.1   The Moon   211
7.1.1   Impact Basins, and Maria   212
7.1.2   The Nature of the Mare Infill   213

	7.1.3	Two Contrasting Hemispheres	215
	7.1.4	Other Lunar Surface Features	215
	7.1.5	Crustal and Mantle Materials	215
	7.1.6	Radiometric Dating of Lunar Events	218
	7.1.7	Lunar Evolution	219
7.2	Mercury		220
	7.2.1	Mercurian Craters and Associated Terrain	220
	7.2.2	Other Surface Features on Mercury	224
	7.2.3	The Evolution of Mercury	225
7.3	Mars		226
	7.3.1	Albedo Features	226
	7.3.2	Two Different Hemispheres	227
	7.3.3	The Northerly Hemisphere	229
	7.3.4	The Southerly Hemisphere	232
	7.3.5	The Polar Regions	233
	7.3.6	Water-related Features	235
	7.3.7	Observations at the Martian Surface	238
	7.3.8	Martian Meteorites	239
	7.3.9	The Evolution of Mars	240
7.4	Icy Surfaces		241
	7.4.1	Pluto and Charon	242
	7.4.2	Titan	242
	7.4.3	Ganymede and Callisto	243
7.5	Summary		245
8	**Surfaces of Planets and Satellites: Active Surfaces**		**247**
8.1	The Earth		247
	8.1.1	The Earth's Lithosphere	247
	8.1.2	Plate Tectonics	249
	8.1.3	The Success of Plate Tectonics	252
	8.1.4	The Causes of Plate Motion	254
	8.1.5	The Evolution of the Earth	255
8.2	Venus		255
	8.2.1	Major Topological Divisions	256
	8.2.2	Radar Reflectivity	256
	8.2.3	Impact Craters, and Possible Global Resurfacing	258
	8.2.4	Volcanic Features	258
	8.2.5	Surface Analyses and Surface Images	261
	8.2.6	Tectonic Features	261
	8.2.7	Tectonic and Volcanic Processes	263
	8.2.8	Internal Energy Loss	264
	8.2.9	The Evolution of Venus	265
8.3	Io		265
8.4	Icy Surfaces: Europa, Enceladus, Triton		267
	8.4.1	Europa	267
	8.4.2	Enceladus	269
	8.4.3	Triton	270

8.5   Summary   272

9 Atmospheres of Planets and Satellites: General Considerations   274
   9.1   Methods of Studying Atmospheres   276
   9.2   General Properties and Processes in Planetary Atmospheres   280
          9.2.1   Energy Gains and Losses   280
          9.2.2   Pressure, Density, and Temperature Versus Altitude   282
          9.2.3   Cloud Formation and Precipitation   288
          9.2.4   The Greenhouse Effect   291
          9.2.5   Atmospheric Reservoirs, Gains, and Losses   292
          9.2.6   Atmospheric Circulation   297
          9.2.7   Climate   300
   9.3   Summary   301

10 Atmospheres of Rocky and Icy–Rocky Bodies   303
   10.1   The Atmosphere of the Earth Today   303
          10.1.1   Vertical Structure, Heating and Cooling   303
          10.1.2   Atmospheric Reservoirs, Gains, and Losses   306
          10.1.3   Atmospheric Circulation   310
          10.1.4   Ice Ages   310
   10.2   The Atmosphere of Mars Today   313
          10.2.1   Vertical Structure, Heating and Cooling   313
          10.2.2   Atmospheric Reservoirs, Gains, and Losses   315
          10.2.3   Atmospheric Circulation   316
          10.2.4   Climate Change   316
   10.3   The Atmosphere of Venus Today   317
          10.3.1   Vertical Structure, Heating and Cooling   317
          10.3.2   Atmospheric Reservoirs, Gains, and Losses   319
          10.3.3   Atmospheric Circulation   319
   10.4   Volatile Inventories for the Terrestrial Planets   321
   10.5   The Origin and Early Evolution of Terrestrial Atmospheres   324
          10.5.1   Inert Gas Evidence   324
          10.5.2   Volatile Acquisition During Planet Formation   326
          10.5.3   Massive Losses   327
          10.5.4   Late Veneers   328
          10.5.5   Outgassing   329
   10.6   The Later Evolution of Terrestrial Atmospheres   330
          10.6.1   Venus   330
          10.6.2   The Earth   332
          10.6.3   Mars   334
          10.6.4   Mercury and the Moon   336
   10.7   Icy–Rocky Body Atmospheres   337
          10.7.1   Titan   337
          10.7.2   Triton and Pluto   338
          10.7.3   The Origin and Evolution of the Atmospheres of Icy–Rocky
                   Bodies   339
   10.8   Summary   340

11 **Atmospheres of the Giant Planets**   **342**
    11.1 The Atmospheres of Jupiter and Saturn Today   342
          11.1.1 Vertical Structure   343
          11.1.2 Composition   345
          11.1.3 Circulation   348
          11.1.4 Coloration   351
    11.2 The Atmospheres of Uranus and Neptune Today   352
          11.2.1 Vertical Structure   352
          11.2.2 Composition   353
          11.2.3 Circulation   354
    11.3 The Origin of the Giant Planets—A Second Look   356
    11.4 The End   359
    11.5 Summary   360

Question Answers and Comments   **362**
Glossary   **388**
Electronic Media   **399**
Further Reading   **401**
Index   **405**

Plate Section   between pages 48 and 49

List of Tables

1.1 Mean orbital elements and some physical properties of the Sun, the planets, and Ceres

1.2 Some properties of planetary satellites

1.3 Some properties of the fifteen largest asteroids

1.4 Some properties of selected comets

1.5 Relative abundances of the fifteen most abundant chemical elements in the Solar System

1.6 Some important constants

2.1 Some broad features of the Solar System today

2.2 The first nine confirmed exoplanets

2.3 A condensation sequence of some substances at 100 Pa nebular pressure

4.1 Some pioneering missions of planetary exploration by spacecraft

4.2 Some physical properties of the planets and larger satellites

4.3 The densities of some important substances

4.4 Radioactive isotopes that are important energy sources

4.5 Mechanisms of heat reaching the surface regions of some planetary bodies today

5.1 Model temperatures, densities, and pressures in the Earth

5.2 Model temperatures, densities, and pressures at the centres of the terrestrial planets and the Moon

5.3 Model pressures at the centres of Pluto and the large satellites of the giant planets, plus some central densities and temperatures

5.4 Model temperatures, densities, and pressures in the giant planets

6.1 Important igneous rocks and minerals in the crusts and mantles of terrestrial bodies

7.1 Ages of some lunar basins and mare infill

7.2 Distinguishing surface features of the inactive intermediate size icy satellites

9.1 Some properties of the substantial planetary atmospheres

9.2 L_{out}/W_{abs}, a, and T_{eff} for some planetary bodies

11.1 The atmospheric composition of the giant planets, given as mixing ratios with respect to H_2

11.2 Elemental mass ratios with respect to hydrogen in the giant planet molecular envelopes and in the young Sun

Preface and Study Guide

In *Discovering the Solar System* you will meet the Sun, the planets, their satellites, and the host of smaller bodies that orbit the Sun. On a cosmic scale the Solar System is on our doorstep, but it is far from fully explored, and there continues to be a flood of new data and new ideas. The science of the Solar System is thus a fast-moving subject, posing a major challenge for authors of textbooks.

A major challenge for the student is the huge range of background science that needs to be brought to bear — geology, physics, chemistry, and biology. I have tried to minimise the amount of assumed background, but as this book is aimed at students of university-level science courses I do assume that you have met Newton's laws of motion and law of gravity, that you know about the structure of the atom and about electromagnetic radiation, and that you have met chemical formulae and chemical equations. Further background science is developed as required, as is the science of the Solar System itself, and it is therefore important that you study the book in the order in which the material is presented. There is some mathematics — simple algebraic equations are used, and there is a small amount of algebraic manipulation. It is assumed that you are familiar with graphs and tables. There is no calculus.

To facilitate your study, there are 'stop and think' questions embedded in the text, denoted by '❏'. The answer follows immediately as part of the development of the material, but it will help you learn if you *do* stop and think, rather than read straight on. There are also numbered questions (Question 1.1, etc.). These are at the end of major sections, and it is important that you attempt them before proceeding — they are intended to test and consolidate your understanding of some of the earlier material. Full answers plus comments are given at the end of the book. Another study aid is the Glossary, which includes the major terms introduced in the book. These terms are emboldened in the text at their first appearance. Each chapter ends with a summary.

The approach is predominantly thematic, with sequences of chapters on the interiors, surfaces, and atmospheres of the major bodies (including the Earth). The first three chapters depart from this scheme, with Chapter 2 on the origin of the Solar System, and Chapter 3 on the small bodies — asteroids, comets, and meteorites. Chapter 1 is an overview of the Solar System, and this is also where most of the material on the Sun is located. Though the Sun is a major body indeed, it is very singular, and it is therefore treated separately. It also gets only very brief coverage, biased towards topics that relate to the Solar System as a whole. There is a significant amount of material on how the Solar System is investigated. The 'discovering' in the title thus has a double meaning — not only can *you* discover the Solar System by studying this book, you will also learn something about how it has been discovered by the scientific community in general.

A large number of people deserve thanks for their assistance with this book. Nick Sleep and Graeme Nash each commented on a whole draft, and Nick Sleep also made a major contribution to generating the figures. Coryn Bailer-Jones, George Cole, Mark Marley, Carl Murray, Peter Read, and Lionel Wilson commented on groups of chapters. Information and comments on specific matters have been received from Mark Bailey, Bruce Bills, Andrew Collier Cameron, Apostolos Christou, Ashley Davies, David Des Marais, Douglas Gough, Tom Haine, Andy Hollis, David Hughes, Don Hunten, Pat Irwin, Rosemary Killen, Jack Lissauer, Mark Littmann, Elaine Moore, Chris Owen, Roger Phillips, Eric Priest, Dave Rothery, Gerald Schubert, Alan Stern, George Wetherill, John Wood, and Ian Wright. Jay Pasachoff supplied some data for the Electronic Media list. Material for some of the figures was made available by Richard McCracken, Dave Richens, and Mark Kesby. John Holbrook loaned me some meteorite samples to photograph.

Good luck with your studies.

1 The Sun and its Family

Imagine that you have travelled far into the depths of space. From your distant vantage point the Sun has become just another star among the multitude, and the Earth, the other planets, and the host of smaller bodies that orbit the Sun are not visible at all to the unaided eye. The Sun is by far the largest and most massive body in the Solar System, and is the only one hot enough to be obviously luminous. This chapter starts with a description of the Sun. We shall then visit the other bodies in the Solar System, but only briefly, the purpose here being to establish their main characteristics — each of these bodies will be explored in much more detail in subsequent chapters. This chapter then continues with an exploration of the orbits of the various bodies. Each body also rotates around an axis through its centre, and we shall look at this too. The chapter concludes with aspects of our view of the Solar System as we see it from the Earth.

1.1 The Sun

This is only a very brief account of the Sun, and it is biased towards topics of importance for the Solar System as a whole. More detailed accounts of the Sun are in books listed in Further Reading.

1.1.1 The Solar Photosphere

The bright surface of the Sun is called the photosphere (Plate 1). Its radius is 6.96×10^5 km, about 100 times the radius of the Earth. It is rather like the 'surface' of a bank of cloud, in that the light reaching us from the photosphere comes from a range of depths, though the range covers only about one thousandth of the solar radius, and so we are not seeing very deep into the Sun. It is important to realise that whereas a bank of cloud scatters light from another source, the photosphere is *emitting* light. It is also emitting electromagnetic radiation at other wavelengths, as the solar spectrum in Figure 1.1 demonstrates. The total power radiated is the area under the solar spectrum, and is 3.85×10^{26} watts (W). This is the solar luminosity.

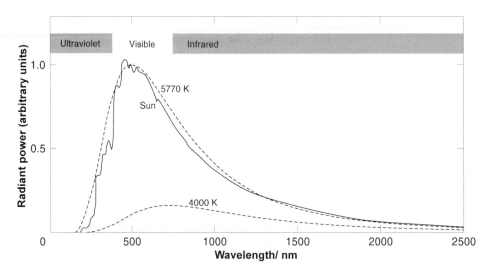

Fig. 1.1   The solar spectrum, and the spectra of ideal thermal sources at 5770 K and 4000 K (1 nm = 10^{-9} m). (Adapted from D. Labs and H. Neckel, *Zeit. für Astrophys.* **69**, 1 (1968))

The photosphere, for all its brilliance, is a tenuous gas, with a density of order 10^{-3} kg m^{-3}, about 1000 times less than that of the air at the Earth's surface.

The spectrum in Figure 1.1 enables us to estimate the mean photospheric temperature. This is done by comparing the spectrum to that of an **ideal thermal source**, sometimes called a black body. The exact nature of such a source need not concern us. The important point is that its spectrum is uniquely determined by its temperature. Turning this around, if we can fit an ideal thermal source spectrum reasonably well to the spectrum of any other body, then we can estimate the other body's temperature. Figure 1.1 shows a good match between the solar spectrum and the spectrum of an ideal thermal source at a temperature of 5770 K. Also shown is the poor match with an ideal thermal source at 4000 K, where the peak of the spectrum is at longer wavelengths. Also, the power emitted by this source is a lot less. The power shown corresponds to the assumption that the 4000 K source has the same physical area as the source at 5770 K, and thus brings out the point that the temperature of an ideal thermal source not only determines the wavelength range of the emission, but the power too. Note that 5770 K is a *representative* temperature of the Sun's photosphere, the local temperature varying from place to place.

At a finer wavelength resolution than in Figure 1.1 the solar spectrum displays numerous narrow dips, called spectral absorption lines. These are the result of the absorption of radiation by various atoms and ions, mainly in the photosphere, and therefore the lines provide information about chemical composition. At 5770 K significant fractions of the atoms of some elements are ionised, and so it is best to define composition in terms of atomic nuclei, rather than neutral atoms. In the photosphere, about 91% of the nuclei are hydrogen, about 9% are helium, and all the other chemical elements account for only about 0.2%.

Plate 1 shows that the most obvious feature of the photosphere is dark spots. These are called (unsurprisingly) **sunspots**. They range in size from less than 300 km across, to around 100 000 km, and their lifetimes range from less than an hour to six months or so. They have central temperatures of typically 4200 K, which is why they look darker than the surrounding photosphere. They are shallow depressions in the photosphere, with magnetic fields playing an important role in their creation. They are a complicated phenomenon, not fully understood. The number of sunspots varies, following a sunspot cycle. The time between successive maxima ranges from about 8 years to about 15 years with a mean value of 11.1 years.

Sunspots provide a ready means of studying the Sun's rotation, and reveal that the rotation period at the equator is 25.4 days, increasing with latitude to about 36 days at the poles. This differential rotation is common in fluid bodies in the Solar System.

1.1.2 The Solar Atmosphere

Above the photosphere there is a thin gas that can be regarded as the solar atmosphere. Because of its very low density, at most wavelengths it emits far less power than the underlying photosphere, and so the atmosphere is not normally visible. During total solar eclipses, the Moon just obscures the photosphere, and the weaker light from the atmosphere then becomes visible. In Plate 2 the long exposure time has emphasised the outer atmosphere, whereas in Plate 3 the short exposure time has emphasised the inner atmosphere. The atmosphere can be studied at other times, either by means of an optical device called a coronagraph that attenuates the radiation from the photosphere or by making observations at wavelengths where the atmosphere is brighter than the photosphere.

Figure 1.2 shows how the temperature and density in the solar atmosphere vary with altitude above the base of the photosphere. A division of the atmosphere into two main layers is apparent, the chromosphere and the corona, separated by a thin transition region.

The chromosphere

The chromosphere lies immediately above the photosphere. It has much the same composition as the photosphere, so hydrogen dominates. The density declines rapidly with altitude, but the temperature *rises*. The red colour that gives the chromosphere its name ('coloured sphere') is a result of the emission by hydrogen atoms of light at 656.3 nm. This wavelength is called Hα.

The data in Figure 1.2 are for 'quiet' parts of the chromosphere. Its properties are different where magnetic forces hold aloft filamentary clouds of cool gas, extending into the lower corona. The filaments are the red prominences above the limb of the photosphere in Plate 3. Prominences are transitory phenomena, lasting for periods from minutes to a couple of months. The chromosphere is also greatly disturbed in regions where a flare occurs. This is a rapid brightening of a small area of the Sun's upper chromosphere or lower corona, usually in regions of the Sun where there are sunspots. The increase in brightness occurs in a few minutes, followed by a decrease taking up to an hour, and the energy release is spread over a very wide range of wavelengths. Flares, like certain prominences, are associated with bursts of ionised gas that escape from the Sun. Magnetic fields are an essential part of the flare process, and it seems probable that the electromagnetic radiation is from electrons

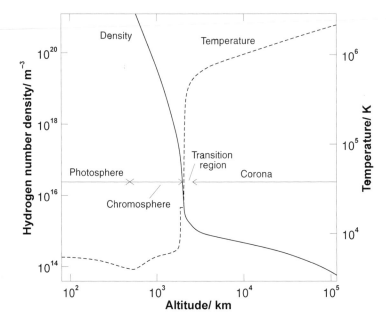

Fig. 1.2 The variation of temperature and density in the Sun's atmosphere with altitude above the base of the photosphere

that are accelerated close to the speed of light by changes in the magnetic field configuration. As with so many solar phenomena, the details are unclear.

The corona

Above the chromosphere the density continues to fall steeply across a thin transition region that separates the chromosphere from the corona (Figure 1.2).

❏ What distinctive feature of the transition region is apparent in Figure 1.2?

A distinctive feature is the enormous temperature gradient. This leads into the corona, where the gradient is not so steep. The corona extends for several solar radii (Plate 2), and within it the density continues to fall with altitude, but the temperature continues to rise, reaching $3–4 \times 10^6$ K, sometimes higher. Conduction, convection and radiation from the photosphere cannot explain such temperatures — these mechanisms would not transfer net energy from a body at *lower* temperature (the photosphere) to a body at *higher* temperature (the corona). The main heating mechanism seems to be magnetic — magnetic fields become reconfigured throughout the corona, and induce local electric currents that then heat the corona.

The corona is highly variable. At times of maximum sunspot number it is irregular, with long streamers in no preferred directions. At times of sunspot minimum, the visible boundary is more symmetrical, with a concentration of streamers extending from the Sun's equator, and short, narrow streamers from the poles. Coronal 'architecture' owes much to solar magnetic field lines. The white colour of the corona is photospheric light scattered by its constituents. Out to two or three solar radii the scattering is mainly from free electrons, ionisation being nearly total at the high temperatures of the corona. Further out, the scattering is dominated by the trace of fine dust in the interplanetary medium.

The solar wind

The solar atmosphere does not really stop at the corona but extends into interplanetary space in a flow of gas called the **solar wind**, which deprives the Sun of about one part in 2.5×10^{-14} of its mass per year. Because of the highly ionised state of the corona, and its predominantly hydrogen composition, the wind consists largely of protons and electrons. The temperature of the corona is so high that if the Sun's gravity were the only force it would not be able to contain the corona, and the wind would blow steadily and uniformly in all directions. But the strong magnetic fields in the corona act on the moving charged particles in a manner that reduces the escape rate. Escape is preferential in directions where the confining effect is least strong, and an important type of location of this sort is the coronal hole. This is a region of exceptionally low density and temperature, where the solar magnetic field lines reach huge distances into interplanetary space. Charged particles travel in helical paths around magnetic field lines, so the outward directed lines facilitate escape. Elsewhere, where the field lines are confined near the Sun, there is an additional outward flow, though at lower speeds.

Solar wind particles (somehow) gain speed as they travel outwards, and at the Earth the speeds range from 200 to $900 \, \text{km s}^{-1}$. The density is extremely low — typically about 4 protons and 4 electrons per cm^3, though with large variations. Particularly large enhancements result from what are called coronal mass ejections, often associated with flares and prominences, and perhaps resulting from the copious opening of magnetic field lines. If the Earth is in the way of a concentrated jet of solar wind, then various effects are produced, as you will see in later chapters. The solar wind is the main source of the extremely tenuous gas that pervades interplanetary space.

Solar activity

Solar activity is the collective term for those solar phenomena that vary with a periodicity of about 11 years.

❐ What two aspects of solar activity were outlined earlier?

You have already met the sunspot cycle, and it was mentioned that the form of the corona is correlated with it. Prominences (filaments) and flares are further aspects of solar activity, both phenomena being more common at sunspot maximum. The solar luminosity also varies with the sunspot cycle, and on average is about 0.15% *higher* at sunspot maximum than at sunspot minimum. This might seem curious, with sunspots being cooler and therefore less luminous than the rest of the photosphere.

However, when there are more sunspots, a greater area of the photosphere is covered in bright luminous patches called faculae.

All the various forms of solar activity are related to solar magnetic fields that ultimately originate deep in the Sun. The origin of these fields will be considered briefly in Section 1.1.3.

1.1.3 The Solar Interior

To investigate the solar interior, we would really like to burrow through to the centre of the Sun, observing and measuring things as we go. Alas! this approach is entirely impracticable. Therefore, the approach adopted, in its broad features, is the same as that used for all inaccessible interiors. A model is constructed and varied until it matches the major properties that we either *can* observe or can obtain fairly directly and reliably from observations. Usually, a *range* of models can be made to fit, so the model is rarely unique. Many features are, however, common to all models, and such features are believed to be correct. This modelling process will be described in detail in Chapter 4, in relation to planetary interiors. Here, I present the *outcome* of the process as applied to the Sun.

A model of the solar interior

Figure 1.3 shows a typical model of the Sun as it is thought to be today. Hydrogen and helium predominate throughout, as observed in the photosphere. Note the enormous increase of pressure with depth, to 10^{16} pascals (Pa) at the Sun's centre — about 10^{11} times atmospheric pressure at sea level on Earth! The central density is more modest, 'only' about 14 times that of solid lead as it occurs on Earth, though the temperatures are so high that the solar interior is everywhere fluid — there are no solids. Another consequence of the high temperatures is that at all but the shallowest depths the atoms are kept fully ionised by the energetic atomic collisions that occur. A highly ionised medium is called a **plasma**. The central temperatures are about 1.5×10^7 K, sufficiently high that nuclear reactions can sustain these temperatures and the solar luminosity, and can have done so for the 4600 million years (Ma) since the Sun formed (an age based on various data to be outlined in Chapter 3). This copious source of internal energy also sustains the pressure gradient that prevents the Sun from contracting.

Though nuclear reactions sustain the central temperatures today, there must have been some other means by which such temperatures were initially attained in order that the nuclear reactions were triggered. This must have been through the gravitational energy released when the Sun contracted from some more dispersed state. With energy being radiated to space only from its outer regions, it would have become hotter in the centre than at the surface. Nuclear reaction rates increase so rapidly with increasing temperature that when the central regions of the young Sun became hot enough for nuclear reaction rates to be significant, there was a fairly sharp boundary between a central core where reaction rates were high and the rest of the Sun where reactions rates were negligible. This has remained the case ever since.

The Sun was initially of uniform composition, all models giving proportions by mass close to 70.7% hydrogen, 27.5% helium, and under 2% for the total of all the

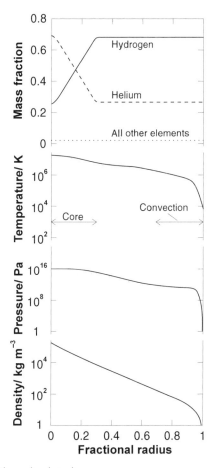

Fig. 1.3 A model of the solar interior

other elements. In such a mixture, at the core temperatures that the Sun has had since its birth, there is only one group of nuclear reactions that is significant — the pp chains. The name arises because the sequence of reactions starts with the interaction of two protons (symbol p) to form a heavier nucleus (deuterium), a single proton being the nucleus of the most abundant isotope of hydrogen (^{1}H). When a heavier nucleus results from the joining of two lighter nuclei, this is called **nuclear fusion**. The details of the pp chains will not concern us, but their net effect is the conversion of four protons into the nucleus of the most abundant isotope of helium (^{4}He), which consists of two protons and two neutrons. Various other subatomic particles are involved, but of central importance are the gamma rays produced — electromagnetic radiation with very short wavelengths.

The gamma rays carry nearly all of the energy liberated by the pp chain reactions. The gamma rays don't get very far before they interact with the plasma of electrons and nuclei that constitutes the solar core. To understand the interaction, it is necessary to recall that although electromagnetic radiation can be regarded as a wave, it can also be regarded as a stream of particles called **photons**. The wave picture is useful for understanding how radiation gets from one place to another; the photon picture is useful for understanding the interaction of radiation with matter. The energy of a photon is related to the frequency f of the wave via

$$e = hf \qquad (1.1)$$

where h is Planck's constant. The frequency of a wave is related to its wavelength via

$$f = c/\lambda \qquad (1.2)$$

where c is the wavespeed. For electromagnetic radiation in space c is the speed of light, $3.00 \times 10^5 \, \text{km s}^{-1}$. Note that Table 1.6 lists physical constants, like c, and h. (For ease of reference, Tables 1.1–1.6 are located at the end of the chapter.)

On average, after only a centimetre or so a gamma ray in the core either bounces off an electron or nucleus, in a process called scattering, or it is absorbed and re-emitted. This maintains the level of random motion of the plasma; in other words, it maintains its high temperature. The gamma ray photons are not all of the same energy. They have a spectrum shaped like that of an ideal thermal source at the temperature of the local plasma. This is true throughout the Sun, so as the photons move outwards their spectrum moves to longer wavelengths, corresponding to the lower temperatures, until at the photosphere the spectrum is that in Figure 1.1 (Section 1.1.1), where the number of photons is greater than in the core, but they are of much lower average energy. From the moment a gamma ray is emitted in the core to the moment its descendants emerge from the photosphere, a time of several million years will have elapsed.

❐ What is the direct travel time?

The direct travel time at the speed of light c across the solar radius of 6.96×10^5 km is $6.96 \times 10^5 \, \text{km}/3.00 \times 10^5 \, \text{km s}^{-1}$, i.e. 2.23 seconds!

The transport of energy by radiation is, unsurprisingly, called **radiative transfer**. This occurs throughout the Sun. Another mechanism of importance in the Sun is convection, the phenomenon familiar in a warmed pan of liquid, where energy is transported by currents of fluid. When the calculations are done for the Sun, then the outcome is as in Figure 1.3. Convection is confined to the outer regions, where it supplements radiative transfer as a means of conveying energy outwards. The tops of the convective cells are seen in the photosphere as transient patterns called granulations. These are typically 1000 km across, and exist for 5–10 minutes. Because convection does not extend to the core in which the nuclear reactions are occurring the core is not being replenished, and so it becomes more and more depleted in hydrogen and correspondingly enriched in helium. The core itself is unmixed, and so with temperature increasing with depth, the nuclear reaction rates increase with depth, and therefore so does the helium enrichment. This feature is apparent in the solar model in Figure 1.3.

The solar magnetic field

The source of any magnetic field is an electric current. If a body contains an electrically conducting fluid, then the motions of the fluid can become organised in a way that constitutes a net circulation of electric current, and a magnetic field results. This is just what we have in the solar interior — the solar plasma is highly conducting, and the convection currents sustain its motion. We shall look more closely at this sort of process in Section 4.2. Detailed studies show that the source of the solar field is concentrated towards the base of the convective zone. The differential rotation of the Sun contorts the field in a manner that goes some way to explaining sunspots and other magnetic phenomena.

The increase of solar luminosity

Evolutionary models of the Sun indicate that the solar luminosity was only about 70% of its present value 4600 Ma ago, that it has gradually increased since, and will continue to increase in the future. This increase is of great importance to planetary atmospheres and surfaces, as you will see in later chapters.

1.1.4 The Solar Neutrino Problem

There is one observed feature of the Sun that solar models have had difficulty in explaining. This is the rate at which solar neutrinos are detected on Earth. Solar neutrinos are so unreactive that most of them escape from the Sun and so provide one of the few direct indicators of conditions deep in the solar interior. A neutrino is an elusive particle that comes in three kinds. The electron neutrino is produced in the pp chains of nuclear reactions that occur in the solar interior. The rates at which electron neutrinos from the Sun are detected by various installations on Earth are significantly below the calculated rates. One possible explanation is a slightly cooler solar core — a slight reduction in temperature results in a substantial reduction in reaction rate. Unfortunately, solar models cannot give such a reduction without introducing features of solar behaviour that seem arbitrary.

Another explanation depends on the observation that neutrinos oscillate between the three different kinds. If, in their 8-minute journey at near the speed of light from the solar core to the terrestrial detectors, they settle into this oscillation, then at any instant only some of the neutrinos arriving here are of the electron type. The neutrino detectors only detect the electron type, so this neatly explains the shortfall.

Question 1.1

The Sun's photospheric temperature, as well as its luminosity, has also increased since its birth. What is the combined effect on the solar spectrum in Figure 1.1?

1.2 The Sun's Family — A Brief Introduction

Within the Solar System we find bodies with a great range of size, as Figure 1.4 shows. The Sun is by far the largest body. Next in size are the four **giant planets**: Jupiter, Saturn, Uranus, and Neptune. We then come to a group of bodies of intermediate size. Prominent are the Earth, Venus, Mars, and Mercury. These four bodies constitute the **terrestrial planets**, so called because they are comparable in size

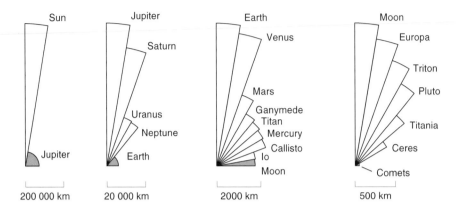

Fig. 1.4 Sizes of bodies in the Solar System

and composition, and are neighbours in space. This intermediate group has an arbitrary lower limit in size which I shall take to be that of the remaining planet Pluto, the ninth and outermost of the planets. Consequently, the group also contains the seven planetary satellites that are larger than Pluto. As their name suggests, planetary satellites are companions of a planet, bound in orbit around it and with a smaller mass. There are plenty of bodies smaller than Pluto: the remaining satellites, of which Titania is the largest, a swarm of **asteroids**, of which Ceres ('series') is easily the largest, a huge number of comets, and a continuous range of even smaller bodies, right down to tiny particles of dust.

Table 1.1 includes the radius, mass, and mean density (globally averaged density) of the nine planets and Ceres. Table 1.2 includes the same information for planetary satellites, and Table 1.3 includes the radius of the fifteen largest asteroids.

Figure 1.5 shows the orbits of the planets. These orbits are roughly circular, and lie more or less in the same plane. The planets move around their orbits at different rates, but in the same direction, anticlockwise as viewed from above the Earth's North Pole — this is called the **prograde direction**. The asteroids are concentrated in the space between Mars and Jupiter, in the asteroid belt. Note that the distances in Figure 1.5 are huge compared even to the solar radius of 6.96×10^5 km. A convenient unit of distance in the Solar System is the average distance of the Earth from the Sun, 1.50×10^8 km, which is given a special name, the **astronomical unit** (AU). Between them, Figures 1.4 and 1.5 provide a map of the Solar System's planetary domain.

1.2.1 The Terrestrial Planets

The terrestrial planets occupy the inner Solar System (Figure 1.5). They consist largely of rocky materials, with iron-rich cores. Mercury's surface is heavily cratered by the accumulated effects of impacts from space (Plate 4), indicating little geological resurfacing since the planet was formed. It has a negligible atmosphere. Venus is the Earth's twin in size and mass, and like the Earth it is geologically active, with volcanic features common (Plate 5), but it differs from the Earth in that it has no oceans. The surface of Venus, at 735 K, is far too hot for liquid water, a consequence

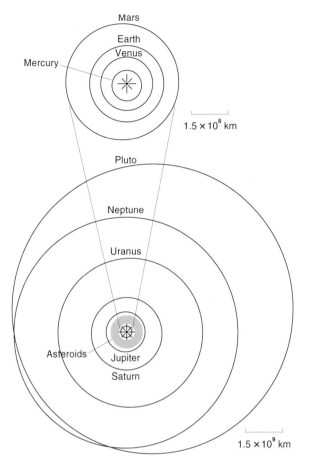

Fig. 1.5 The orbits of the planets as they would appear from a distant viewpoint perpendicular to the plane of the Earth's orbit, from above the North Pole

of its proximity to the Sun, and its massive, carbon dioxide atmosphere. The Earth is further from the Sun and has an atmosphere about a hundred times less massive, mainly nitrogen and oxygen. It is thus cool enough to have oceans, but not so cold that they are frozen (Plate 6). Unlike Mercury and Venus, the Earth has a satellite — the Moon. It is a considerable world, larger than Pluto. It is devoid of an appreciable atmosphere and has a heavily cratered surface (Plate 7).

Beyond the Earth we come to Mars, smaller than the Earth but larger than Mercury. It has a thin carbon dioxide atmosphere through which its cool surface is readily visible (Plate 8). About half of the surface is heavily cratered. The other half is less cratered, and shows evidence of the corresponding past geological activity. Plate 9 is a view at the surface in the less cratered hemisphere. Mars has two tiny satellites, Phobos and Deimos (Table 1.2). These orbit very close to the planet, and might be captured asteroids.

It is the domain of the asteroids — the asteroid belt — that we cross in the large gulf of space that separates Mars from Jupiter. Asteroids are rocky bodies of which

Ceres is by far the largest (Table 1.3), though it is still a good deal smaller than Pluto (Figure 1.4). It is thought that there are about 10^9 asteroids larger than 1 km, and Plate 10 shows what is thought to be a fairly typical one. At one metre there is a switch in terminology, with smaller bodies being called **meteoroids**, and these are even more numerous. Below about 0.01 mm there is another switch in terminology — smaller particles are called dust, and this is widely distributed within and beyond the asteroid belt, and is predominantly submicron in size (less that 10^{-6} m across). The asteroids are sometimes called minor planets, and consequently the nine planets are sometimes called the **major planets**.

1.2.2 The Giant Planets

The giant planets are very different from the terrestrial planets, not just in size (Figure 1.4) but also in composition. Whereas the terrestrial planets are dominated by rocky materials, including iron, the giants are dominated by hydrogen and helium, and by icy materials, notably water. The icy materials tend to concentrate towards the centres of the giant planets, where it is so hot, typically 10^4 K, that the icy materials are liquids not solids. Rocky materials make up only a small fraction of the mass of each giant, and they are also concentrated towards the centres. The giant planets are fluid throughout their interiors.

❐ What other body in the Solar System is also dominated by hydrogen and helium, and is fluid throughout?

The Sun is also a fluid body, dominated by hydrogen and helium (Section 1.1).

Jupiter is the largest and most massive of the planets. Plate 11 shows the richly structured uppermost layer of cloud, which consists mainly of ammonia particles, coloured by traces of a wide variety of substances, and patterned by atmospheric motions. The prominent banding is parallel to the equator.

Jupiter has a large and richly varied family of satellites. Figure 1.6 is a plan view, drawn to scale, of the orbits of the four largest by far of Jupiter's satellites — Io, Europa, Ganymede, Callisto. They are called the **Galilean satellites**, after the Italian astronomer Galileo Galilei (1564–1642), who discovered them in 1610 when he made some of the very first observations of the heavens with the newly invented telescope. They orbit the planet close to its equatorial plane. These remarkable bodies are shown in Plates 12–15. They range in size from Ganymede, which is somewhat larger than Mercury and is the largest of all planetary satellites, to Europa, which is somewhat smaller than the Moon. Io is a rocky body. The other three contain increasing amounts of water (mainly as ice) with increasing distance from Jupiter. Table 1.2 includes all the named satellites of Jupiter.

We move on to Saturn, which is somewhat smaller than Jupiter, but is otherwise not so very different (Plate 16), and so I will say no more about the planet in this chapter, but turn to its family of satellites, and in particular to its largest satellite Titan, an icy-rocky body larger than Mercury, and second only to Ganymede among the satellites. The most remarkable thing about Titan is that it has a massive atmosphere. Indeed, per unit area of surface, it has about ten times more mass of atmosphere than the Earth. The atmosphere is mainly nitrogen, but

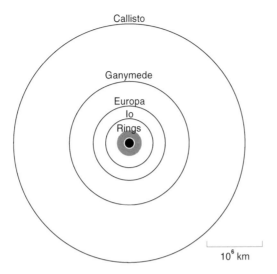

Fig. 1.6   The orbits of the Galilean satellites of Jupiter

contains so much hydrocarbon cloud and haze that the surface is almost invisible (Plate 17).

Saturn is most famous for its rings (Plate 18). These lie in the planet's equatorial plane, and consist of small solid particles. The rings are extremely thin, probably no more than a few hundred metres. They are, however, so extensive that they were observed by Galileo in 1610, though it was the Dutch physicist Christiaan Huyghens (1629–1693) who, in 1655, was first to realise that they are rings encircling the planet. Plate 18 shows that each main ring is broken up into many ringlets, to form a structure of exquisite complexity. The other three giant planets also have ring systems, but they are far less substantial.

Beyond Saturn we head off across another of the increasingly large gulfs of space that separate the planets as we move out from the Sun. We come to Uranus, a good deal smaller than Saturn, and with a smaller proportion of hydrogen and helium. In spite of its size it was unknown until 1781 when it was discovered accidentally by the Germano-British astronomer William Herschel (1738–1822) during a systematic survey of the stars. This was the first planet to be discovered in recorded history. It had escaped earlier detection because it is at the very threshold of unaided eye visibility, owing to its great distance from us. It lacks the strong bands of Jupiter and Saturn (Plate 19).

Neptune, like Uranus, was discovered in recorded history, but the circumstances were very different. Whereas Uranus was found accidentally, Neptune was discovered as a result of predictions made by two astronomers in order to explain slight departures of Uranus from its expected orbit. The British astronomer John Couch Adams (1819–1892) and the French astronomer Urbain Jean Joseph Le Verrier (1811–1877) independently predicted that the cause was a previously unknown planet orbiting beyond Uranus, and in 1846 Neptune was discovered by the German astronomer Johann Gottfried Galle (1812–1910) close to its predicted positions. Neptune, the last of the giants, is not so very different from Uranus (Plate

20), and so in the spirit of this quick tour I will say no more here about the planet itself.

Uranus and Neptune have many satellites. The largest among them by far, Neptune's satellite Triton, is an icy–rocky body slightly larger than Pluto, and it is the only satellite other than Titan that has a significant atmosphere, though it is fairly tenuous, and allows the icy surface of Triton to be seen (Plate 21). Among Neptune's other satellites, Nereid has a huge and extraordinarily eccentric orbit (Table 1.2). The orbit of Triton is curious in a different way — though it is nearly circular it is **retrograde**, which is the opposite direction to the prograde orbital motion of the planets and all other large satellites.

1.2.3 Pluto, and Beyond

Pluto is the furthest known planet, in an orbit where sunlight is 1600 times weaker than at the Earth. Pluto was discovered in 1930 by the American astronomer Clyde William Tombaugh (1906–1997) during a systematic search of a band of sky straddling the orbital planes of the known planets. It is a small world (Figure 1.4) and it is the only planet in the Solar System not yet visited by a spacecraft. Consequently we know rather little about Pluto and its only known satellite Charon. Pluto is an icy world, with about half of its volume consisting of frozen water and other icy substances, and the remainder consisting of rock. Charon probably has a broadly similar composition.

We have now reached the edge of the planetary domain. The orbits of Uranus, Neptune, and Pluto, and the paths of distant spacecraft, indicate no large masses beyond Pluto. But the space beyond is not empty, and we have certainly not come to the edge of the Solar System. One type of body is abundant beyond Pluto — the **comets**. These are small icy bodies that, through the effect of the Sun, develop huge fuzzy heads and spectacular tails when their orbits carry them into the inner Solar System (Plate 22). It is thought that beyond Pluto there is a swarm of 10^{12}–10^{13} comets in prograde orbits concentrated towards the ecliptic plane, mainly between the orbit of Pluto and about 10^3 AU from the Sun. This is the **Edgeworth–Kuiper belt**, and several of the larger members have been detected.

The Edgeworth–Kuiper belt probably blends into a comparably numerous swarm of comets in a thick spherical shell surrounding the Solar System, extending from about 10^3 to 10^5 AU. This is the **Oort cloud** (also called the Öpik–Oort cloud). Its outer boundary is at the extremities of the Solar System, where passing stars can exert a gravitational force comparable to that of the Sun. The Oort cloud has not been observed directly, but its existence is inferred from the comets that we see in the inner Solar System. These are a small sample of the Oort cloud and also of the Edgeworth–Kuiper belt, but in orbits that have been greatly modified. Table 1.4 lists some properties of selected comets.

Question 1.2

In about 100 words, discuss whether there is any correlation between the size of a planet and its distance from the Sun.

1.3 Chemical Elements in the Solar System

With most of the mass in the Solar System in the Sun, and the Sun composed almost entirely of hydrogen and helium, the chemical composition of the Solar System is dominated by these two elements. Hydrogen is the lightest element. Its most common isotope (by far) has a nucleus consisting of a single proton. Helium is the next lightest element, with the nucleus of its most common isotope (again by far) consisting of two protons and two neutrons. Recall that an element is defined by the number of protons in its nucleus — this is the atomic number — and that the isotopes are distinguished by different numbers of neutrons. To denote a particular isotope the number of neutrons plus protons is included with the chemical symbol, e.g. ^{4}He for helium's common isotope.

Most of the mass outside the Sun is in Jupiter and Saturn, and these are also composed largely of hydrogen and helium, though they contain larger proportions of the other elements — the so-called **heavy elements**. For the Solar System as a whole, Table 1.5 gives the relative abundances of the fifteen most abundant of the 110 or so chemical elements. Note that the value for helium is for the observable part of the Sun. These outer regions have not been modified by the conversion of hydrogen to helium by nuclear fusion, such as occurs in the core of the Sun.

Except in very high temperature regions, most of the atoms of most elements are combined with one or more other atoms, either of itself or of other elements. The important exceptions are helium, neon, argon, krypton, and xenon, which are so chemically unreactive that they remain monatomic and have been given the name **inert gases**. If an element is combined with itself, as in H_2, then we have the element in molecular form, whereas if it is combined with other elements, then we have it as a chemical compound.

Water (H_2O) is the most abundant hydrogen compound in the Solar System. Table 1.5 provides a reason.

❐ What is the reason?

This is because oxygen has a high abundance. But hydrogen is so overwhelmingly abundant that there is plenty left over after the formation of compounds. Most of the uncompounded hydrogen outside of the Sun is in the giant planets, as H_2, whereas water is the main repository of hydrogen in most of the other bodies.

1.4 Orbits of Solar System Bodies

1.4.1 Kepler's Laws of Planetary Motion

Each planet orbits the Sun as shown in plan view in Figure 1.5. As a crude approximation, the planetary orbits can be represented as circles centred on the Sun, with all the circles in the same plane, and each planet moving around its orbit at a constant speed; the larger the orbit, the slower the speed. A far better approximation is encapsulated in three empirical rules called **Kepler's laws of planetary motion**. These were announced by the German astronomer Johannes Kepler (1571–1630), the first two in 1609, the third in 1619.

Kepler's first law   Each planet moves around the Sun in an ellipse, with the Sun at one focus of the ellipse.

Kepler's second law   As the planet moves around its orbit, the straight line from the Sun to the planet sweeps out equal areas in equal intervals of time.

We will come to the third law shortly.

Figure 1.7 shows an **ellipse**. The shape is that of a circle viewed obliquely, the more oblique the view the greater the departure from circular form. The important features of an ellipse are marked in Figure 1.7, and are that

- It has a major axis of length $2a$, and a minor axis of length $2b$ — unsurprisingly, a and b are called, respectively, the **semimajor axis**, and the semiminor axis
- There are two foci that lie on the major axis, each a distance ae from the centre of the ellipse, where e is the **eccentricity** of the ellipse; note that the foci are in the plane of the ellipse, and that $e=\sqrt{(1 - b^2/a^2)}$.

The eccentricity is a measure of the departure from circular form. If e is zero, then the foci coalesce at the centre, a equals b, and the ellipse has become a circle of radius a. If e approaches 1 then the ellipse becomes extremely elongated.

Kepler's first law tells us that the shape of a planetary orbit is an ellipse, and that the Sun is at one focus. Figure 1.8 shows the orbit of Pluto, which among planetary orbits has the greatest eccentricity, $e = 0.249$. Note that the shape is very close to a circle but that the Sun, which is at one of the foci, is distinctly off-centre. Note also that the semimajor axis is less than the greatest distance of a body from the Sun, but is greater than the minimum distance, and it is therefore some sort of average distance. At its greatest distance from the Sun the body is at a point in its orbit called **aphelion**, the closest point is called **perihelion**. These terms are derived from the Greek words Helios for the Sun and peri- and apo- which in this context mean 'in the vicinity of' and 'away from' respectively. The length of the semimajor axis of the Earth's orbit is called the astronomical unit (AU), mentioned earlier.

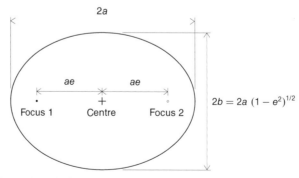

Fig. 1.7   An ellipse, though far more eccentric than the orbit of any planet. This is the shape of the orbit of the comet Giacobini–Zinner (Table 1.4)

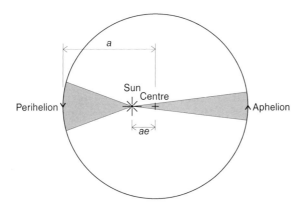

Fig. 1.8 The orbit of Pluto

Kepler's laws don't apply just to planets. Figure 1.7 is in fact the shape of the orbit of the comet Giacobini–Zinner.

❏ Where should the Sun be marked in Figure 1.7?

The Sun should be shown at either one of the two foci. This is an orbit of fairly high eccentricity, $e = 0.7065$. The non-circular form is now very clear, and the foci are greatly displaced from the centre.

Kepler's *second* law tells us how a planet (or comet) moves around its orbit. For the case of Pluto the shaded areas within the orbit in Figure 1.8 are equal in area, and so by Kepler's second law these are swept out in equal intervals of time. Thus, around aphelion the body is moving slowest, and around perihelion it is moving fastest. The difference in these two speeds is larger, the greater the eccentricity.

❏ What are the speeds at different positions in a *circular* orbit?

In a circular orbit the equal areas correspond to equal length arcs around the circle, so the body moves at a constant speed around its orbit.

So far, Kepler's laws have described the orbital motion around the Sun of an *individual* body. The third and final law compares the motion of one body to another:

Kepler's third law   If P is the time taken by a planet to orbit the Sun once, and a is the semimajor axis of the orbit, then

$$P = ka^{3/2} \qquad (1.3)$$

where k has the same value for each planet.

P is called the orbital period, or the period of revolution. It is the period as observed from a non-rotating viewpoint, which, for practical purposes, is any viewpoint fixed with respect to the distant stars. This leads to the term **sidereal**

(= 'star related') **orbital period** for P. For the Earth this period is called the sidereal year. Therefore, with $P = 1$ (sidereal) year and $a = 1$ AU, $k = 1$ year AU$^{-3/2}$. According to Kepler's third law, this is the value of k for all the planets.

Equation (1.3) tells us that the larger the orbit, the longer the orbital period. This is partly because the planet has to travel further, and partly because the planet moves more slowly. We can see that the planet moves more slowly from the simple case of a circular orbit of radius a. The circumference of the orbit is $2\pi a$, so if the orbital speed were independent of a then P would be proportional to a, not to $a^{3/2}$. Therefore, the orbital speed must be proportional to $a^{-1/2}$. In an elliptical orbit the circumference still increases as a increases, and now it is the *average* speed that decreases.

Kepler's third law enables us to obtain relative distances in the Solar System. If we measure the orbital periods of bodies A and B, then the ratio of the semimajor axes of their orbits is obtained from equation (1.3)

$$\frac{a_A}{a_B} = \left(\frac{P_A}{P_B}\right)^{2/3}$$

If one of the two bodies is the Earth, then we can express the other semimajor axis in astronomical units. This can be repeated for all orbits. Moreover, from the shape and orientation of the orbits, we can draw a scale plan of the Solar System, and at any instant we can show where the various planets are. At any instant we can thus express in astronomical units the distance between any two bodies. If at the same instant we can measure the distance between any two bodies in metres, we can then obtain the value of the astronomical unit in metres.

Today, the astronomical unit is best measured using radar reflections. Radar pulses travel at the speed of light c, which is known very accurately. Time intervals can also be measured very accurately, so if we measure the time interval Δt between sending a radar pulse from the Earth to a planet and receiving its echo, then the distance from the Earth to the planet is $c\,\Delta t/2$. Accurate measurements of distances in the Solar System have revealed that the semimajor axis of the Earth's orbit is subject to very slight variations. As a consequence the AU is now defined as *exactly* equal to $1.495\,978\,7 \times 10^8$ km. The Earth's semimajor axis is currently $1.000\,000\,03$ AU.

Question 1.3

The asteroid Fortuna is in an orbit with a period of 3.81 years. Show that the semimajor axis of its orbit is 2.44 AU.

Question 1.4

Suppose that when the Earth is at perihelion Venus lies on the straight line between the Earth and the Sun. The time interval between sending a radar pulse from the Earth to Venus and receiving its echo is 264 s. If the speed of light in space is 3.00×10^5 km s^{-1}, calculate to two significant figures the astronomical unit in metres. Proceed as follows.

For the instant of measurement

- From the orbital details calculate the distance between the Earth and Venus in AU
- From the radar data calculate the distance in km between the Earth and Venus. Hence calculate the number of metres in 1 AU.

Note: *For two significant figures accuracy Venus is sufficiently close to perihelion when the Earth is at perihelion for you to use the perihelion distance of Venus.*

1.4.2 Orbital Elements

The quantities a and e are two of the five quantities — of the five **orbital elements** — that are needed to specify the elliptical orbit of a body. Note that P is *not* among the three remaining elements.

❏ Why is P redundant?

The orbital period is redundant because it can be obtained from a via Kepler's third law. The need for three further elements is illustrated in Figure 1.9, which shows the plane of the Earth's orbit plus the orbit of another body. Note that, for clarity, the orbit of the Earth is not shown, though the direction of the Earth's orbital motion is indicated by an arrow. The plane of the Earth's orbit acts as a reference plane for all other orbits and is called the **ecliptic plane**. The position of the Earth in its orbit at a certain moment in the year provides a reference direction. The direction chosen is that from the Earth to the Sun when the Earth is at the **vernal (March) equinox**. The direction points to the stars at a location called the **first point of Aries**. The direction (and the location) has the symbol ♈. The basis of these names will be given later.

For the other body in Figure 1.9, its orbital plane intersects the ecliptic plane to form a line. The Sun lies on this line at the point S — the Sun must lie in both orbital planes (Kepler's first law). Another point on the line is marked N, and this is where the body crosses the ecliptic plane in going from the south side to the north side, north and south referring to the sides of the ecliptic plane on which the Earth's north

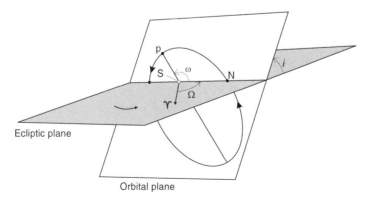

Ecliptic plane

Orbital plane

Fig. 1.9 The three orbital elements i, ω, Ω, used to specify the orientation of an elliptical orbit with respect to the ecliptic plane

and south poles lie. N is called the ascending node of the body's orbit. The angle Ω is measured in the direction of the Earth's motion, from γ to the line SN. This is the orbital element called the longitude of the ascending node. It can range from $0°$ to $360°$. The orbital plane of the planet makes an angle i with respect to the ecliptic plane, and this is the element called the **orbital inclination**. It can range from $0°$ to $180°$ — values greater than $90°$ correspond to retrograde orbital motion.

❐ What is the inclination of the Earth's orbit, and why is the longitude of the ascending node an inapplicable notion?

The Earth's orbit lies in the ecliptic plane. With the ecliptic plane as the reference plane, the inclination of the Earth's orbit is therefore zero. An ascending node is one of the two points where an orbit intersects the ecliptic plane. The Earth's orbit lies in this plane and therefore the ascending node is undefined.

The last of the five elements that are needed to specify the elliptical orbit of a body is the angle ω, measured from SN to the line Sp, where p (Figure 1.9) is the perihelion position of the body. The angle ω is measured in the direction of motion of the body, and can range from $0°$ to $360°$. It is called the argument of perihelion. However, it is somewhat more common to give as the fifth element the angle $(\Omega + \omega)$. This is called the longitude of perihelion. It is a curious angle, being the sum of two angles that are not in the same plane. Note that if the sum exceeds $360°$, then $360°$ is subtracted.

To specify exactly where a body will be in its orbit at some instant we need to know when it was at some specified point at some earlier time. For example, we could specify one of the times at which the body was at perihelion. This sort of specification is a sixth orbital element. Table 1.1 lists the values of the orbital elements for each planet, and for Ceres. Note that

- The orbital inclinations are small; the planets' orbital planes are almost coincident, Pluto's inclination of $17.1°$ being by far the greatest
- Except for Pluto and Mercury, and to a lesser extent Mars, the orbital eccentricities are also small, and the exceptions are not dramatic.

Question 1.5

(a) The comet Kopff has the following orbital elements: inclination $4.7°$, eccentricity 0.54, argument of perihelion $163°$, longitude of the ascending node $121°$. Sketch the orbit with respect to the ecliptic plane and the direction γ. (An accurate drawing is *not* required.)

(b) The distance of comet Kopff from the Sun at its perihelion on 2 July 1996 was 1.58 AU. Calculate the semimajor axis of the orbit, and hence calculate: its aphelion distance, its orbital period, and the month and year of the first perihelion in the 21st century (given that there are 365.24 days per year).

(c) The perihelion and aphelion distances of Mars are 1.38 AU and 1.67 AU, and yet the orbits of Mars and comet Kopff do *not* intersect. In a few sentences, state why not. (A proof is *not* required.)

1.4.3 Asteroids, and the Titius–Bode Rule

Nearly all the asteroids are in the asteroid belt between Mars and Jupiter, and though their orbital inclinations and eccentricities are more diverse than for the planets (Table 1.3), the asteroids in the asteroid belt do, by and large, partake in the nearly circular swirl of prograde motion near to the ecliptic plane.

If we compare the semimajor axes of the planets, and include the asteroids, then something curious emerges. One way of making this comparison is shown in Figure 1.10. The planets have been numbered in order from the Sun: Mercury is numbered 1, Venus 2, Earth 3, Mars 4, the asteroids 5, Jupiter 6, and so on. The semimajor axes of the orbits have been plotted versus the planets' numbers. For the asteroids the dot is Ceres and the bar represents the range of semimajor axes in the **main belt**, a concentration within the broader asteroid belt. The curious thing is that, with a logarithmic scale on the 'vertical' axis, the data in Figure 1.10 lie close to a straight line. This means that the semimajor axes increase by about the same factor each time we go from one planet to the next one out. This is one of several ways of expressing the **Titius–Bode rule**, named after the German astronomers Johann Daniel Titius

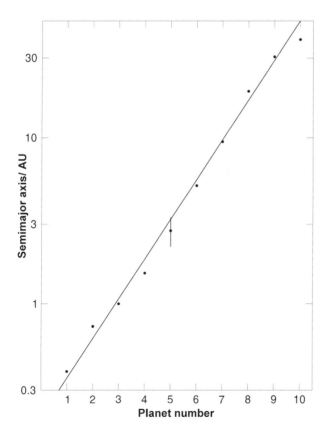

Fig. 1.10 The semimajor axes of the planets versus the number of the planet in order from the Sun: 1 = Mercury, 2 = Venus, etc. The bar at 5 is the asteroid belt, and so on, until 10 = Pluto

(1729–1796) who formulated a version of the rule in 1766, and Johann Elert Bode (1747–1826) who published it in 1772. Theories of the formation of the Solar System (Chapter 2) do give rise to an increase in spacing of planetary orbits as we go out from the Sun, so the Titius–Bode rule is an expression of this feature of the theories.

1.4.4 A Theory of Orbits

Kepler's laws are empirical rules that describe pretty well the motion of the planets around the Sun. One of the many achievements of the British scientist Isaac Newton (1642–1727) was that he was able to explain the rules in terms of two universal theories that he had developed. One theory is encapsulated in **Newton's laws of motion**, and the other in **Newton's law of gravity**. I state these laws here on the assumption that you have met them before, and will concentrate on using them to explore motion in the Solar System.

Newton's first law of motion   An object remains at rest or moves at constant speed in a straight line unless it is acted on by an unbalanced force. In other words, an unbalanced force causes acceleration (i.e. either a change of speed, or a change of direction, or a change of both speed and direction).

Newton's second law of motion   If an unbalanced force of magnitude F acts on a body of mass m, then the acceleration of the body has a magnitude given by

$$a = F/m \qquad (1.4)$$

and the direction of the acceleration is in the direction of the unbalanced force.

Newton's third law of motion   If body A exerts a force of magnitude F on body B, then body B will exert a force of the same magnitude on body A but in the opposite direction.

Newton's law of gravity   If two point masses M and m are separated by a distance r then there is a gravitational force of attraction between them with a magnitude given by

$$F = GMm/r^2 \qquad (1.5)$$

where G is the universal gravitational constant (its value is given in Table 1.6).

A point mass has a spatial extent that is negligible compared to r. For extended bodies the net gravitational force is the sum of the gravitational forces between all the points in one body, and all the points in the other.

To derive Kepler's laws from Newton's laws three conditions have to be met.

(1) The only force on a body is the gravitational force of the Sun.
(2) The Sun and the body are **spherically symmetrical**. This means that their densities vary only with radius from the centre to the (spherical) surface. In this case they interact gravitationally like point masses with all the mass of each body concentrated at its centre.

(3) The mass of the orbiting body is negligible compared to the Sun's mass.

The detailed derivation can be found in books on celestial mechanics, and will not be repeated here, but we can illustrate some links between Kepler's laws and Newton's laws.

Kepler's first and second laws

Let's take the first and second laws together and examine a body A in an elliptical orbit such as orbit 1 in Figure 1.11. Newton's law of gravity tells us that the Sun attracts A. Thus, from the second law of motion, A accelerates towards the Sun, its speed increasing as its distance from the Sun decreases. Because of its 'sideways' motion it does not fall directly towards the Sun. It therefore misses the Sun and swings through perihelion (p) at its maximum speed. It is then slowed down by the Sun's gravity as it climbs away from the Sun, and has its minimum speed as it passes through aphelion (a). The mathematical details show that under conditions 1 and 2 the precise shape of the orbit is elliptical with the Sun at a focus (Kepler's first law) and that the increase in speed with decreasing solar distance gives the equal areas law (Kepler's second law).

Consider the body now in the circular orbit 2 in Figure 1.11. This orbit has the same perihelion distance as orbit 1, but the body is now moving more slowly at p than it was in orbit 2, and so it does not climb away from the Sun. It still accelerates towards the Sun in that its motion is always curving towards the Sun, but the 'sideways' motion is just right to keep it at the same distance from the Sun. Consequently its speed in its orbit is constant, and its acceleration is entirely in its change of direction. If the body had *no* sideways motion then it would accelerate straight into the Sun.

Non-elliptical orbits

Now consider the body in an orbit with a speed at perihelion greater than in orbit 1.

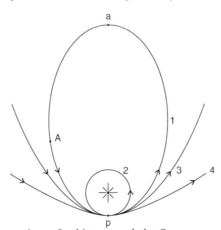

Fig. 1.11   A body in a variety of orbits around the Sun

❐ What would be the orbit were the speed at p only slightly greater?

In this case the body would climb slightly further away at aphelion — the semimajor axis would be greater, as would be the eccentricity. If we increase the speed further then Newton's laws predict that there comes a point when the body climbs right away from the Sun, never to return. This threshold is met in orbit 3 in Figure 1.11. This is a parabolic orbit. It is not a closed curve — the two arms become parallel at infinity. Orbits with even greater perihelion speeds are even more opened out, and one example is orbit 4. These are **hyperbolic orbits**. At infinity, the two arms of a hyperbola become tangents to diverging straight lines, the greater the perihelion speed, the greater the angle between the lines. Parabolic and hyperbolic orbits are called unbound orbits, whereas an elliptical orbit is a bound orbit.

Are there any Solar System bodies in unbound orbits? Yes there are. Table 1.4 shows that the orbital eccentricities of some comets are indistinguishable from 1, a value that corresponds to a parabolic orbit. Some comets might be in hyperbolic orbits. If a comet is in an unbound orbit then, unless its orbit is suitably modified to become bound, for example by a close encounter with a planet, it will leave the Solar System. Also, unless its orbit has been modified on its way inwards, it must have come from beyond the Solar System. Comets are a major topic in Chapter 3.

Kepler's third law

For Kepler's third law ($P = ka^{3/2}$) we have to consider bodies in orbits with different semimajor axes. You saw earlier that the $a^{3/2}$ dependence is the combined result of an increase in the distance around the larger orbit, and a lower orbital speed. This lower speed is explained by the decrease of gravitational force with distance (Newton's law of gravity, equation (1.5)) and the corresponding decrease in acceleration, a result derived in detail in standard texts. Such texts also show that, under the conditions 1 and 2, Newton's laws give

$$ P = \left(\frac{4\pi^2}{G(M_\odot + m)} \right)^{1/2} a^{3/2} \qquad (1.6) $$

where $M_\odot$ is the mass of the Sun and m is the mass of the other body. This is not quite Kepler's third law.

❐ What further condition is needed?

To get Kepler's third law $4\pi^2/G(M_\odot + m)$ must be a constant for the Solar System. With m being the property of the non-solar body, this condition is met if m is negligible compared to the Sun's mass. This is condition 3. In the Solar System Jupiter is by some way the most massive planet, but even so is only 0.1% the mass of the Sun. Therefore, condition 3 is met to a good approximation, and Kepler's third law is explained satisfactorily by Newton's laws.

Question 1.6

From the orbital data for the Earth in Table 1.1, calculate the mass of the Sun. Work in SI units, and note that 1 year $= 3.156 \times 10^7$ s. Repeat the calculation using the data for Jupiter's orbit. State any approximations you make, and whether your calculated masses seem to bear them out.

1.4.5 Orbital Complications

Conditions 1–3 in Section 1.4.4 are met only approximately in the Solar System, and because of this, complications arise, as follows.

The mass of the orbiting body is NOT negligible compared to the Sun's mass

Consider a single planet and the Sun, as in Figure 1.12(a). You can see that they each orbit a point on a line between them. This point is called the **centre of mass** of the system comprising the Sun and the planet. For any system of masses the centre of mass is the point that accelerates under the action of a force external to the system *as if* all the mass in the system were concentrated at that point. Thus if the external forces are negligible then the centre of mass is unaccelerated. By contrast, both the Sun and the planet accelerate the whole time because of their orbital motions with respect to the centre of mass. In Figure 1.12(b) the same planet is shown in its orbit

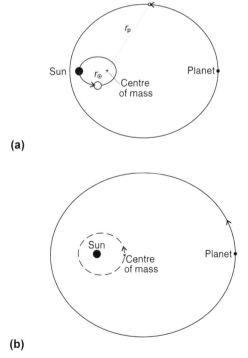

Fig. 1.12 A planet in orbit around the Sun (a) motion with respect to the centre of mass (b) motion of the planet with respect to the Sun

with respect to the Sun. This orbit is bigger than the two in Figure 1.12(a) but all three orbits have the same eccentricity and orbital period. Kepler's first two laws apply to the planetary orbit with respect to the Sun, as in Figure 1.12(b), and are *not* invalidated by the non negligible planet's mass.

For two spherically symmetrical bodies, such as the Sun and planet in Figure 1.12, the centre of mass is at a position such that

$$\frac{r_\odot}{r_p} = \frac{m_p}{M_\odot} \tag{1.7}$$

where $r_\odot$ and r_p are the *simultaneous* distances of the Sun and planet from the centre of mass at any point in the orbits, and $M_\odot$ and m_p are the masses. Though I shall not prove this equation, it has reasonable features. For example, the greater the value of $m_p/M_\odot$, the further the centre of mass is from the centre of the Sun. In Figure 1.12 $m_p/M_\odot = 1/4$, corresponding to a planet far more massive than any in the Solar System!

❐ Where is the centre of mass if the mass of the planet is negligible compared to the solar mass?

It is then at the centre of the Sun.

Jupiter, the most massive planet, has a mass 0.0955% of that of the Sun. Jupiter is in an approximately circular orbit with a semimajor axis of 7.78×10^8 km, and so, from equation (1.7) we can calculate that the centre of mass of the Jupiter–Sun system is 740 000 km from the Sun's centre. Thus, if Jupiter were the only planet in the Solar System the Sun's centre would move around a nearly circular orbit of radius 740 000 km — not much more than the solar radius. The effects of the other planets is to make the Sun's motion complicated, though the excursions of the Sun's centre are confined to within a radius of about 1.5×10^6 km.

There are forces on a body ADDITIONAL to the gravitational force of the Sun

❐ List some forces on a planet other than the gravitational force of the Sun.

Most obviously there is the gravitational force exerted by the *other* planets. The planets have much smaller masses than the Sun, and are relatively well separated. Therefore, from Newton's law of gravity (equation (1.5)), it is clear that the combined gravitational force of the other planets is small, giving only slight effects on the planet's orbit. In contrast, a comet can approach a planet fairly closely, in which case the comet's orbit will be greatly modified. Planetary satellites also have an effect — it is the centre of mass of a planet–satellite system that follows an elliptical orbit around the Sun, in accord with Kepler's laws. The planet and each satellite thus follow a slightly wavy path.

As well as other *gravitational* forces there are non-gravitational forces. For example, when a comet approaches the Sun, icy materials are vaporised — it is these that give rise to the head and the tails. But they also exert forces on the comet, rather in the manner of rocket engines, and considerable orbital changes can result.

The Sun and the body are NOT spherically symmetrical

Though the Sun and the planetary bodies are close to spherical symmetry, they are not perfectly so. One cause is the rotation of the body. No body is rigid and so the rotation causes the equatorial region to bulge, as in Figure 1.13(a), to give a tangerine-shape. The rotational distortion of Saturn is clear in Plate 16. Another cause of departure from spherical symmetry is a gravitational force that varies in magnitude across a body. From Newton's law of gravity (equation (1.5)) we can see that the surface of a planet closest to the Sun experiences a slightly larger gravitational force than the surface furthest from the Sun, and so the planet distorts under this differential force in the (exaggerated) manner of Figure 1.13(b) — a shape like a rugby ball, or an American football. The differential force is called a **tidal force**, and the distortion is called a tide. The Sun produces a tide in the body of the Earth, and a larger tide in the oceans. The Moon also produces tides in the Earth and actually raises greater tides than the Sun does, in spite of the Moon's far lower mass. This is because it is so much closer than the Sun that the *difference* in the force it exerts across the Earth is greater than the *difference* in the force exerted by the Sun;

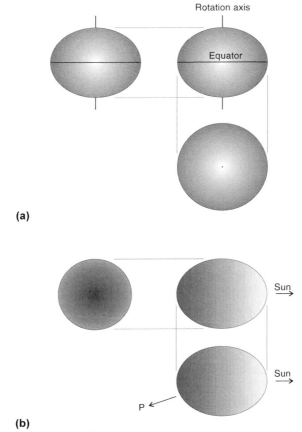

Fig. 1.13 Departures from spherical symmetry in a planet due to (a) rotation (b) the tidal force of the Sun

the gravitational force of the Sun is almost uniform across the Earth, whereas that of the Moon is less so.

The importance of departures from spherical symmetry, however caused, is that they enable one body to exert a **torque** — a twisting — on another body. For example, a planet in Figure 1.13(b) in the direction P is slightly closer to the left end of the distorted planet than to the right end. It therefore exerts a greater overall gravitational force to the left than to the right, and so there is a torque. It can be shown that orbital changes result from such torques.

Because of additional forces and a lack of spherical symmetry the planetary orbits are therefore not quite as described by Kepler's three laws. However, the departures from the laws are sufficiently slight that we can regard the orbits as ellipses in which the orbital elements change, usually slowly, and often chaotically, i.e. without pattern, though the semimajor axes, eccentricities, and inclinations are usually confined to narrow ranges of values. The values given in Table 1.1 apply around the year 2000, and, except for Pluto, will have the quoted values to the precision given for many decades. Pluto is so much less massive than the outer giants that their gravitational forces on Pluto cause somewhat more rapid changes in Pluto's elements.

Question 1.7

Explain briefly why the orbital elements of Venus would be subject to greater variation than at present, if
(a) The Sun rotated more rapidly
(b) The mass of Jupiter were doubled
(c) The Sun entered a dense interstellar cloud of gas and dust.

1.4.6 The Orbit of Mercury

As for all the planetary orbits, the orbit of Mercury is not *quite* an ellipse fixed in space. An important departure is a slow rotation of the ellipse in the prograde direction, around an axis through the Sun, and perpendicular to the orbital plane. This is the **precession of the perihelion** of Mercury, and it is illustrated in Figure 1.14. The actual precession (with respect to a coordinate system fixed with respect to the distant stars) is through an angle of 5600.73 arc seconds per century (3600 arc seconds = 1 degree of arc). The effect of all the other planets, and of the slight departure of the Sun from spherical symmetry, accounts for 5557.62 arc seconds per century, leaving a discrepancy of 43.11 arc seconds per century. This discrepancy (at rather less precision) was a great puzzle when it was identified in the nineteenth century, and it was not accounted for until 1915 when the German–Swiss physicist Albert Einstein (1879–1955) applied his newly developed theory of **general relativity** to the problem. General relativity is *not* a modification of Newton's laws, but a very different sort of theory. Fortunately, for most purposes in the Solar System, the far simpler theory of Newton suffices. Einstein's theory accounts for the observed rate of precession of the perihelion of Mercury to within the measurement uncertainties.

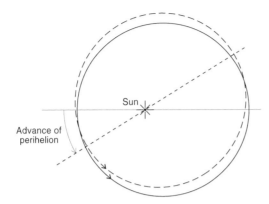

Fig. 1.14 Precession of the perihelion of the orbit of Mercury. The two orbits are separated by 2000 years

1.5 Planetary Rotation

Each planet rotates around an axis that passes through its centre of mass. In the case of the Earth this rotation axis is shown in Figure 1.15. It intersects the Earth's surface at the North and South Poles, and the Equator is the line half-way between the Poles. You can see that the rotation axis is not perpendicular to the Earth's orbital plane (the ecliptic plane) but has an **axial inclination** of 23.4° from the perpendicular.

As the Earth moves around its orbit the rotation axis remains (very nearly) fixed with respect to the distant stars. This is shown (from an oblique viewpoint) in Figure 1.16. The axis is *not* fixed with respect to the Sun, and so the aspect varies around the orbit. At A the North Pole is maximally tilted towards the Sun. This is called the June solstice, and it occurs on or near 21 June each year. Six months later, at C, the North Pole is maximally tilted away from the Sun. At B and D we have the only two

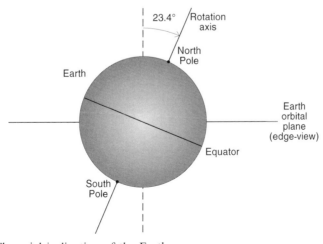

Fig. 1.15 The axial inclination of the Earth

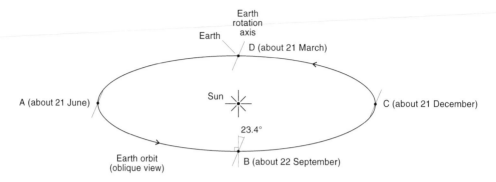

Fig. 1.16 The Earth's rotation axis as the Earth orbits the Sun. This is an oblique view of the orbit, which is nearly circular

moments in the year when the Earth's rotation axis is perpendicular to the line from the Earth to the Sun. Over the whole Earth, day and night are of equal length, which gives us the name for these two configurations — the **equinoxes**. The direction from the Earth to the Sun at the vernal (March) equinox is used as the reference direction in the ecliptic plane that you met in Section 1.4.2.

   We now turn to the *period* of rotation. Figure 1.17 shows the Earth moving around a segment of its orbit. As it does so it also rotates, and the arrow extending from a fixed point on the Earth's surface enables us to monitor this rotation. Between positions 1 and 2 the Earth has rotated just once with respect to a distant star. This is the **sidereal rotation period**. The distant stars, to sufficient accuracy, provide a non-rotating frame of reference (just as for the sidereal orbital period in Section 1.4.1). For the Earth, the sidereal rotation period is actually called the mean rotation period — astronomical terminology can be perverse. However, the Earth has not yet rotated once with respect to the Sun. The Earth has to rotate further to complete this rotation, and in the extra time taken it moves further around its orbit, to position 3. The period of rotation of the Earth with respect to the Sun is called the **solar day**. It is clearly longer than the mean rotation period, though only by a few minutes.

❐ State in what way the motions in Figure 1.17 are *not* shown to scale.

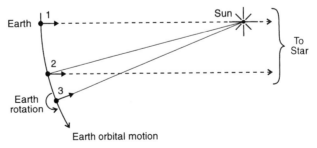

Fig. 1.17 The rotation of the Earth with respect to the Sun, and with respect to the distant stars. (*Not* to scale)

In Figure 1.17 the Earth moves too far around its orbit between positions 1, 2, and 3. As there are 365 days in a year, the Earth should only proceed about 1° around its orbit in the time it takes the Earth to rotate once.

The mean rotation period does not vary significantly through the year, but the solar day does. This is a consequence of the eccentricity of the Earth's orbit, and the inclination of its rotation axis (I won't go into details). By contrast, the mean solar day is defined to be fixed in duration, and has the mean length of the solar days averaged over a year. If solar time and mean solar time coincide at some instant, they will coincide again a year later, but in-between, differences develop, sometimes solar time being ahead of mean solar time, and sometimes behind. The maximum differences are about 15 minutes ahead or behind. The day that we use in our everyday lives, as marked by our clocks, is the mean solar day. Even this varies in length, *very* slightly, and so for scientific purposes a standard day is defined, very nearly the same as the current length of the mean solar day. It is this standard **day** that appears in Tables 1.1–1.4, and elsewhere. It is exactly $24 \times 60 \times 60$ seconds in length, and thus consists of exactly 24 hours of 60 minutes, with each minute consisting of 60 seconds.

On this basis, the mean rotation period is 23 h 56 min 4 s, i.e. 3 min 56 s shorter than the mean solar day. Over one sidereal year, this difference must add up to one extra rotation of the Earth with respect to the distant stars. You can convince yourself that there is a difference of one rotation by considering a planet that is rotating as in Figure 1.18. In this case there are three rotations per orbit with respect to the Sun and four with respect to the stars. For the Earth, during the sidereal year there are 365.26 mean solar days, and 366.26 mean rotation periods.

Table 1.1 gives the axial inclination and sidereal rotation period of each planet, and also of the Sun. The inclination of each planet is with respect to the plane of its orbit, whereas in the case of the Sun it is with respect to the ecliptic plane. Note that, with three exceptions, the inclinations are fairly small. This means that the prograde swirl of motion of the orbits, almost in one plane, is shared by planetary and solar rotation. The exceptions are Venus, Uranus, and Pluto. The inclination of Venus is not far short of 180°.

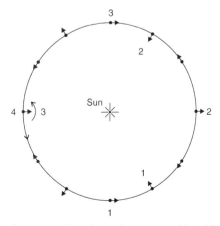

Fig. 1.18 A fictitious planet rotating three times per orbit with respect to the Sun

❑ What is the difference between an axial inclination of 180° and 0°?

The difference is that 0° is prograde rotation whereas 180° is retrograde rotation, in each case with the rotation axis perpendicular to the orbital plane. Any inclination greater than 90° is retrograde, and so Pluto and Uranus are also in retrograde rotation, though Uranus's inclination of 97.9° means that its rotation axis is almost in its orbital plane. We shall return to these oddities when we discuss the origin of the Solar System in Chapter 2.

As in the case of the orbital elements, the axial inclinations and rotation periods of a body are subject to changes, and for the same basic reason — the forces applied by the other bodies in the Solar System. For example, the sidereal rotation period of the Earth is currently increasing, somewhat erratically, by 1.4×10^{-3} seconds per century, largely because of the torque exerted by the Moon on the Earth's tidal distortion. The Earth has had a similar effect on the Moon, and has slowed down the Moon so that it is now locked into a rotation period that keeps it facing the Earth. When one body rotates so that it keeps one face to the body it orbits, it is said to be in **synchronous rotation**.

Seasons

Figure 1.19 is an edge-view of the Earth's orbit with the positions A and C in Figure 1.16 marked, and the size of the Earth *greatly* exaggerated. When the North Pole of the Earth is maximally tilted towards the Sun, as at A, there is summer in the northern hemisphere because the surface there is receiving its greatest solar power. This is not only because the Sun reaches high in the sky, but also because of the long duration of daylight. By contrast, the southern hemisphere is maximally tilted *away* from the Sun.

❑ What season is this hemisphere experiencing?

It is winter in this hemisphere, because solar radiation is thinly spread over the surface and daylight is short. Six months later, at C, the December solstice, the seasons are reversed. It is thus the axial inclination that is responsible for seasonal changes. The eccentricity of the Earth's orbit has only a secondary effect. The Earth is at perihelion in early January, with the northern hemisphere in the depths of winter, and so, as a result of the orbital eccentricity, the seasonal contrasts are reduced in the northern hemisphere, and increased in the southern hemisphere.

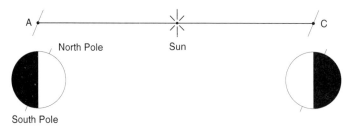

Fig. 1.19 Seasonal changes in the solar radiation at the Earth's surface

Question 1.8

Discuss whether you would expect seasonal changes on Venus.

1.5.1 Precession of the Rotation Axis

So far, the direction of the Earth's rotation axis has been regarded as fixed with respect to the distant stars. This is not quite the case. In fact, it cones around in the manner of Figure 1.20, a motion called the **precession of the rotation axis**. It is a result of the torques exerted by other bodies in the Solar System on the slightly non-spherical form of the Earth. The Moon and the Sun account for almost the whole effect. All planets are slightly non-spherical, so all of them are subject to precession. For the Earth, one complete coning takes 25 800 years, an interval called the precession period of the Earth.

One consequence of precession is that the positions of the equinoxes and solstices move around the orbit, giving rise to the term precession of the equinoxes. In the case of the Earth this motion is in a retrograde direction, taking 25 800 years to move around once. Figure 1.21 compares the present configuration (dashed lines) with the configuration 12 900 years from now (solid lines) — each equinox and solstice has moved half-way around the orbit. Recall that the reference direction in the ecliptic plane is the line from the Earth to the Sun when the Earth is at the vernal equinox. Therefore, with respect to the distant stars, this reference direction has moved through 180° in Figure 1.21. At present, when the Earth is at the vernal equinox, the direction is to a point in the constellation Aquarius, but about 2000 years ago, when precession became widely recognised, it was in the constellation Aries, when its location was called the first point of Aries. The name sticks, even though the point has moved on to other constellations.

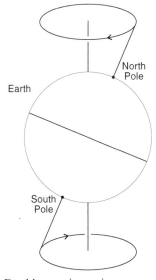

Fig. 1.20 Precession of the Earth's rotation axis

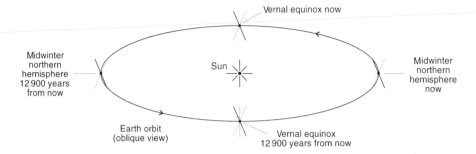

Fig. 1.21   The effect of the precession of the Earth's rotation axis on the position of the equinoxes and solstices. The dashed line is the Earth's rotation axis now, and the solid line the axis 12 900 years from now

The slow retrograde motion of the vernal equinox around the Earth's orbit means that the time taken for the Earth to traverse its orbit from one vernal equinox to the next is very slightly less than the sidereal year. The time interval between vernal equinoxes is called the tropical year, and it is the year on which our calendars are based. Its duration is 365.2422 days, whereas the sidereal year is 365.2564 days. From now on, the term '**year**' will mean the tropical year. It is this year that is the unit of time measurement in Tables 1.1, 1.3, and 1.4, and elsewhere. It is denoted by the symbol 'a', from the Latin word for year, 'annus'.

Question 1.9

What would be the problem with basing our calendar on the *sidereal* year?

1.6 The View from the Earth

1.6.1 The Other Planets

The way that the planets appear in our skies depends on whether the orbit of the planet is larger or smaller than the orbit of the Earth. Figure 1.22 shows Venus representing the two planets (Venus and Mercury) with smaller orbits, and Mars representing those with larger orbits. The planets are shown at two instances. In position 1 all three planets are lined up with the Sun, a very rare occurrence but a useful one for describing the view from the Earth. The planets move at different rates around their orbits, so this alignment lasts only for an instant.

In position 1, Venus is between the Earth and the Sun. It is then at what is called inferior conjunction. The alignment is rarely exact, because of the inclination of the orbit of Venus. Exact or not, our view of Venus is drowned by the overwhelming light of the Sun. The greater angular speed of Venus in its orbit then causes it to draw ahead of the Earth and we start to see part of the hemisphere illuminated by the Sun as an ever-thickening crescent. At position 2, Venus has reached its greatest angle from the Sun and is at what is called its maximum western elongation. It is now relatively easy to see (before sunrise) and half of its illuminated hemisphere is visible.

As it moves on we see even more of its sunlit hemisphere, but it is getting further away from the Earth, and closer in direction to the Sun, until at superior conjunction it is pretty well in the same direction as the Sun again, but now Venus is on the far side of the Sun. Subsequently, it moves towards maximum eastern elongation, then again to inferior conjunction, and the whole cycle is repeated.

For planets beyond the Earth, such as Mars in Figure 1.22, the sequence of events is different. The line-up with Mars and Earth on the same side of the Sun does not result in an inferior conjunction, but in what is called an **opposition**, Mars being in the opposite direction in the sky from the Sun, as viewed from the Earth. Mars is then well seen, with the illuminated hemisphere facing us, and the separation between the planets being comparatively small—though this distance is different from opposition to opposition.

❐ When will the opposition distance be a minimum?

It will be a minimum when opposition occurs with the Earth near aphelion, and Mars near perihelion. After opposition the greater angular orbital speed of the Earth causes it to overtake Mars 'on the inside track' as the configuration moves towards superior conjunction, with Mars on the far side of the Sun as seen from Earth.

The time interval between similar configurations of the Earth and another planet is called the **synodic period** of the planet. Opposition and inferior conjunction are important examples of types of configuration. For any type of configuration the synodic period varies slightly, mainly because of the variations in the rate at which the Earth and the planet move around their respective orbits, as described by Kepler's second law. It is thus the mean value of the synodic period that is normally quoted, as in Table 1.1. For a given planet this mean synodic period is the

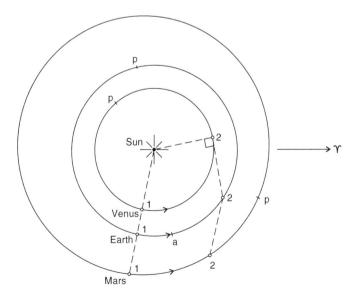

Fig. 1.22 The motion of Venus and Mars with respect to the Earth. Perihelia are denoted by 'p', and Earth's aphelion by 'a'

same for all types of configuration. Note that these periods are not simple multiples of the sidereal year, and so successive configurations have the Earth at different points in its orbit.

Question 1.10

Discuss why the opposition distance to Mars is least when oppositions occur in mid-August.

1.6.2 Solar and Lunar Eclipses

Figure 1.23 shows an oblique view of the nearly circular orbit of the Moon around the Earth, and part of the orbit of the Earth around the Sun (strictly, the orbit of the centre of mass of the Earth and Moon around the Sun). The size and inclination of the lunar orbit, and the sizes of the Sun, Earth and Moon, have all been exaggerated. When the Moon is at A (Figure 1.23(a)) its unilluminated hemisphere faces the Earth, and we have a new Moon. A quarter of an orbit later, at B, half of its illuminated hemisphere is facing us, and we see a half Moon, also called first quarter. At C the fully lit hemisphere faces the Earth, and the Moon is full. At D, three quarters of the way around its orbit from A we see another half Moon, the third quarter. We then get another new Moon at A, on average 29.53 days after the previous new Moon.

The plane of the Moon's orbit is inclined by 5.15° to the ecliptic plane, which it crosses at the two nodes labelled B and D in Figure 1.23(a). This orbital inclination means that for this configuration the Moon, as seen from the Earth, cannot pass in

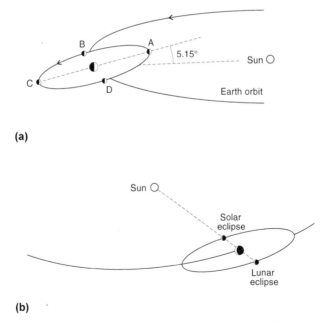

(a)

(b)

Fig. 1.23   The Moon's motion around the Earth, as the Earth orbits the Sun. (*Not* to scale)

front of the Sun, nor can the Moon be touched by the Earth's shadow. However, as the Earth moves around the Sun, the Moon's orbit stays (almost) fixed with respect to the distant stars, so that about a quarter of an Earth orbit later (3 months) the nodes lie on or near the line that joins the Earth and the Sun, as in Figure 1.23(b). If, at this time, the Moon is sufficiently near the node between the Earth and the Sun, then, as seen from Earth, part or all of the Moon will pass in front of the Sun, and we get a solar **eclipse**. If the Moon is at or near the other node then the Earth's shadow will fall on part or all of the Moon, and we get a lunar eclipse. The nodes line up twice a year, and usually the Moon is sufficiently near a node for there to be an eclipse of some sort.

There are different types of solar eclipses. Figure 1.24(a) shows umbral and penumbral shadows of the Moon on the Earth. If we are at a point on the Earth's surface within the umbral shadow then the photosphere of the Sun is completely obscured and we see a total solar eclipse. With the photosphere obscured, we see the pearly white solar corona (Plate 2), the chromosphere and prominences (Plate 3). It is worth making a considerable effort to see a total solar eclipse, which is a most magnificent spectacle. If we are in the penumbral shadow the Sun is only partly obscured and we see a partial solar eclipse. If the Moon is just too far from the node, then the umbral shadow misses the Earth completely, and nowhere on Earth can we see a total eclipse.

Even with the Moon at the node, not all solar eclipses are total. The Sun is about 400 times the diameter of the Moon, but it is also about 400 times further away, so the angular diameters of the two bodies are about the same, about 0.5 degrees of arc. Because of the eccentricity of the orbits of the Earth and the Moon, this coincidence means that sometimes an eclipse occurs with the Moon's angular diameter a bit smaller than that of the Sun, as in Figure 1.24(b). The umbral shadow does not reach the Earth, and from the Earth's surface, at the centre of the penumbral shadow, a thin ring of the photosphere is still exposed. This is called an annular solar eclipse. Largely because of tidal interactions with the Moon, the distance between the Moon and the Earth is currently increasing at a rate of about 25 mm per year, and so, from about 1000 Ma in the future, the Moon will never be close enough to the Earth to produce a total solar eclipse, and all solar eclipses will then be partial or annular.

Figure 1.25 shows the umbral shadow paths for the years 1998–2020. Within these narrow paths a total solar eclipse occurs. The paths are determined by the line-up of the Earth, Moon, and Sun, and by the combined effect of the orbital motion of the Moon and the rotation of the Earth, which together sweep the umbral shadow across the Earth. The duration of totality is longest at the centre of the path, and varies from eclipse to eclipse. The longest durations, a little over 7 minutes, occur when the Earth is at aphelion, and the Moon is closest to the Earth, at what is called perigee.

Figure 1.24(c) shows a total lunar eclipse, which occurs when the Moon is entirely within the umbral shadow of the Earth. Where the umbral shadow falls on the Moon, the lunar surface is not completely dark. Sunlight is refracted by the Earth's atmosphere, red light more than the other visible wavelengths, which can give the Moon a coppery tint. At any moment the eclipsed Moon can be seen from half the Earth's surface. The view *from* the Moon would be of the black, night side of the Earth, surrounded by a thin red ring of sunlight refracted by the Earth's atmosphere.

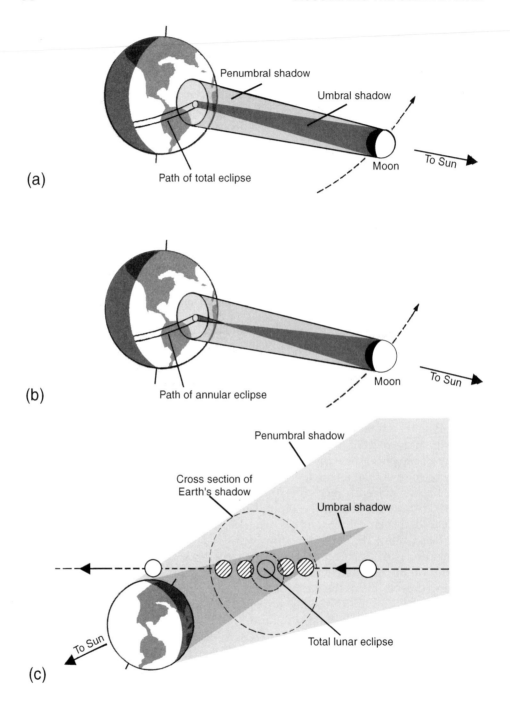

Fig. 1.24 Different sorts of eclipses (not to scale). (a) Total and partial solar eclipses. (b) An annular solar eclipse. (c) A lunar eclipse. (Adapted with permission from Figures 3–9, 3–12, and 3–17 of *Foundations of Astronomy*, M. A. Seeds, Wadsworth, 1994, © Wadsworth Inc.)

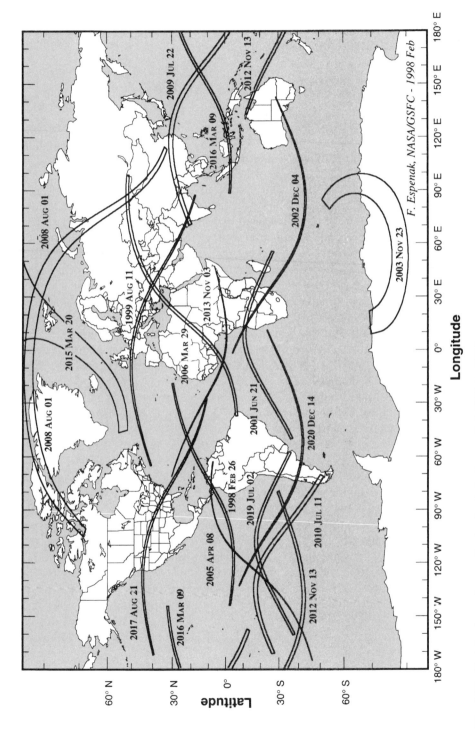

Fig. 1.25 Total solar eclipse paths (umbral shadow paths) 1998 to 2020

Question 1.11

The nodes of the lunar orbit are not quite fixed but move around the lunar orbit in the retrograde direction in 18.6 years. How does this explain why eclipses are not confined to particular months?

1.7 Summary

The Solar System consists of the Sun, nine planets with their satellites and rings, many asteroids (about 10^9 greater than 1 km across), of order 10^{13} comets, and an interplanetary medium of tenuous gas and small solid bodies ranging in size down to less than 10^{-6} m.

The planets orbit the Sun in one direction—the prograde direction—in approximately circular, coplanar orbits with the Sun near the centre. The orbital planes of the asteroids have a wider range of inclinations and eccentricities. The rotation of the Sun is prograde, as is that of most of the planets. If the inclination of the rotation axis to the orbital plane is more than a few degrees then the surface of the planet will experience seasonal changes. Rotation axes are subject to precession.

Some comet orbits reach to within a few AU of the Sun, but the great majority spend most or all of their time at far greater distances. The Edgeworth–Kuiper belt lies immediately beyond the planetary domain, and contains comets in orbits that are predominantly prograde and that are concentrated towards the ecliptic plane. The Oort cloud is more far flung and consists of comets in a spherical distribution around the Sun, reaching out to the edge of interstellar space, about 10^5 AU from the Sun.

The Sun is by far the largest, the most massive, and the most luminous body in the Solar System. It is fluid throughout, and consists largely of hydrogen and helium. Its luminosity is sustained by the nuclear fusion of hydrogen deep in its interior where temperatures reach 1.5×10^7 K.

The four planets closest to the Sun—Mercury, Venus, the Earth, and Mars—are the terrestrial planets. They are comparable to the Earth in size, and consist of iron-rich cores overlain by rocky materials. The Earth is the largest of these bodies. The asteroids are rocky bodies concentrated between Mars and Jupiter.

The giant planets—Jupiter, Saturn, Uranus, and Neptune—are considerably larger and more massive than the terrestrial planets, Jupiter by some margin being the most massive planet of all. The giants consist largely of hydrogen, helium, and icy–rocky materials, and (like the Sun) are fluid throughout. The giants have richly varied families of satellites, and they all have rings, those of Saturn being by far the most substantial. Beyond Neptune we come to the outermost planet, Pluto, smaller in size than the terrestrial planets, and more icy in its composition. The comets are also icy bodies.

The orbits of the planets are described to a very good approximation by Kepler's laws of planetary motion.

First law
Each planet moves around the Sun in an ellipse, with the Sun at one focus of the ellipse.

Second law
As the planet moves around its orbit, the straight line from the Sun to the planet sweeps out equal areas in equal intervals of time.

Third law
If P is the sidereal period of a planet, and a is the semimajor axis of the orbit, then

$$P = ka^{3/2} \tag{1.3}$$

where $k = 1$ year 1 $\text{AU}^{-3/2}$.

Elliptical orbits are characterised by five orbital elements: the semimajor axis a, the eccentricity e, the inclination i, the longitude of the ascending node Ω, and the longitude of perihelion $(\Omega + \omega)$. To calculate the position of a body in its orbit we need a sixth element — a single position at any known time.

Newton's laws of motion and law of gravity account for Kepler's laws, and go further by accounting for the motion of comets and of other bodies, and for slight departures from Kepler's laws that have various causes. The precession of the perihelion of Mercury shows that at the highest level of precision Einstein's theory of general relativity is superior to Newton's laws.

Our view from the Earth of the apparent motion of a planet depends on whether it is in a smaller or larger orbit than our own. Solar and lunar eclipses result when the Moon, Sun, and the Earth line up. Figure 1.25 shows the umbral tracks of forthcoming total solar eclipses.

Tables 1.1–1.6 list basic data on the Solar System.

Table 1.1 Mean orbital elements and some physical properties of the Sun, the planets, and Ceres

Object	Mean orbital elements[a]							From the Earth		Physical properties				
	Semimajor axis		Sidereal period/ years	Eccen-tricity	Inclin-ation/°	Long of ascend-ing node/°	Long of perihe-lion/°	Mean syn. period/ days	Date, oppn or inf. conj.	Axial inclin-ation/°[b]	Sid. rotn period/ days[c]	Equat. radius/ km[d]	Mass/kg	Mean density/ kg m^{-3}
	AU	10^6 km												
Sun	—	—	—	—	—	—	—	—	—	7.2	25.4	696 265	1.9891×10^{30}	1410
Mercury	0.387	57.9	0.2408	0.206	7.0	48.3	77.5	116	1 Dec. 98	0.0	58.646	2 440	3.302×10^{23}	5430
Venus	0.723	108.2	0.6152	0.007	3.4	76.7	131.6	584	16 Jan. 98	177.4	243.019	6 052	4.869×10^{24}	5240
Earth	1.000	149.6	1.0000	0.017	0	—	102.9	—	—	23.4	0.9973	6 378	5.974×10^{24}	5520
Mars	1.524	227.9	1.8808	0.093	1.8	49.6	336.1	780	24 Apr. 99	25.2	1.0260	3 397	6.419×10^{23}	3940
Ceres	2.768	414.1	4.6052	0.077	10.6	78.7	147.8	467	28 Nov. 98	54	0.378	457	1.1×10^{21}	2700
Jupiter	5.203	778.3	11.86	0.048	1.3	100.5	14.3	399	16 Sep. 98	3.1	0.4135	71 490	1.8987×10^{27}	1330
Saturn	9.555	1429.4	29.42	0.056	2.5	113.7	93.1	378	23 Oct. 98	25.3	0.4440	60 270	5.685×10^{26}	700
Uranus	19.218	2875.0	83.75	0.046	0.8	74.0	173.0	370	3 Aug. 98	97.9	0.7183	25 560	8.662×10^{25}	1300
Neptune	30.110	4504.4	163.73	0.009	1.8	131.8	48.1	367	23 Jul. 98	28.3	0.6712	24 765	1.0278×10^{26}	1760
Pluto	39.545	5915.8	248.03	0.249	17.1	110.3	224.1	367	28 May 98	123	6.3872	1 150	1.50×10^{22}	2100

[a]Another orbital element is the time of any perihelion passage. For the Earth in 1998 this was at 21 UT on 4 January.
[b]The Sun's axial inclination is with respect to the ecliptic plane.
[c]The rotation period of the Sun is at the equator; it increases with latitude, reaching about 36 days near the poles of the Sun.
[d]For the giants the equatorial radius is to the atmospheric altitude where the pressure is 10^5 Pa (1 bar).

Table 1.2 Some properties of planetary satellites[a]

Object[b]	Some mean orbital elements				Size, mass, mean density		
	Semimajor axis/10^3 km	Sidereal period/ days	Eccen- tricity	Inclin- ation/$^{\circ c}$	Radius/ km[d]	Mass/ 10^{18} kg	Mean density/ kg m^{-3}
Earth							
Moon	384.5	27.322	0.0549	5.15	1738	73 490	3340
Mars							
Phobos	9.4	0.319	0.015	1.1	11	0.013	~2000
Deimos	23.5	1.263	0.0005	1.8 var	6	0.0018	~2000
Jupiter							
Metis	128	0.294	0.0	small	23	?	?
Adrastea	129	0.297	0.0	small	7	?	?
Amalthea	180	0.498	0.003	0.4	85	?	?
Thebe	222	0.674	0.013	small	48	?	?
Io	422	1.769	0.004	0.0	1821	89 320	3530
Europa	671	3.551	0.010	0.5	1565	48 700	3020
Ganymede	1 070	7.155	0.001	0.2	2635	149 000	1940
Callisto	1 885	16.689	0.007	0.2	2405	107 500	1850
Leda	11 110	240	0.147	27	~8	?	?
Himalia	11 470	251	0.158	28	93	?	?
Lysithea	11 710	260	0.107	29	~18	?	?
Elara	11 740	260	0.207	28	38	?	?
Ananke	21 200	631	0.17	147	~15	?	?
Carme	22 350	692	0.21	164	~20	?	?
Pasiphae	23 330	735	0.38	148	~25	?	?
Sinope	23 370	758	0.28	153	~18	?	?
Saturn							
Pan	134	0.577	~0	~0	~10	?	?
Atlas	137	0.601	0.002	0.3	15	?	?
Prometheus	139	0.613	0.002	0.0	50	?	?
Pandora	142	0.628	0.004	0.1	45	?	?
Janus	151	0.695	0.009	0.3	95	?	?
Epimetheus	151	0.695	0.007	0.1	60	?	?
Mimas	187	0.942	0.020	1.5	195	38	1200
Enceladus	238	1.370	0.004	0.02	250	80	1100
Tethys	295	1.888	0.000	1.1	530	760	1200
Telesto	295	1.888	~0	~0	13	?	?
Calypso	295	1.888	~0	~0	13	?	?
Dione	378	2.737	0.002	0.02	560	1 050	1400
Helene	378	2.737	0.005	0.2	15	?	?
Rhea	526	4.517	0.001	0.4	765	2 490	1300
Titan	1 221	15.945	0.029	0.3	2575	134 570	1880
Hyperion	1 481	21.276	0.104	0.4	128	?	?
Iapetus	3 561	79.331	0.028	14.7	730	1 880	1200
Phoebe	12 960	550.46	0.163	150	110	?	?
Uranus							
Cordelia	49.8	0.333	~0	~0	~13	?	?
Ophelia	53.8	0.375	~0	~0	~15	?	?

(*continued*)

Table 1.2   (*continued*)

Object[b]	Some mean orbital elements				Size, mass, mean density		
	Semimajor axis/10^3 km	Sidereal period/ days	Eccen- tricity	Inclin- ation/$^{\circ c}$	Radius/ km[d]	Mass/ 10^{18} kg	Mean density/ $kg\,m^{-3}$
Bianca	59.2	0.433	~0	~0	~23	?	?
Cressida	61.8	0.463	~0	~0	~33	?	?
Desdemona	62.6	0.475	~0	~0	~30	?	?
Juliet	64.4	0.492	~0	~0	~43	?	?
Portia	66.1	0.513	~0	~0	~55	?	?
Rosalind	70.0	0.558	~0	~0	~30	?	?
Belinda	75.3	0.621	~0	~0	~34	?	?
Puck	86.0	0.763	~0	~0	78	?	?
Miranda	129.9	1.413	0.017	3.4	243	75	1300
Ariel	190.9	2.521	0.0028	0.0	580	1 340	1600
Umbriel	266.0	4.146	0.0035	0.0	595	1 270	1400
Titania	436.3	8.704	0.0024	0.0	805	3 470	1600
Oberon	583.4	13.463	0.0007	0.0	775	2 920	1500
S/1997 U1	7 170	579	0.082	139.7	~40	?	?
S/1997 U2	12 214	1289	0.509	152.7	~80	?	?
Neptune							
Naiad	48.0	0.30	~0	4.5	30	?	?
Thalassa	50.0	0.31	~0	0.4	40	?	?
Despina	52.5	0.33	~0	<1	75	?	?
Galatea	62.0	0.43	~0	<1	80	?	?
Larissa	73.6	0.55	~0	<1	95	?	?
Proteus	117.6	1.12	~0	<1	210	?	?
Triton	354	5.877	<0.0005	160.0	1350	21 400	2070
Nereid	5510	365.21	0.76	27.6	170	?	?
Pluto							
Charon	19.1	6.387	~0	~0	600	~1 100	~1200

[a]The inclinations of the rotation axes of the satellites, and the rotation periods are not given, but in most cases the inclinations are small. Many of the satellites, like the Moon, are in synchronous rotation around their planet. Note that '?' denotes 'mass unknown', and consequently the density is also unknown.
[b]Some very small satellites of Jupiter and Saturn are not included.
[c]Note that in most cases the orbital inclination is with respect to the *equatorial* plane of the planet. The exceptions are the Moon, Phoebe, and Nereid, and for Jupiter all the satellites beyond Callisto. In these cases the inclination is with respect to the *orbital* plane of the planet. This is because the inclination with respect to the equatorial plane changes periodically through a fairly large range of values. Inclinations greater than 90° indicate retrograde orbital motion, i.e. opposite to the direction of rotation of the planet.
[d]Values less than a few hundred km are average radii of irregularly shaped bodies.

Table 1.3 Some properties of the fifteen largest asteroids[a]

| | Some orbital elements[a] | | | | | Rotation and size | |
| | Semimajor axis | | Sidereal period/ years | Eccentricity | Inclination/° | Sid. rotn period/ hours | Radius[b]/ km[b] |
Object	AU	10^6 km					
Ceres	2.768	414.1	4.61	0.077	10.58	9.08	457
Pallas	2.774	415.0	4.62	0.232	34.82	7.81	262
Vesta	2.361	353.2	3.63	0.090	7.13	5.34	251
Hygiea	3.136	469.1	5.55	0.120	3.84	~18.4	215
Davida	3.172	474.5	5.65	0.182	15.94	5.13	169
Interamnia	3.064	458.4	5.36	0.146	17.32	8.73	167
Europa	3.099	463.6	5.46	0.101	7.48	5.63	156
Eunomia	2.644	395.5	4.30	0.187	11.75	6.08	136
Sylvia	3.499	523.5	6.55	0.082	10.84	5.18	136
Psyche	2.922	437.1	4.99	0.138	3.09	4.20	132
Euphrosyne	3.145	470.5	5.58	0.228	26.34	5.53	124
Cybele	3.432	513.4	6.36	0.104	3.43	6.07	123
Juno	2.669	399.3	4.36	0.258	12.97	7.21	122
Bamberga	2.686	401.8	4.40	0.337	11.10	29.43	121
Camilla	3.495	522.9	6.53	0.080	10.02	4.84	119

[a]In December 1997.

Table 1.4 Some properties of selected comets

Comet	Some orbital properties				Perihelion distance/AU	Date, last perihelion passage	Associated meteor shower(s)
	Semi-major axis/AU	Sidereal period/years	Eccentricity	Inclination/°			
Short-period							
Giacobini–Zinner	3.52	6.59	0.7065	31.86	1.0337	Nov. 1998	Giacobinids
Biela	3.53	6.62	0.756	12.55	0.861	1852 (lost)	Andromedids
Halley	17.95	76.08	0.9673	162.24	0.5871	Feb. 1986	Eta Aquarids
Swift–Tuttle	25.32	135.1	0.9636	113.43	0.9582	Dec. 1992	Perseids
Encke	2.21	3.28	0.8500	11.93	0.3314	May 1997	Some Taurids?
Whipple	4.17	8.52	0.2587	9.93	3.0939	Dec. 1994	—
Schwassmann–Wachmann 1	5.99	14.65	0.0442	9.39	5.7236	Oct. 1989	—
Oterma	7.24	19.5	0.2446	1.94	5.4707	Jun. 1983	—
Brorsen–Metcalfe	17.1	70.7	0.9720	19.33	0.4788	Sep. 1989	—
Long-period (bright)							
Great Comet of 1843	640	16 000	0.9999914	114.35	0.005527	1843	—
Donati	~150	~1700	0.996	116.96	0.578	1858	—
Arend–Roland	Large	Long	1.000	119.95	0.316	1957	—
Mrkos	Large	Long	0.999	93.94	0.355	1957	—
Seki–Lines	Large	Long	1.000	65.01	0.031397	~1962/3	—
Ikeya–Seki	92	880	0.999915	141.86	0.007786	1965	—
Bennett	~130	~1600	0.996	90.04	0.538	~1970	—
Kohoutek	Large	Long	1.000	14.31	0.142	1973	—
West	Large	Long	1.000	43.07	0.197	1976	—
Hyakutake[a]	1000	32 000	0.9997705	124.92	0.2302234	1996	—
Hale–Bopp[a]	188	2 600	0.9951315	89.43	0.9141639	1997	—

[a]The orbital elements apply after the last perihelion passage.

Table 1.5 Relative abundances of the fifteen most abundant chemical elements in the Solar System

Chemical element				Relative abundance	
Atomic number	Name	Symbol	Relative atomic mass ($^{12}C \equiv 12$)	By number of atoms	By mass
1	Hydrogen	H	1.0080	1 000 000	1 000 000
2	Helium[a]	He	4.0026	97 000	390 000
6	Carbon	C	12.0111	360	4300
7	Nitrogen	N	14.0067	110	1500
8	Oxygen	O	15.9994	940	15 000
10	Neon	Ne	20.179	130	2 600
11	Sodium	Na	22.9898	2	44
12	Magnesium	Mg	24.305	32	780
13	Aluminium	Al	26.9815	3	89
14	Silicon	Si	28.086	45	1 300
16	Sulphur	S	32.06	16	510
18	Argon	Ar	39.948	1	40
20	Calcium	Ca	40.08	2	88
26	Iron	Fe	55.847	32	1 800
28	Nickel	Ni	58.71	2	110

[a]The helium values correspond to those before the conversion of some of the hydrogen in the Sun's core to helium, i.e. to the Sun at its formation.

Table 1.6 Some important constants

Name	Symbol	Value
Speed of light (in a vacuum)[a]	c	$2.997\,924\,58 \times 10^8 \, \text{m s}^{-1}$
Gravitational constant[b]	G	$6.672 \times 10^{-11} \, \text{N m}^2 \, \text{kg}^{-2}$
Boltzmann's constant	k	$1.381 \times 10^{-23} \, \text{J K}^{-1}$
Planck's constant	h	$6.6261 \times 10^{-34} \, \text{J s}$
Stefan's constant	σ	$5.67 \times 10^{-8} \, \text{W m}^{-2} \, \text{K}^{-4}$
Astronomical unit	AU	$1.495\,978\,7 \times 10^{11} \, \text{m}$
Light year[c]	ly	$9.460\,536 \times 10^{15} \, \text{m}$
Parsec	pc	$3.085\,678 \times 10^{16} \, \text{m}$
Solar luminosity	$L_\odot$	$3.85 \times 10^{26} \, \text{W}$
Day[d]	d	86 400 s exactly
Tropical year	a	365.2422 d
pi	π	3.14159.....

[a]This is an exact value. The second (s) is now defined in terms of atomic vibrations, and the metre (m) as the distance travelled by light in a vacuum in $1/(2.997\,924\,58 \times 10^8)$ s.
[b]The kilogram (kg) is still defined as the mass of a metal cylinder at the International Bureau of Weights and Measures, Sèvres, France.
[c]This is the distance travelled by light in a vacuum in one year (of 365.2425 days).
[d]The mean solar day is presently 86 400.002 s.

2 The Origin of the Solar System

In Chapter 1 you met the broad features of the Solar System. It is these broad features that any theory of the origin must explain, and this chapter presents the type of theory that is very widely accepted. This is the solar nebular theory, in which the planets form from a disc of gas and dust around the Sun. Such a type of theory also accounts for many of the *details* of the Solar System, as you will see in subsequent chapters.

You might think that we could *deduce* the origin of the Solar System by working back from the state in which we observe the Solar System to be today. This cannot be done, for several reasons. First, our knowledge of the present state of the Solar System is incomplete. Second, there are areas of ignorance about the way the Solar System has interacted with its interstellar environment. Third, our understanding of the fundamental physical and chemical processes that operate on all matter, though extensive and deep, is incomplete. Fourth, and most profoundly, even if these three areas of ignorance were eliminated, it would still not be possible to 'reverse time' and deduce the origin. This is because an infinitesimal adjustment in the present state of the Solar System would lead to a very different result; it is not possible to have sufficiently accurate knowledge to deduce the origin. This is an example of the scientific phenomenon of chaos, and it is a barrier in principle, not just a barrier in practice.

Astronomers must therefore construct theories as best they can, guided by the broad features of the Solar System and by the sparse but growing observations of other planetary systems. Observations of star formation and of young stars are also important, because these increase our understanding of the formation of the Sun, an event that was surely intimately involved in the formation of the rest of the Solar System.

2.1 The Observational Basis

2.1.1 The Solar System

Table 2.1 lists some of the broad features of the Solar System, most of which you met in Chapter 1. Any theory worthy of serious consideration really has to be able to account for most of these features, and for some others too. But it doesn't necessarily

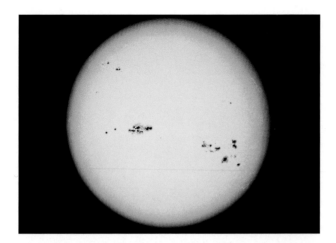

Plate 1 The Sun's photosphere, showing sunspots. (Reproduced by permission of Akira Fujii)

Plate 2 The corona of the Sun, imaged during the total solar eclipse of 11 July 1991. The bright sparkle is a tiny chink of photosphere. (Reproduced by permission of Akira Fujii)

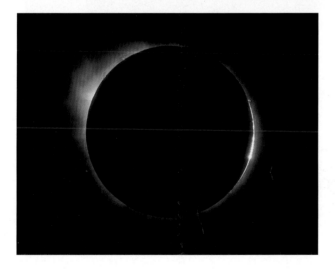

Plate 3 The inner corona and chromosphere of the Sun, imaged during the total solar eclipse of 11 June 1983. Also visible are prominences (red), huge, transient clouds of gas arching above the chromosphere. (Reproduced by permission of Akira Fujii)

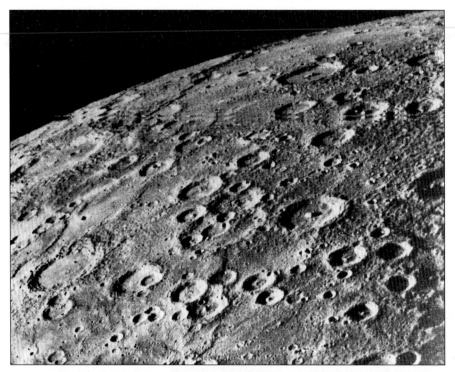

Plate 4 Mercury—part of the surface imaged by the flyby spacecraft Mariner 10 in 1974. (NASA/NSSDC P14679)

Plate 5 The Venusian volcano Maat Mons (on the horizon), altitude 6 km, though the vertical scale is exaggerated 10 times. The sky should not be black but totally overcast by a high cloud layer. This is a radar image by the Magellan Orbiter. (NASA/NSSDC P40175)

Plate 6 Earth, showing oceans, continents, polar caps, and clouds. This view was obtained by the Apollo 17 astronauts in 1972. (NASA/NSSDC AS17-148-22727)

Plate 7 The Moon—part of the surface that faces the Earth. (Reproduced by permission of Akira Fujii)

Plate 8   Mars, showing light and dark features, and the north polar cap. This image was obtained by the Hubble Space Telescope. (AURA/STScI PRC97-09a)

Plate 9   Mars: a view from the surface, obtained by Mars Pathfinder. The two peaks on the horizon—Twin Peaks—are 30–35 m tall and about 1 km to the west. (NASA/JPL P48984)

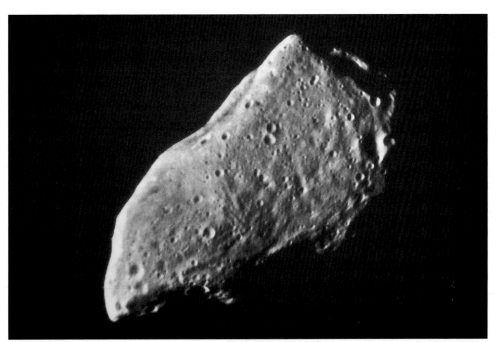

Plate 10 The asteroid Gaspra, imaged by the Galileo spacecraft *en route* to Jupiter. It is about 10 km across. (NASA/NSSDC P40449)

Plate 11 Jupiter—a general view showing the richly structured clouds, including the Great Red Spot, one of the largest and most persistent atmospheric features. This image was obtained by the Hubble Space Telescope. (AURA/STScI WFPC91-13)

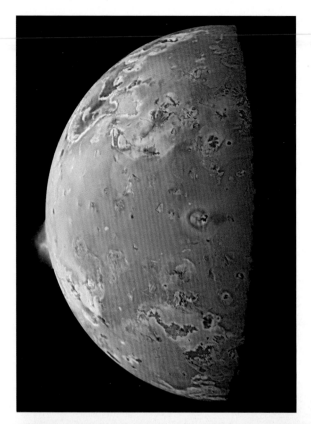

Plate 12 Io, imaged by the Galileo Orbiter. (NASA/JPL MRPS85377)

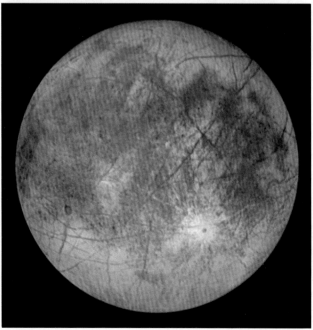

Plate 13 Europa, imaged by the Galileo Orbiter. (NASA/JPL P48040)

Plate 14 Ganymede, imaged by the Galileo Orbiter. (NASA/JPL P47970)

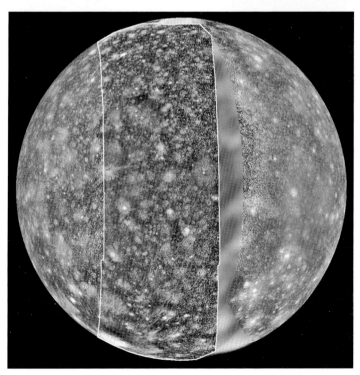

Plate 15 Callisto, imaged by Voyager 1 (left), Galileo Orbiter (centre), Voyager 2 (right). (NASA/JPL MRPS77654)

Plate 16 Saturn — a general view showing the richly structured clouds (and the satellites Tethys, Dione, and Rhea). This image was obtained by one of the two Voyagers (it is not known which one). (NASA/NSSDC P23887)

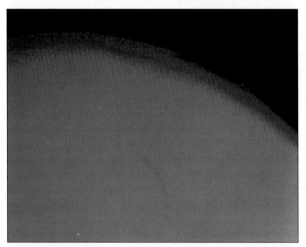

Plate 17 Titan, Saturn's largest satellite, imaged by Voyager 1. (NASA/NSSDC P23108)

Plate 18 The rings of Saturn, imaged by Voyager 1. This is a false colour image. (NASA/NSSDC P23953)

Plate 19 Uranus, imaged by Voyager 2. (NASA/NSSDC P29478)

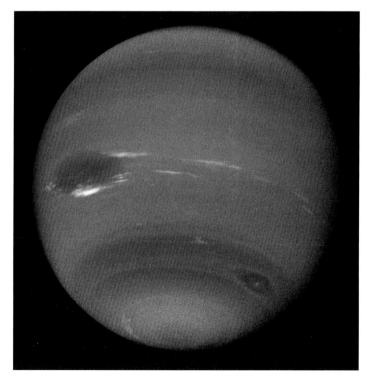

Plate 20 Neptune, imaged by Voyager 2. (NASA/JPL P34611)

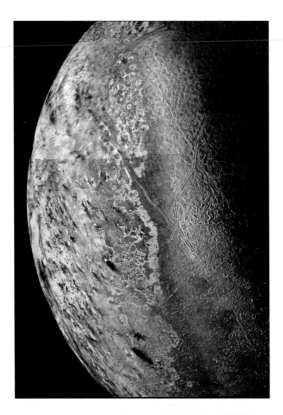

Plate 21 Triton, imaged by Voyager 2. (NASA/NSSDC P34764)

Plate 22 Comet Hale–Bopp, which passed through the inner Solar System in 1997. (Reproduced by permission of Akira Fujii)

Plate 23 The star cluster NGC 2024, about 5 light years across, and so young that the dense cloud that gave it birth is still evident. This is a near-infrared image, in which the longer wavelengths are red, the shorter, blue. Dense clouds are less opaque at infrared than at visible wavelengths. (Reproduced by permission of I. S. McLean)

Plate 24 The dust component in a disc around the star Beta Pictoris. The disc is presented nearly edge-on, and is nearly 2000 AU across in this red light image. The star image is blocked in the telescope. (Reproduced by permission of P. Kalas and D. Jewitt)

Plate 26 An aurora borealis over northern Sweden. (Reproduced by permission of D. H. Rooks)

Plate 25 Various meteorites. The sizes in (a)-(c) are the longest dimensions.
(a) Fragments showing fusion crusts (20 mm). (The author)

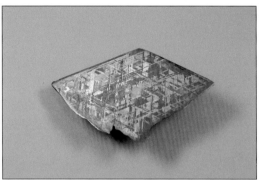

Plate 25 (b) An iron, showing the Widmanstätten pattern (115 mm). (The author)

Plate 25 (c) A cut-open stony-iron (123 mm). (The author)

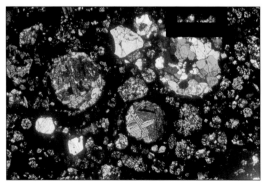

Plate 25 (d) A thin section of a carbonaceous chondrite. The largest circular features (chondrules) are about 2 mm diameter. (Reproduced by permission of Arabelle Sexton and the OUPSRI)

Table 2.1 Some broad features of the Solar System today

1	The Sun consists almost entirely of hydrogen and helium.
2	The orbits of the planets lie in almost the same plane, and the Sun lies near the centre of the orbits.
3	The planets all move around the Sun in the same direction that the Sun rotates (called the prograde direction).
4	The rotation axis of the Sun has a small but significant inclination, 7.2° with respect to the ecliptic plane (the Earth's orbital plane).
5	Whereas the Sun has 99.8% of the mass of the Solar System it has only about 0.5% of its total angular momentum.
6	The axial rotations of six of the major planets are prograde with small or modest axial inclinations. The rotations of Venus, Uranus, and Pluto are retrograde.
7	The inner planets are of low mass and consist of rocky materials, including iron or iron-rich compounds, the closer to the Sun the more refractory the composition.
8	The giants lie beyond the inner planets, are of high mass, and are dominated by hydrogen, helium, and icy materials, with a decreasing mass from Jupiter to Saturn to Uranus/Neptune.
9	The asteroids are numerous small rocky bodies concentrated between Mars and Jupiter.
10	The comets are even more numerous small icy bodies concentrated beyond Neptune in two populations, the Edgeworth–Kuiper belt and the Oort cloud.
11	The giant planets have large families of satellites, that are rocky or icy–rocky bodies.

have to be able to account for them all. If there are any features that a theory can't account for this would not necessarily rule the theory out. For example, it might be that the theory has not yet been worked out in sufficient detail, perhaps because a physical process is insufficiently well understood, or because we don't know enough about the state of the substances from which the Solar System formed. It is however, fatal for a theory if it unavoidably produces features that are clearly unlike those observed. For example, if a theory predicts that roughly half the planets should be in retrograde orbits then we can rule the theory out.

❒ What about a theory that predicts that there are no giant planets?

We can rule this out too!

2.1.2 Other Planetary Systems

Though our knowledge of other planetary systems is sparse, it is beginning to give us some guidance about the origin of our own system. Direct detection of planets is beyond present instrumental capabilities, because a planet is a very faint object with a very small angular separation from a far brighter object — its star. Therefore, up to now the successful techniques have been indirect, in that astronomers have detected planets through their influence on the star they orbit, or on any disc of gas and dust around the star.

In our Solar System the planets cause the Sun to follow a small (complicated) orbit around the centre of mass of the system (Section 1.4.5). Therefore, if small orbital motion of other stars can be detected we can infer the presence of one or more planets even if they are too faint to be seen. One way is to repeatedly measure the position of the star with respect to much more distant stars. This is called the

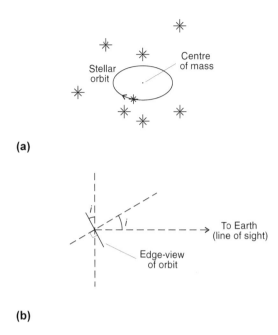

(a)

(b)

Fig. 2.1 (a) The orbit of a star due to a single planet in orbit around it, in the simple case when the centre of mass of the system is stationary with respect to the distant stars. (b) The angle of inclination i of the normal to the plane of the stellar orbit with respect to our line of sight

astrometric technique. An outcome is shown in Figure 2.1(a), where, for simplicity, it has been assumed that the centre of mass is fixed against the more distant stellar background. In reality, the motion of the centre of mass would add to that in Figure 2.1(a) to give a wavy stellar path. A second technique is possible if the angle i in Figure 2.1(b) is greater than zero. In this case the orbit is *not* presented face-on to us, and therefore as the star moves around its orbit its speed along the direction to the Earth varies, i.e. its line-of-sight speed varies. These speed variations cause variations in the wavelengths of the spectral lines of the star. This is due to the **Doppler effect**, whereby the observed wavelength depends on the speed of the radiation source with respect to the observer. The technique based on this effect is called the radial velocity technique.

From either technique we can obtain the mass of the planet(s) and some of the orbital elements. The details will not concern us except to note that whereas the astrometric technique gives the mass of the planet m_p, the radial velocity technique gives $m_p \times \sin(i)$. Thus, if i is unknown we obtain only a lower limit on the mass of the planet corresponding to $i = 90°$, as if we had an edge-on view of the orbit.

The first exoplanets (as planets in orbit around other stars are called) were discovered in 1992 in orbit around a pulsar. A pulsar is the remnant of a star that has suffered a catastrophic explosion — a supernova explosion. Such an explosion would surely have destroyed any planetary system, and so the planets are presumed to have formed subsequently. But interesting though these pulsar planets are, pulsars are rare objects, quite unlike the Sun. It was in 1995 that the first exoplanet was detected in orbit around a star other than a pulsar—the star was 51 Pegasi. Table 2.2 summarizes the first nine (non-pulsar) exoplanets to be discovered, with the Solar System included for comparison. The estimates of planetary masses are lower limits, because these discoveries have been made by the radial velocity technique. The masses are in units of Jupiter's mass M_J, and so in all cases the planets are giants. Note that the 'planet' of HD114762 has such a large lower limit to its mass that it might not be a planet at all, but an object called a brown dwarf, and the likelihood of this would be greater if its actual mass were considerably more than 11.6 M_J. A brown dwarf is a feeble star, and forms rather like the Sun did (Section 2.1.3), and not like the planets (Section 2.2).

Table 2.2 shows that in all but two cases the giants are very close to the star, a striking difference from the Solar System. Some of these giants are also in very eccentric orbits. However, we cannot conclude that it is the *Solar System* that is unusual. This is because the easiest planets to detect with the radial velocity technique are those that induce the greatest orbital speed of the star, and these are massive planets close to the star—an example of an observational selection effect. It might be that planetary systems like ours, with giants far from the star, are more common than Table 2.2 indicates. Indeed, recent discoveries have redressed the balance somewhat. However, astronomers do need to explain how giants can be in such very different orbits from those of the giant planets in our system.

All the stars in Table 2.2 are nearby (1 light year = 63 240 AU), and because most searches have been directed towards solar-type stars, none is very different from the Sun. The number of stars in Table 2.2 is several per cent of those so far examined, and this proportion is a minimum because further observations could reveal low-mass planets, or planets in large orbits. Therefore, we reach the important conclusion that planetary systems are fairly common, at least around solar-type stars.

Table 2.2 The first nine confirmed exoplanets

Stellar characteristics[a]			Planetary characteristics[b]			
Name	Mass/$M_\odot$	Distance/ light years	Orbital period/ years	Semi-major axis/AU	Eccentri-city	Mass/M_J
51 Pegasi	1.0	50.2	0.01158	0.051	0.01	>0.47
Rho[1] 55 Cancri	0.85	43.7	0.04013	0.11	0.05	>0.84
Upsilon Andromedae	1.25	53.8	0.01265	0.053	0.03	>0.63
Tau Bootis	1.25	49	0.009069	0.042	0.0	>3.64
Rho Corona Borealis	1.0	54.4	0.109	0.23	0.028	>1.1
70 Virginis	0.95	59.0	0.3195	0.47	0.40	>6.84
16 Cygni B	1.0	72	2.198	1.70	0.68	>1.74
HD114762	1.15	91	0.2301	0.36	0.34	>11.6
47 Ursae Majoris	1.1	46.0	2.992	2.08	0.09	>2.42
Sun	1	—	11.86 etc.	5.203 etc.	0.048 etc.	1 (Jupiter) etc.

[a]All these stars are solar-type main sequence stars. Planets around pulsars are excluded. Note that the number of stars known to have planets is growing rapidly.
[b]The top group are in small, low-eccentricity orbits, the middle group are in small, high-eccentricity orbits, the bottom group are in large, low-eccentricity orbits. Note that the 'planet' of HD114762 might be a brown dwarf.

Question 2.1

Discuss why, in the astrometric technique

(a) Planets with large mass will be easier to detect than planets with small mass
(b) It will be easier to detect planets around nearby stars than around distant ones.

2.1.3 Star Formation

Observations of star formation provide further insight into the origin of the Solar System. Star formation is a relatively rapid process by astronomical standards, but it still takes many millions of years, and therefore the process has been pieced together by observing it at different stages in different locations, linking the observations together by physical theory. This is rather like observing a large number of people at a particular moment — they are seen at all stages of their lives, and it is therefore possible to use general biological principles to construct a theory of the complete human lifecycle from an observation that occupied only a small fraction of the human lifespan.

From dense clouds to cloud fragments

Stars form from the interstellar medium (ISM) — the thin gas with a trace of dust that pervades interstellar space. Its chemical composition everywhere is dominated by hydrogen and helium, with hydrogen typically accounting for about 71% of the mass, helium for about 27% and all the other chemical elements (the 'heavy' elements) for only about 2%. Almost all of the hydrogen and helium is in the form of gases, but a significant fraction of most of the heavy elements is condensed in the dust in a variety of compounds. Dust accounts for roughly 1% of the mass of the ISM.

The density and temperature of the ISM vary considerably from place to place. Star formation occurs in the cooler, denser parts of the ISM, because low temperatures and high densities each favour the gravitational contraction that must occur to produce a star from diffuse material. Low temperatures favour contraction because the random thermal motions of the gas that promotes spreading is then comparatively weak. High densities favour contraction because the gravitational attraction between the particles is then relatively strong. The cooler, denser parts of the ISM are called, unsurprisingly, dense clouds. They are often components of giant molecular clouds, 'molecular' because the predominant form of hydrogen throughout them is the molecular form, H_2.

Dense cloud temperatures are of order 10 K. They must not, however, be thought of as chunky things — a typical density at the high end of a wide range is only of order $10^{-14}\,\text{kg}\,\text{m}^{-3}$, rather less than the density in a typical laboratory vacuum! A typical size is, however, a few light years across, and therefore most dense clouds are massive enough to form many hundreds of stars. They are also large enough for the dust content to make them opaque at visible wavelengths.

Though the conditions for gravitational contraction are best met in dense clouds, it is likely that in most cases they will contract only if they are subject to some external compression, particularly because magnetic fields and gas flows within the cloud hinder contraction. Compression can occur in one or more of a variety of ways, such as in a collision between two clouds, or by the impact of a shock wave from an exploding star, or by the action of a so-called spiral density wave that sweeps through the whole Galaxy (thereby sustaining its spiral arms). One way or another, a dense cloud, or a good part of it, becomes dense enough to become gravitationally unstable, and it starts to contract. As it contracts it becomes denser, to the point where the denser parts of the cloud, called dense cores, each contract independently. Many of these fragments are each destined to give birth to a star. This leads us to expect stars to form in clusters, and indeed the great majority of young stars *are* found in clusters (though a few form in isolation, from *small* dense clouds). Typically, a cluster contains a few hundred stars, and Plate 23 shows an example. Star clusters gradually disperse, and therefore older stars, like the Sun, are no longer in clusters.

From a cloud fragment to a star

Let's now follow the fate of a typical cloud fragment as it contracts. The gas molecules and dust particles gain speed as they fall inwards, and when they collide there is an increase in the random element of their motion. Temperature is a measure of the random motion of an assemblage of microscopic particles, and therefore the temperature rises. However, the rise is initially small because when the gas molecules collide they are raised into higher energy states of vibration or rotation. When the molecules return to lower energy states they get rid of their excess energy by emitting photons, usually at infrared wavelengths. Initially the density of the cloud fragment is so low that most of these photons escape. This loss of energy by the fragment retards the temperature rise.

The fragment continues to contract, and its density rises further. Detailed calculations show that the central regions of the fragment contract the most rapidly. It is therefore these central regions that become opaque to the photons emitted by the molecules. The temperature rise is then rapid and if the central object is more than a few percent of a solar mass it becomes a protostar. Contraction continues, now more slowly, and a few million years after the fragment separated from the dense cloud the temperature in the core of the protostar has become high enough for nuclear fusion to occur—about 10^7 K. This fusion releases energy and creates a pressure gradient that halts the contraction of the protostar. At this point the protostar has become a star—a compact body sustained by nuclear fusion.

The fusion that dominates the nuclear reactions in the core of the star depends on its composition.

❐ What element accounts for most of the mass of the star?

As in the dense cloud, typically 71% of the mass is hydrogen. It is also the case that nuclear fusion involving hydrogen occurs at a lower temperature than fusion involving helium and the other elements. Therefore it is the fusion of hydrogen nuclei that is by far the dominant source of energy. This fusion results in the creation of nuclei of helium (Section 1.1.3).

The hydrogen fusion phase of a star's lifetime is called the main sequence phase, and the star itself is called a **main sequence star**. It lasts longer the less massive the star, and for a star of solar mass it lasts about 10^{10} years. The Sun itself is nearly half-way through its main sequence phase. In all stars it is a period of relative stability, but it is immediately preceded by a well-observed period of instability that is of considerable importance to the formation of any planetary system. This is the **T Tauri phase**, named after the protostar that was the first to be observed in this phase, and for a protostar of solar mass it is thought to last for the order of 10^7 years ($= 10$ Ma). It is marked by a considerable outflow of gas, called a T Tauri wind, a protostar of solar mass losing the order 10% of its mass in this way, and by a high level of ultraviolet (UV) radiation from the protostar. The root causes of T Tauri activity are the final stages of infall of matter to the protostar, plus its strong interior convection and rapid rotation.

After the onset of hydrogen fusion the T Tauri activity quickly subsides, the UV radiation falling to a much lower level, and the wind declining to a much smaller rate of mass loss, called a stellar wind in general, and the solar wind in the case of the Sun (Section 1.1.2).

2.1.4 Circumstellar Discs

Meanwhile, that part of the fragment that has remained outside the protostar has also been evolving. As it contracts, it becomes denser towards its central regions. But the fragment is rotating, and so it is to be expected that only the material on or near the rotation axis falls fairly freely towards the centre; the infall of the remainder is moderated by its rotation. A circumstellar disc should thus form in the plane perpendicular to the axis of rotation. We have noted that such discs are thought to be the birthplace of a planetary system.

In recent decades circumstellar discs have been detected around many protostars. The discs have masses ranging from a few times the mass of Jupiter to hundreds of times Jupiter's mass. The gas component in the discs is readily imaged through its emission at radio and millimetre wavelengths. Discs have also been detected around over 100 young main sequence stars through the infrared emission from the dust in the discs. By this stage, the disc masses are considerably less than around protostars. There is a growing number of images of these dust discs, some utilising the dust emission at infrared and submillimetre wavelengths, others utilising infrared and visible wavelengths with the disc in silhouette against a bright background. In Plate 24 the dust component in the disc around the young main sequence star Beta Pictoris is imaged through the light from the star that the dust scatters. There is good evidence that the dust in this disc is replenished by collisions of cometary bodies. A disc of dust around the star Rho[1] 55 Cancri, that has at least one planet (Table 2.2), is thought to be sustained in the same way. The Beta Pictoris disc does not extend inwards of about 20 AU from the star — could the hole have been hollowed out by the formation of planets? This possibility is supported by a warping of the inner disc that could be caused by a giant planet just within the hole, in an orbit inclined at about $3°$ to the plane of the disc. Other discs have similar holes.

Thus, around protostars we have discs of material that could form a planetary system, and around young main sequence stars we have discs that seem to indicate that planetary formation has actually occurred. These observations lend strong support to solar nebular theories, to which we now turn.

Question 2.2

Identify the feature of the Solar System in Table 2.1 that is already present in the existence of circumstellar discs around protostars.

2.2 Solar Nebular Theories

Over the centuries there have been several different types of theory of the origin of the Solar System, but in recent decades one type, with antecedents in the eighteenth century, has emerged as the firm favourite. This is the solar nebular theory. Theories of this type are characterised by the formation of the planetary system from a disc of gas and dust encircling the young Sun — the **solar nebula**. This is clearly in accord with the relatively recent observations of circumstellar discs around protostars and young stars. Overall, such theories fit the observational data better than any other type of theory, and there are certainly no observations that rule them out. Within the general type there have been many variants, though there has been some degree of convergence so that most variants now differ only in relatively minor details. I shall concentrate on the *typical* features of solar nebular theories, pointing out where the variants differ significantly.

We pick up the story at the point where the protoSun is surrounded by a disc of gas and dust of order 100 AU across — the solar nebula. This is shown edge-on in Figure 2.2(a). The plane of the disc coincides with the equatorial plane of the protoSun, and the disc and protoSun rotate in the same direction. The disc rotates differentially, the orbital period increasing with distance from the centre, as in the Solar System today. The elemental composition of the disc is much the same as the outer part of the present Sun — by mass, 71% hydrogen, 27% helium, 2% the rest. The gas in the disc is predominantly hydrogen and helium, and a significant fraction of the other elements is in the various compounds that constitute the dust. Clearly the formation of planets in this disc will lead to planets orbiting in the same plane, all in the same direction, the direction of solar rotation. These are features of the Solar System that any acceptable theory of its origin has to explain (Table 2.1).

The starting point is, however, ill defined in one important respect; we do not know the initial mass of the disc. In one subtype of solar nebula theories the mass of the disc is about 1% of the present solar mass $M_\odot$, just sufficient for the mass of the planetary system today plus sufficient hydrogen and helium to top up the composition of the disc to be the same as the protoSun. This is the minimum mass subtype. At the other extreme is the subtype in which the initial mass of the disc is comparable to $M_\odot$. Some indication of an appropriate choice of mass is obtained by considering the angular momentum in the Solar System.

2.2.1 Angular Momentum in the Solar System

The magnitude l of the **angular momentum** of a body with mass m moving at a speed v with respect to an axis, as in Figure 2.3, is given by

$$l = mvr \qquad (2.1)$$

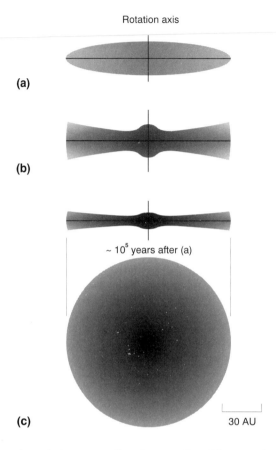

Fig. 2.2   The solar nebula surrounding the protoSun. The protoSun is too small to show on this scale

where *r* is the perpendicular distance from the path of the body to the axis. The angular momentum of *m* is with respect to this axis. In the solar nebula a natural choice of axis is the rotation axis in Figure 2.2 — through the centre of the protoSun and perpendicular to the plane of the disc. In the Solar System today the natural

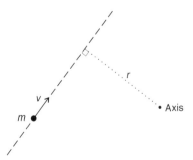

Fig. 2.3   A body of mass *m* moving at a speed *v* at a perpendicular distance *r* from an axis perpendicular to the page

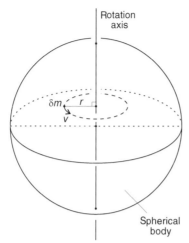

Fig. 2.4 The rotation of an element δm of a spherical body

choice is for the axis to go through the centre of mass of the Solar System and to be perpendicular to the ecliptic plane. It is the angular momenta with respect to these natural choices of axes that are of concern here.

Equation (2.1) applies to the mass m when its dimensions are small compared to r so that the whole of m can be regarded as being the same distance from the axis. This is closely approximated by a planet in orbit around the centre of mass of the Solar System, and the quantity is called the orbital angular momentum. If this condition is not met then the body is notionally subdivided into many small masses δm and the magnitude of its angular momentum is then a combination of the quantities $\delta m v r$. A simple case is when the angular momentum of a rotating planet or the Sun is calculated, as in Figure 2.4. The natural choice of axis is again the rotation axis, and because the paths of the δm around this axis are all in parallel planes, the combination is simply the sum of $\delta m v r$. The quantity in this case is called the rotational (or spin) angular momentum.

In the Solar System today about 85% of the angular momentum is in the orbital motion of Jupiter and Saturn, and only about 0.5% is in the rotation of the Sun. Less than 0.5% is in the orbital motion of the Sun around the centre of mass of the Solar System. This is in sharp contrast to the Sun having about 99.8% of the mass of the Solar System. Thus today, 'where the mass is, the angular momentum isn't'. The Sun's rotational angular momentum is small because it rotates slowly, about once every 26 days. Its orbital angular momentum is small for two reasons. First, the centre of mass of the Solar System is just outside the Sun, so r in equation (2.1) is small (call it $r_\odot$). Second, the orbital period $P_\odot$ for its small orbit is about 12 years, so its speed $v_\odot (= 2\pi r_\odot / P_\odot)$ is very small. For a planet, the average orbital angular momentum is well approximated by mva, where m is the mass of the planet, v is its average orbital speed, and a is the semimajor axis of the orbit. By combining Kepler's third law

$$P = ka^{3/2} \qquad (1.3)$$

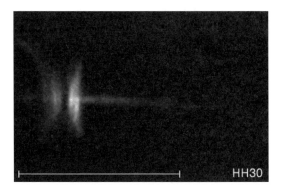

Fig. 2.5 Bipolar outflow from the protostar in an object called HH-30. A disc (edge-on) is also apparent. The scale bar is 1000 AU long. (AURA/STScI ID Team, ESA, C. Burrow)

with equation (2.1), and using $2\pi a/P$ for the average speed, where P is the planet's orbital period, we get

$$l_{\text{orb}} = \frac{2\pi}{k} m a^{1/2} \tag{2.2}$$

where l_{orb} is the average orbital angular momentum of the planet.

❐ So, why are the orbital angular momenta of Jupiter and Saturn large?

They have large orbits, and they are by far the most massive of the planets.

This distribution of angular momentum today is in sharp contrast with that calculated for a contracting cloud fragment. The protoSun rotates rapidly, and has a correspondingly large fraction of the total angular momentum. Therefore we need to explain how most of the angular momentum of the protoSun could have been lost. One of the more convincing explanations involves turbulence in the disc at the time it still blended with the outer protoSun. Turbulence is the random motion of parcels of gas and dust and is expected to have been a feature of the contracting nebula. (Note that though the parcels can be small, they are much larger than atomic scale — this is not the random thermal motion that occurs at the atomic level.) Turbulent motions are superimposed on the orderly swirl of circular orbital motion around the protoSun. Turbulence transfers parcels radially, and it can be shown that the *net* transfer of disc mass is outwards in the outer part of the disc and inwards in the inner part of the disc. The associated net transfer of angular momentum is from the protoSun to the disc, and it is carried by a small fraction of the disc mass.

Note that the transfer of disc mass leads to the loss of mass from the disc. In the outer disc this loss is to interstellar space. In the inner disc it is to the protoSun, and also to interstellar space via an outflow along the rotation axis of the disc, perhaps enhanced by mass loss from the polar regions of the protoSun. Such bipolar outflow is observed from protostars, as in Figure 2.5, where the disc is also apparent edge-on. Why these outflows are so tightly collimated is not well understood, but the magnetic field of the protostar or of the disc itself, acting on electrically charged particles in the flow, might be important. Bipolar outflow would have carried off only a small proportion of the rotational angular momentum of the protoSun. But this is not the whole outflow, particularly in the strong T Tauri phase (Section 2.1.3), when an additional proportion of the protoSun's angular momentum could have been carried off.

How does the distribution of angular momentum cast light on the initial mass of the disc? If the initial disc mass were only about 1% of $M_\odot$ the angular momentum transfer is weak, to the extent that it is difficult to explain the necessary loss of angular momentum by the protoSun. At the other extreme, an initial disc mass comparable to $M_\odot$ is considerably more than is necessary and requires a huge proportion of disc mass to be lost to space. Therefore, many astronomers favour an intermediate value, say 10% of $M_\odot$. The broad features of the following story are not sensitive to the exact mass at these intermediate values.

Question 2.3

(a) Derive equation (2.2).
(b) Use equation (2.1) to calculate the average orbital angular momentum of the Earth, and then use equation (2.2) to calculate the average orbital angular momenta of Jupiter and Neptune.

2.2.2 The Evaporation and Condensation of Dust in the Solar Nebula

In nebular theories the formation of planets and other bodies occurs in a number of stages, the first of which is the evaporation of some of the initial complement of the dust in the solar nebula.

Evaporation of dust

You have seen (Section 2.1.3) that the protoSun only begins to increase greatly in temperature when it becomes dense enough to be opaque to its own radiation. The disc never becomes as dense as the protoSun and therefore the tendency for its temperature to rise as it contracts is more strongly moderated. Nevertheless the disc does rise in temperature, particularly in its inner regions where it is denser and thus more opaque, and where infall to the protoSun causes greater frictional heating.

❐ What additional source of energy heats the inner disc more than the outer disc?

The protoSun when it becomes luminous is such an additional source.

In the inner disc, out to perhaps 1 AU, the calculated temperatures exceed about 2000 K, high enough to evaporate practically all of the dust in the disc — only those substances with very high sublimation temperatures escape evaporation. In **sublimation**, a substance goes directly from solid to gas, like carbon dioxide ice (dry ice) does at the Earth's surface. It occurs when the pressure is too low to sustain a liquid, and in the disc the pressures are well below this threshold. Above the sublimation temperature the substance is a gas, and below it, it is a solid. For any particular substance the sublimation temperature depends on the pressure, the higher the pressure, the higher the sublimation temperature. The value of the sublimation temperature for a particular substance at a given pressure is one measure of the

volatility of the substance. The most volatile substances, such as hydrogen (H_2) and helium (He) have *extremely low* sublimation temperatures, whereas substances such as corundum (Al_2O_3) have *extremely high* sublimation temperatures, and are said to be **refractory**. With increasing distance from the protoSun the disc temperature decreases, and therefore increasingly more volatile substances survive. Beyond the order of 10 AU even quite volatile substances escape evaporation, such as water ice.

Condensation of dust

So far the disc has evolved in completely the wrong direction to make planets — it has gained gas at the expense of solid material! However, at some point the contraction of the disc slows. Moreover the luminosity of the protoSun declines as it contracts, its surface area decreasing greatly while its surface temperature increases only slightly. (In contrast, the protosolar core temperature is increasing enormously, because of the lower rate of energy transfer across the outer layers of the protoSun.) Heat generation within the disc also declines and so the disc's temperature begins to fall because it continues to emit infrared radiation. At some point fresh dust begins to condense, its composition depending on the composition of the gas and on the local temperature. Because of the low pressures, solids rather than liquids appear.

Table 2.3 gives the condensation temperatures of representative substances that would have condensed. The pressure for the data is 100 Pa, 0.1% of the atmospheric pressure at the Earth's surface, and this is a theoretical value for the total gas pressure in the disc. The temperature at which a substance condenses will depend on this total pressure and also on the proportion of the disc accounted for by the substance. Therefore, the data in Table 4.3 are not definitive. Indeed, the pressure might have been rather lower than 100 Pa, though the condensation temperatures are only slightly lower even at 10 Pa.

The disc temperatures are generally lower, the further we are from the Sun. Therefore as the disc cools a substance condenses rather in the manner of a wave

Table 2.3 A condensation sequence of some substances at 100 Pa nebular pressure

Temperature/K	Substance	Chemical formula
1758	Corundum	Al_2O_3
1471	Iron–nickel	Fe plus $\sim$6% Ni by mass
1450	Diopside[a]	$CaMgSi_2O_6$
1444	Forsterite[b]	Mg_2SiO_4
<1000	Alkali feldspars	$(Na, K)AlSi_3O_8$
700	Troilite	FeS
550–330	Hydrated minerals[c]	$X(H_2O)_n$ or $X(OH)_n$
190	Water	H_2O
135	Hydrated ammonia	$NH_3.H_2O$
77	Hydrated methane	$CH_4.7H_2O$
70	Hydrated nitrogen	$N_2.6H_2O$
37	Methane	CH_4
$\sim$8	Hydrogen	H_2
$\sim$1	Helium	He

[a]A particular form of pyroxene, $(Ca, Fe, Mg)_2Si_2O_6$.
[b]A particular form of olivine, $(Mg, Fe)_2SiO_4$.
[c]X can be a molecule of a variety of rocky minerals, and *n* is greater than or equal to 1.

spreading inwards to some minimum distance within which the temperature is always too high.

In the innermost part of the disc the temperatures are probably always too high at the dust-condensation stage for anything much less refractory than iron–nickel to condense. At greater distances less refractory dust components appear, including an important range of substances exemplified in Table 2.3 by diopside, forsterite, and alkali feldspars. These are examples of silicates. A **silicate** is a chemical compound that has a basic unit consisting of atoms of one or more metallic elements and atoms of the abundant elements silicon and oxygen. For example, olivine has the chemical formula $(Fe, Mg)_2SiO_4$. Therefore, in the basic unit there is one atom of silicon (Si), four of oxygen (O), and two atoms of iron or magnesium — either two iron atoms, or two magnesium atoms, or one of each. A particle of dust consists of very many units, and so the proportion of iron to magnesium in the particle as a whole can vary almost continuously over the range 0–100%. The particular version of olivine in Table 2.3 (forsterite) has no iron at all. Particular versions are called **minerals**, naturally occurring substances with a basic unit that has a particular chemical composition and structure. Olivine is thus the name for a range of closely related minerals. Silicates have a far wider range of compositions.

Silicates are by far the most common refractory substances in the Solar System after iron–nickel, and they are common in rocks, a **rock** being an assemblage of one or more minerals, in solid form. Together with iron–nickel, silicates account for most of the common **rocky materials**. Note that this is the name of a group of refractory substances and not an implication that they are solid.

An extremely important boundary in the disc is the distance beyond which water condenses. This is important because there is a considerable mass of water in the disc, and where it condenses it becomes the dominant constituent of the dust grains. Water must have been abundant because oxygen, among the heavy elements, is particularly abundant (Table 1.5), and in a hydrogen-rich gas, at all but very high temperatures, most of the oxygen combines with hydrogen to form water. Figure 2.6 shows the column mass of the disc versus distance from the protoSun at a time well into the dust-condensation stage, when the disc probably resembled Figure 2.2(c). The **column mass** is the total mass in a cylinder of unit cross-sectional area running perpendicular to the plane of the disc. The increase in dust column mass at about 5 AU from the protoSun is due to the condensation of water beyond this distance. This distance is sometimes called the **ice line**. Note that the values in Figure 2.6 are illustrative, and not definitive.

Though water as H_2O condenses beyond about 5 AU, the dust closer in is not devoid of water; **hydrated minerals** (Table 2.3) can have higher condensation temperatures. These are substances that have one or more water molecules (H_2O) attached to their basic unit, or one or more hydroxyl molecules (OH) which are fragments of the water molecule. Water is one of a group of substances called **icy materials**. As in the case of rocky materials this is the name of a group of volatile substances with no implication that they are present as solids. The solid form is called an ice. Other important icy materials include ammonia (NH_3) and nitrogen (N_2), which are shown in Table 2.3 in their hydrated forms, methane (CH_4), shown in hydrated and non-hydrated form, carbon monoxide (CO) and carbon dioxide (CO_2).

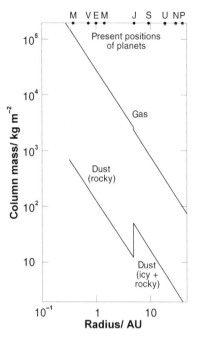

Fig. 2.6 The column masses of gas and dust in the solar nebula disc well into the stage of dust condensation

All of these are more volatile than water and so have minimum condensation distances that are greater than that of water.

Some evidence for extensive though incomplete evaporation of dust, followed by recondensation, is provided by meteorites, where a proportion of their refractory and not so refractory substances have isotope ratios that differ markedly from the Solar System average. This would be the outcome if these proportions had survived evaporation — material that has not been recondensed from a nebular gas retains an imprint of its origin beyond the Solar System.

Question 2.4

What indications are there already that we will get two zones in the Solar System, with terrestrial planets in the inner zone, and giant planets in the outer zone?

2.2.3 From Dust to Planetesimals

At the stage we have reached, the dust grains throughout the nebula are tiny, with sizes in the range 1–30 μm (1 μm $= 10^{-6}$ m), as observed in circumstellar discs (Section 2.1.4). The grains now grow slightly by acquiring atoms and molecules from the gas, rather as raindrops grow by acquisition of water vapour. This slow growth phase is accompanied by an increasing tendency for grains to settle to the mid-plane of the disc, a result of the net gravitational field and gas drag. This tendency

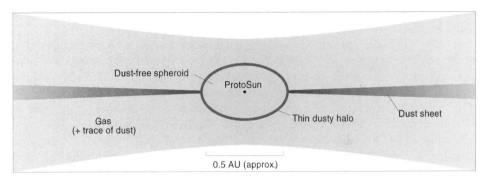

Fig. 2.7   Edge-view of the gas and dust around the protoSun after the dust has settled towards the mid-plane. The thickness of the dust sheet has been greatly exaggerated

increases as the turbulence in the disc dies away and as the grains grow in size. There is thus an increasing concentration of dust around the mid-plane of the nebula, forming a very thin sheet of order 10^4 km thick. The rest of the nebula is much thicker and much more massive, and consists of gas, mainly hydrogen (H_2) and helium, and of remnant dust. This is shown in Figure 2.7 for the inner part of the dust sheet, where the thickness of the dust sheet has been greatly exaggerated. Note that the sheet gets thicker with increasing distance from the Sun, i.e. with increasing heliocentric distance.

❐ So how (except for the step-up at the ice line) can the column mass of the dust decrease as in Figure 2.6?

As the heliocentric distance increases, the mass of dust per unit volume of the sheet must decrease. This is mainly because the dust grains are further apart, and not because they are smaller.

The concentration of the dust into a sheet leads to a greatly increased chance of a collision between two grains. Neighbouring grains tend to be in similar orbits and therefore a significant fraction of the collisions is at sufficiently low relative speeds for the grains to stick together in a process called coagulation. Coagulation is more likely when one or both grains have a fluffy structure, and it is aided when the two grains have opposite electric charges, or when they contain magnetised particles.

The outcome of coagulation is the gradual build-up of bodies of order 10 mm across. The time required for this to happen depends on the relative speeds of grains in slightly different orbits, the lower the relative speeds, the lower the collision rate and the slower the coagulation. The relative speeds are lower, the smaller the orbital speeds, and so there is a tendency for the coagulation time to increase with increasing heliocentric distance. But the coagulation time also depends on the average spacing of the grains, the greater the spacing, the slower the coagulation. This spacing increases as the column mass of the disc decreases, and so the coagulation time is further increased with increasing heliocentric distance, except at the ice line, where the step-up in column mass causes a significant reduction in the coagulation time in the Jupiter region. Broadly speaking, the coagulation times are of the order of thousands of years out to about 5 AU, increasing to many hundreds of thousands of years at 30 AU. The times for dust condensation and settling were shorter.

By the time bodies 10 mm or so in size are appearing at 30 AU, the bodies out to about 5 AU have grown to 0.1–10 km across. These are called **planetesimals** — 'little planets', rocky within the ice line, icy–rocky mixtures beyond it. There are at least two means of producing planetesimals, both of which might have been significant. The first is a continuation of coagulation, promoted by the continuing thinning of the dust sheet with the corresponding increase in its density. The second is a different consequence of this density increase. At a sheet thickness of order 100 km it is possible that the gravitational attraction between the bodies constituting the sheet leads to gravitational instability, the sheet breaking up into numerous fragments, each fragment forming a planetesimal.

Though the formation of planetesimals is a considerable step towards bodies of planetary size, there is clearly some way still to go. The theory of the remaining stages of planetary formation indicates that the process was rather different in the inner Solar System — within the ice line — than in the outer Solar System.

Question 2.5

(a) Use Figure 2.6 to estimate the total mass in the planetesimals between 0.8 AU and 1.2 AU from the protoSun. Compare your result with the mass of the Earth, and comment on the significance of the comparison.

(b) If the mean density of a planetesimal around 1 AU from the protoSun is 2500 kg m^{-3}, calculate the number of planetesimals corresponding to the mass you calculated in part (a).

In both parts, state any assumptions that you make.

2.2.4 From Planetesimals to Planets in the Inner Solar System

A planetesimal about 10 km across has sufficient mass for it to exert a significant gravitational attraction on neighbouring planetesimals. This increases the collision rate between planetesimals, and models show the net effect to be growth of the larger planetesimals at the expense of the smaller ones. An essential condition for net growth is that the collisions are at low speed, thus requiring neighbouring planetesimals to be in low-eccentricity, low-inclination orbits. Such orbits could have been promoted by nebular gas drag. The acquisition by a larger body of smaller bodies is called **accretion**.

As a planetesimal gets more massive its accretional power increases, and consequently there is a strong tendency for a dominant planetesimal to emerge that ultimately accretes most of the mass in its neighbourhood. (This is an example of the Matthew effect, 'For unto every one that hath shall be given, and he shall have abundance: but from him that hath not, shall be taken away even that which he hath', *The Gospel according to St Matthew XXV, 29*. The Matthew effect is also familiar to players of Monopoly.) The outcome is runaway growth, in which the population of planetesimals in a neighbourhood evolves to yield a single massive planetesimal called an **embryo**, that accounts for over 90% of the original planetesimal mass in the neighbourhood, plus a swarm of far less massive planetesimals, the largest being perhaps a million times less massive than the embryo. The neighbourhood of an

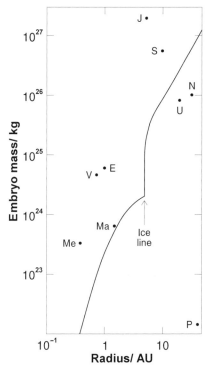

Fig. 2.8 Embryo mass versus heliocentric distance as calculated in one model. The masses of the planets are also shown, in their present positions

embryo is an annular strip covering a small range of heliocentric distances, and so we get a set of embryos each at a different heliocentric distance.

Figure 2.8 shows the embryo mass versus distance calculated from one model, the results from which must be regarded as illustrative and not definitive — for example, the step at the ice line is smaller in some other models. In this model the orbits of the embryos within the ice line are spaced by about 0.02 AU, and the time taken for the full development of a single embryo from a swarm of planetesimals is about 0.05 Ma at 1 AU, and increases with heliocentric distance. These times are *very* uncertain, though a general increase in time with increasing heliocentric distance emerges in all models, largely because of the decrease in column mass. Another common feature is an increase in embryo spacing with increasing solar distance.

❐ In this model, how many embryos are there between 0.3 AU and 5 AU?

There are about $(5 - 0.3)/0.02$ embryos in this region, i.e. 200 or so. This is the region presently occupied by the terrestrial planets and the asteroids. From Figure 2.8 and the 0.02 AU spacing, their total mass can be estimated to be of order 10 times the mass of the Earth.

Figure 2.8 shows that (except perhaps for Mars) we have to put embryos together to form the terrestrial planets. However, assembly is a slow process because of the small number of embryos and their consequent large spacings. We have to rely on

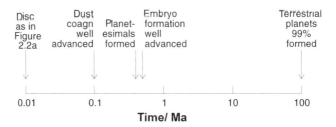

Fig. 2.9 Time line of the formation of a terrestrial planet. The relative durations of each stage are more reliable than the absolute durations, which vary considerably from model to model

modest orbital eccentricities to produce collisions. Such collisions would produce fragmentation, but the gravitational field would be sufficient to assemble most of the fragments into a body with nearly the combined mass of the two colliding embryos. At some intermediate stage there could have been a few dozen Mars-sized embryos, and a host of less massive bodies. Collisions would usually have been off-centre, and so even if many of the embryos initially had small inclinations of their rotation axes, larger inclinations could readily be imparted to some planets through the arrival of large embryos. This is in accord with point 6 in Table 2.1.

The time occupied by the transition from embryos to a terrestrial planet increases with increasing heliocentric distance. It is estimated that for the Earth the time was of order 100 Ma. This is by far the slowest stage in the formation of the terrestrial planets, though it is short compared to the 4500 Ma or so that have elapsed since. Figure 2.9 is a time-line summary of the formation of a terrestrial planet. The relative durations of each stage are more reliable than the absolute durations, which vary considerably from model to model. Note the logarithmic scale.

After the last embryo collision we are left with planetesimals bombarding an essentially complete planet. There is widespread evidence that the terrestrial planets and the Moon suffered such a **heavy bombardment**, and that it tailed off about 3900 Ma ago, to be followed by a light bombardment by small bodies that persists to the present day.

If the terrestrial planets did indeed form as proposed here, then they should consist largely of substances more refractory than the hydrated minerals in Table 2.3, and the closer to the Sun, the more refractory the composition should be. Note, however, that the compositional differences between the planets will have been moderated by embryos and planetesimals arriving from different regions of formation. This is in accord with what we know about terrestrial planet composition. Nebular gas would also have been captured, and though only to the extent of a tiny fraction of the planetary mass it would have provided the planets with atmospheres rich in hydrogen and helium. In the theory such atmospheres are removed by the T Tauri activity of the protoSun.

The T Tauri activity also removes the remnant nebular gas that pervades interplanetary space. The drag of this gas has been causing a slow inward spiralling of the terrestrial planets, and so this is halted. Conversely, the mass lost by the protoSun in its T Tauri phase reduces its gravitational field, resulting in some orbital enlargement. Subsequently the semimajor axes of the orbits generally change very little, though changes in the other orbital elements are greater (Section 1.4.5).

The effects of Jupiter

Figure 2.8 shows that around 3 AU from the protoSun, there are expected to have been several embryos with masses of the order of 10^{24} kg. Today this region is occupied by the asteroids, with a total mass only of order 10^{22} kg, the most massive being Ceres at 1.1×10^{21} kg. The answer to this seeming contradiction is the effect of Jupiter, the most massive planet, which is in orbit just beyond this region. In the theory, as Jupiter grows, its gravitational field 'stirs' the orbits of the planetesimals and embryos, producing a range of eccentricities, inclinations, and semimajor axes, so that most collisions occur between two bodies occupying substantially different orbits, with huge relative speed, as in Figure 2.10. The result is fragmentation and dispersal of the fragments, rather than accretion. Some of the dispersed fragments are flung into huge orbits, some are lost from the Solar System, and some are captured by the Sun and planets. Only a small fraction remains around 3 AU. This population has continued to evolve, and we see the survivors today as the asteroids, with perhaps only 0.1% of the original mass in this region.

The growth of Mars must also have been stunted by the stirring of the planetesimal and embryo orbits by Jupiter. Nearer the Sun, the effect of Jupiter might have been to speed up the final stages of growth of the Mercury, Venus, and the Earth, partly through the provision of material from outside the terrestrial zone, and partly by increasing the eccentricity of the embryo orbits, thus increasing their collision rate without producing the huge relative speeds of the asteroid region.

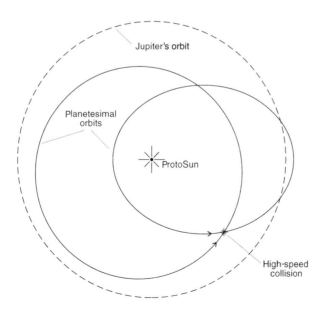

Fig. 2.10   High relative speeds when two bodies in substantially different orbits collide

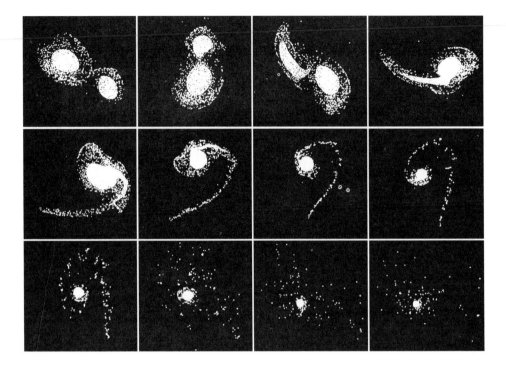

Fig. 2.11　One way in which the Moon could have formed from a grazing embryo impact on the Earth. Note the decreasing scale from frame to frame — the Earth (the larger object) is about the same size in all frames. (Reproduced by permission of A. G. W. Cameron)

Formation of the Moon

Of the terrestrial planets only the Earth and Mars have satellites. The two tiny satellites of Mars are probably captured asteroids. The Moon, at 1.2% of the mass of the Earth, is far too massive for capture to be likely. In recent years widespread support has grown for the view that the Moon is the result of an embryo at least twice the mass of Mars colliding with the nearly formed Earth at a grazing angle. All but the core of the embryo, and some of the outer part of the Earth, is scattered along an arc, predominantly as a gas produced by vaporisation during the impact. Much of this material returns to Earth, some escapes, but a small fraction goes into orbit around the Earth, from where, in only a year or so, it forms the Moon. You will see in later chapters that the detailed composition and structure of the Earth and the Moon provide a good deal of support for this theory. Recent simulations of the process are illustrated in Figure 2.11.

Question 2.6

List the features of the Solar System in Table 2.1 that apply to the terrestrial planets. For each feature in your list state whether it can be explained by solar nebular theories.

2.2.5 From Planetesimals to Planets in the Outer Solar System

Figure 2.8 shows a huge increase in the embryo mass beyond the ice line. You might think this is simply the result of the huge mass of condensed water, but comparison with Figure 2.6 shows that this is not the only important factor. In Figure 2.6 there is certainly an important step-up in the column mass of the solar nebula at about 5 AU, i.e. at the ice line, but there is also an underlying decrease with heliocentric distance. As a result, the column mass in the neighbourhood of Jupiter is less than in the neighbourhood of the terrestrial planets. Figure 2.8 therefore indicates that, in the models, as the heliocentric distance increases, planetary embryos not only become more massive, they also become fewer in number. In other words, the 'feeding zone' of each embryo covers a wider annular strip of the disc. Consequently, masses of order 10^{26} kg are typical in the models for the Jovian region, i.e. of order 10 times the Earth's final mass!

At greater heliocentric distances it also takes longer to form the embryos from planetesimals — about 0.5 Ma at 5 AU, and even longer further out. However, the time required is shorter for smaller planetesimals, and so if there is a trend whereby the greater the heliocentric distance, the smaller the planetesimals, then this would partly offset the increasing embryo formation times. In any case, many planetesimals are left over after the embryos have formed.

The embryos are so few and far between beyond the ice line that embryo collisions are unlikely, and so the slow embryo-to-final-planet phase that operates in the terrestrial region does not occur. Instead the embryos are massive enough to act as *kernels* that gravitationally capture large quantities of the considerable mass of gas that still dominates the solar nebula.

❐ What two substances make up nearly all of the mass of this gas?

Hydrogen (as H_2) and helium (as He) together account for about 98% of the mass of the nebula, and for nearly all of the gas component. At first, the rate of capture of gas by the kernels is low, and it is estimated that it takes several times the kernel-formation time for the capture of a mass of gas equal to the initial kernel mass. At this point, the capture rate is much higher and it is rising rapidly with further mass increase — there is a runaway.

As nebular gas is captured it undergoes self-compression, to yield an envelope with an average density that grows as its mass increases. As well as gas, the growing giants also capture a small but significant proportion of the surviving planetesimals, which still account for nearly 2% of the mass of the nebula. These icy–rocky bodies partially or wholly dissolve in the envelope, particularly in its later, denser stages. Icy materials dissolve more readily than rocky materials, so some preferential accretion of rocky materials onto the kernel might occur. On the other hand, convection in the envelope opposes core growth, so the further central concentration of icy–rocky materials might be slight. The (runaway) capture of gas is halted by the T Tauri phase of the protoSun, when the high radiation and particle fluxes sweep the remaining nebular gas into interstellar space.

We can thus account for the presence of giants in the outer Solar System. However, some critical timing is seen to be essential when we look at the *differences*

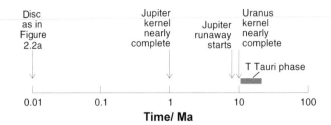

Fig. 2.12 A possible time line for the formation of the giant planets. The times are far more uncertain than the sequence of events

between the giants, notably the decrease in mass with increasing heliocentric distance (item 8 in Table 2.1). In the models, the key to understanding this trend is the increasing time it takes to reach the runaway stage with increasing heliocentric distance. If the T Tauri phase occurs *after* the onset of runaway at Jupiter and Saturn, but *before* it starts at Uranus and Neptune, then we can account qualitatively for the giants' different masses. Figure 2.12 is one possible time line for the formation of the giant planets. Again, this is illustrative, not definitive.

❐ What would have been the consequence of a much later T Tauri phase?

If the T Tauri phase had been much later, then all the giants would now be more massive than Jupiter.

After the T Tauri phase the giants must have captured further icy–rocky planetesimals. These will have added only very slightly to the total mass, but could have significantly enriched the envelopes in icy and rocky materials.

The axial inclinations of the giants are presumably the result of the off-centre accumulation of mass — the same sort of explanation that can account for the axial inclinations of the terrestrial planets. The inclination of Uranus is remarkable (Table 1.1), though this too can be explained by the accumulated effects of icy–rocky planetesimals that just happened to nudge the rotation axis predominantly in one direction — a catastrophe is not necessary.

A minority view

The scenario just outlined for the formation of the giant planets is not without problems, notably our poor understanding of the runaway growth of Jupiter and Saturn, and the possible lack of time for this to happen before the T Tauri phase of the protoSun dissipates the nebula. Consequently, some astronomers believe that each giant planet formed by the gravitational contraction of a fragment of the nebula, rather in the manner that the Sun formed at the centre of the nebula. In this single-stage process the fragment contracts to form a protoplanet, that further contracts to form the giant. This is different from the two stage process in which an icy–rocky embryo (kernel) forms first, and then captures nebular material.

In protoplanet theories there is a tendency for all four giants to acquire similar compositions. You will see in Chapter 5 that we have good reasons to believe that there are substantial differences among the giants' compositions. Also, it is difficult to provoke models to produce protoplanets rather than other types of structure. Therefore we have some reason to favour the embryo model over the protoplanet model, in spite of our incomplete understanding of the former.

Comets, and Pluto

The formation of giant embryos leaves behind large numbers of icy–rocky planetesimals, most of them less massive than the rocky planetesimals in the asteroid zone. The gravitational fields of the giant planets leads to the capture of some of these planetesimals by each of the planets and by the Sun, but these fields cause most of the planetesimals to be flung far outwards. Most of these escape from the Solar System, but a few per cent do not. The effects of Uranus and Neptune are less energetic than those of Jupiter and Saturn, and so the Uranus–Neptune region is a particularly copious source of icy–rocky planetesimals that are flung out but do not escape. Their perihelion distances are subsequently increased by the overall gravitational force of the stars and interstellar matter that constitute our Galaxy. The force has this effect because it varies across a planetesimal orbit, i.e. it is a differential force.

❐ What is a suitable name for this force?

A differential force is a tidal force (Section 1.4.5), and so a suitable name is the Galactic tidal force, though it is usually called the **Galactic tide**. The planetesimal orbits are subsequently randomised by this tidal force and also by passing stars and by giant molecular clouds. The Oort cloud of comets is thus emplaced (Section 1.2.3), perhaps on a time scale as short as 10^3 Ma. In Section 3.2.5 you will see how the Oort cloud can account for some of the comets observed today in the inner Solar System.

The increasing difficulty of forming embryos at increasing heliocentric distance makes it quite feasible that beyond Neptune the planetesimal population is relatively unevolved. This would be a second population of icy–rocky bodies, one that has remained more or less where it formed, and not as far out as the Oort cloud. This postulated population is called the Edgeworth–Kuiper (E–K) belt (Section 1.2.3), and about 60 objects that are thought to be its larger innermost members have now been detected. It is possible that Pluto and Charon are the largest members of the E–K belt. The E–K belt is a second source of observed comets, as will be described in Section 3.2.5.

Question 2.7

If the protoSun went through its T Tauri phase much *earlier* than in Figure 2.12, what might the planets in the outer Solar System be like today?

2.3 Formation of the Satellites and Rings of the Giant Planets

2.3.1 Formation of the Satellites of the Giant Planets

With the exception of Triton, all of the massive satellites of the giants, and nearly all of their less massive satellites, orbit the planet in the same direction as the planet rotates, and in a plane tilted at only a small angle with respect to the equatorial plane of the planet (Table 1.2). This orderly arrangement is strong evidence against separate formation and capture, and strong evidence for formation in a disc of dust and gas around each planet, called a protosatellite disc. In some ways this mimics the formation of the planets from the disc of gas and dust around the protoSun, but it differs in one important respect — most of the angular momentum in the giant planet system is in the rotation of the planet and not in the orbits of the satellites. Therefore, there is no need to transfer angular momentum away from the planet, in contrast to the protoSun.

In the models, the protosatellite disc is composed of material attracted to the growing giant, but that fails to be incorporated into it. The material forms a cloud of gas, dust, and planetesimals. Interactions within the cloud and between the cloud and the planet cause the cloud to evolve into a thin disc in the equatorial plane of the planet, and orbiting in the same direction as the planet is rotating. Though much of the icy–rocky material in the disc is lost to interplanetary space, coagulation and accretion occur, building up the satellites. Remnant gas is lost during the T Tauri phase of the protoSun.

The surface temperature of Jupiter reaches about 1000 K — the result of infall of material from the nebula to Jupiter's outer envelope. The luminosity is then high enough for a significant rise in the temperature of the inner part of the protosatellite disc, which is also heated directly by infall. Therefore, the satellites that form close to Jupiter are expected to be more depleted in icy materials than those that form further away. In particular they should be depleted in the proportion of *water* that they contain, the temperatures in the solar nebula at Jupiter always being too high for appreciable condensation of the more volatile substances.

❐ Given that water is less dense than rocky materials, why are the densities of the Galilean satellites (Table 1.2) in accord with this prediction of a decreasing proportion of water with decreasing distance from Jupiter?

Table 1.2 shows that the densities of the Galilean satellites increase with decreasing distance from Jupiter, which is consistent with a decrease in water content.

The known densities of the satellites of Saturn, Uranus and Neptune show no trend with distance from the giant, and none is high enough to suggest a lack of icy materials. This could be the result of the lower masses of these giants and the correspondingly reduced heating of the inner protosatellite disc. An additional reason is suggested by the relatively small sizes of most of their satellites — most are much smaller than the Galileans. Small satellites are prone to collisional fragmentation, particularly the inner ones where, in the theories, the space is crowded with planetesimals gravitationally attracted to the giant and accelerated to high speeds. Therefore, any newly formed ice-poor satellite of modest mass in the inner region is readily disrupted. Satellites subsequently forming in this region are built of substances from across a wide range of distances from the giant, and compositional differences are consequently diluted. Among the smaller satellites of the giants there are some that seem to be collisional fragments, and some of the medium-sized satellites display evidence that they were indeed once disrupted.

A few other small satellites are probably captured asteroids or icy–rocky planetesimals. Any satellite in a peculiar orbit might be a captured body, particularly if it is in a large orbit, capture being easier into large orbits than into small orbits — large orbits require the captured body to lose a smaller fraction of its orbital energy than do small orbits. A retrograde orbit (one in the opposite direction to the rotation of the planet) is also suggestive of capture.

Just one *large* satellite might have been captured. This is Triton, one of the largest satellites, and by far the largest satellite of Neptune. It is unique among the large satellites in that it orbits its planet in the retrograde direction (Table 1.2). Indeed, there are very few satellites of any size that orbit in this manner, and none of the other satellites of Neptune does so. This is strong evidence that Triton was indeed captured. Initially its orbit would have been eccentric and perhaps inclined at a large angle with respect to Neptune's equatorial plane. Once captured, its gravitational interaction with Neptune would have reduced the eccentricity and inclination.

The capture of Triton would have wreaked havoc with any emerging or fully formed satellite system. The orbit of another satellite of Neptune, Nereid, might bear witness to this. Its large eccentricity and large semimajor axis could be the result of the capture of Triton. Nereid's orbit would have remained peculiar because of its large average distance from Neptune. If Triton was captured, then its broad similarity in size and density to Pluto suggests that it might initially have been a large member of the E–K belt.

Question 2.8

Examine Table 1.2 and list the satellites of the giant planets, additional to Triton, that are likely candidates for a capture origin. Justify your choices.

2.3.2 Formation and Evolution of the Rings of the Giant Planets

All four giants have rings (Figure 2.13), and these are particularly extensive in the case of Saturn (Plate 18). For all giants the rings are close to the planet and lie in the equatorial plane. The rings consist of small bodies called ring particles. Very few are more than 1 m across, and the great majority are far smaller, down to less than 1 μm. The rings are very thin — even in the case of Saturn they are no more than about 100 m thick, and their total mass is only of order 10^{-5} times the mass of the Earth!

Rings and the Roche limit

An important concept relating to the origin of the rings is that of the Roche limit, named after the French scientist Edouard Albert Roche (1820–1883) who in 1847 derived the eponymous limit. It arises from the tidal force that one body exerts on

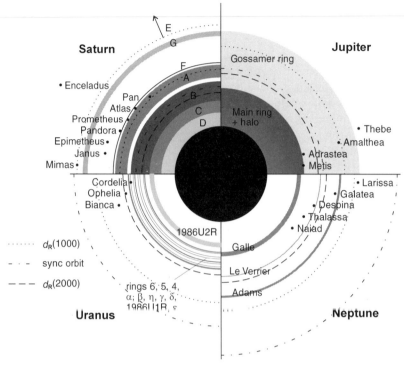

Fig. 2.13 Simplified diagram of the ring systems of the giant planets, scaled so that all four giants are the same radius. $d_R(1000)$ and $d_R(2000)$ are, respectively, the Roche limits of bodies with densities $1000\,\mathrm{kg\,m^{-3}}$ and $2000\,\mathrm{kg\,m^{-3}}$

another. Figure 2.14 shows the tidal distortion of a body of mass m at distance d from a body of mass M. If m is held together only by gravitational forces, and if both bodies are of uniform density, then it can be shown that m is torn apart if the distance d is less than d_R, where

$$d_R = 2.44 \times R_M \left(\frac{\rho_M}{\rho_m}\right)^{1/3} = 1.51 \times \left(\frac{M}{\rho_m}\right)^{1/3} \qquad (2.3)$$

R_M is the radius of M, and ρ_m and ρ_M are respectively the densities of m and M. The second form is obtained from the first form using $M = \rho_M \times (4\pi/3)R_M^3$. The

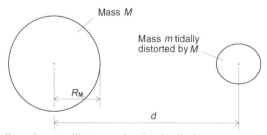

Fig. 2.14 Tidal distortion, to illustrate the Roche limit

quantity d_R is the **Roche limit**. As one might expect, d_R increases as M increases and as ρ_m decreases.

Equation (2.3) applies to an initially uniform spherical body held together only by the gravitational attraction of one part on another. Bodies are also held together by non-gravitational forces. These operate at short range, binding molecule to molecule. By contrast, gravity, being proportional to $1/r^2$ (equation (1.5), Section 1.4.4) is a long-range force, and it is therefore the dominant cohesive force in large bodies. Therefore, the Roche limit applies only to bodies that are sufficiently large for gravitational cohesion to dominate — this explains why astronauts and satellites in Earth orbit are not torn apart by the tidal force of the Earth. For non-porous solid bodies made of icy or rocky materials, gravitational cohesion dominates only when they are more than a few hundred kilometres in radius. For poorly consolidated bodies, such as comets and other loose aggregates, gravity dominates down to far smaller sizes.

Thus any sufficiently large body that strays within the Roche limit will be torn apart. The resulting fragments will be numerous and in similar orbits, and therefore collisions among them will be frequent, resulting in further fragmentation. Any tendency to reassemble is counteracted by tidal disruption. After hundreds of millions of years the material evolves to a population of bodies that are predominantly smaller than a metre. This process is a plausible source of ring particles. But the Roche limit also provides a second source — the disruption of bodies already within this limit that are growing by accretion.

❐ How can this happen?

If such a body grows larger than the size at which gravitational cohesion predominates, it is then torn apart by tidal forces. Both processes continue today.

The importance of the Roche limit is illustrated by Figure 2.13, which shows that very little ring material exists outside the limits. The small quantity that does is readily explained by the inward spiralling of dust produced outside the limit.

Ring particles and ring structures

Observations show that Jupiter's ring particles seem to be largely devoid of volatiles, whereas those of Saturn seem to be 'dirty snowballs' — mainly water ice mixed with a trace of less volatile substances, including rocky materials. This can be explained by the higher temperatures expected in the inner protosatellite disc of Jupiter than in Saturn's disc. This left Jupiter with ice-poor materials for its initial and subsequent populations of ring particles. The ring particles of Jupiter and Saturn also differ in size, with most of Saturn's being in the range 0.01–1 m, and most of Jupiter's being far smaller.

❐ How can the greater average size of Saturn's ring particles be explained?

This can be put down to the survival of water ice, which is an abundant substance.

Little is known about the composition of the ring particles of Uranus and Neptune. They are very dark and for an unknown reason seem to be less icy than

Saturn's particles. Their low reflectivity might be the result of solar wind action on **hydrocarbons** (compounds of carbon and hydrogen).

Different-sized ring particles are affected differently by a variety of processes acting on them. One of several gravitational processes arises from the slight departure from spherical symmetry of the giant planet's gravitational field. The outcome depends on whether the orbital period of the particle is greater or less than the giant planet's rotation period. If these two periods are equal then the particle (or any other orbiting body) is said to be in a **synchronous orbit** (Figure 2.13). In such an orbit there is zero effect. In a closer orbit the outcome is a slow spiralling towards the giant, whereas in a larger orbit the outcome is a slow spiralling *outwards*. This effect tends to clear the rings of bodies of all sizes, but the replenishment rate is higher for small particles, and so the net effect is a downward trend in the size distribution.

Another gravitational effect occurs in close encounters between particles in nearly identical orbits. After the encounter is over the inner particle is in an even smaller orbit, and the outer one is in an even larger one. This effect is greater, the larger the mass of the particles, and thus it also causes a downward trend in the size distribution of the ring particles. The observed scarcity of ring bodies larger than a metre or so can be explained by these two gravitational effects.

Two other effects are greater, the *smaller* the body. As a body is swept by solar radiation it encounters the photons rather in the manner that you encounter raindrops when you are running — the front of you collects more raindrops than your back. The effect of the extra photon bombardment on the leading face of a body is to decelerate it. This is the Poynting–Robertson effect, and for a ring particle the effect is to cause it to spiral towards the giant. The effect is greater, the smaller the particle because the magnitude F of the net force exerted by the bombardment is proportional to the surface area of the particle, whereas the magnitude of the deceleration (or acceleration) is given by F/m where m is the mass of the particle (equation (1.4), Section 1.4.4). The area, and hence F, is proportional to the square of the particle's mean radius r_m, and m is proportional to its cube, and so F/m is proportional to r_m^{-1}. The Poynting–Robertson effect explains the sparseness of ring particles within the inner edge of the rings — particles of sizes that typify the rings traverse this inner region rapidly in their downward spiral.

The second effect is really a group of effects involving electromagnetic forces. A proportion of ring particles is electrically charged through the action on them of electrons and ions in the vicinity of the giant planet. These charged ring particles are then susceptible to electromagnetic forces exerted not only by the planet's magnetic field but also by the electric and magnetic forces exerted by the ions and electrons that charged the ring particles. As for the Poynting–Robertson effect, small bodies suffer greater accelerations, and therefore electromagnetic forces are particularly important at micrometre sizes and below. Additional removal mechanisms that affect small particles are collisions, including collisions with micrometeorites sweeping in from interplanetary space. Collisions fragment or remove small particles. Bodies of all sizes are removed by the gravitational effects of satellites.

Ring particle lifetimes of the order of 10^2 Ma have been estimated, much shorter than the 4600 Ma age of the Solar System. Persistent sources of ring particles are therefore needed, and several have been identified. As well as disruption within the Roche limit these include: the (rare) collisions of satellites and other modest-sized bodies within and beyond the rings, the erosion of such bodies by micrometeorite bombardment, the capture of micrometeorites themselves, and in the case of Jupiter, the volcanic outpourings of Io.

The ring systems are structures of exquisite complexity (Plate 18). Electromagnetic forces and gravitational forces are responsible for this fine structure too. Of particular note are the gravitational effects of satellites, not only the large satellites well outside the rings but small satellites embedded within the rings. Their gravitational effects sustain much of the fine structure of gaps and numerous ringlets.

Question 2.9

Discuss whether, at some time in the future, compared to today

(a) The ring system of Saturn could be much *less* extensive
(b) The ring system of Jupiter could be much *more* extensive.

2.4 Successes and Shortcomings of Solar Nebular Theories

The solar nebular theory outlined in this chapter accounts for many of the features in Table 2.1. It accounts for other features too. Overall, it is a successful theory. Perhaps the most worrying aspect left unexplained is the 7.2° tilt of the solar rotation axis with respect to the ecliptic plane. Whereas it is not difficult to account for the axial tilts of the planets' axes by the effect of material added asymmetrically, it is less easy to understand how the addition of material to the protoSun could have been sufficiently asymmetrical. A possible explanation is that there was a close encounter between the protoSun and another young star in the cluster in which the Sun was born.

Another possible problem is posed by the occurrence of giant planets close to the star in other planetary systems (Table 2.2). These lie well *within* the ice line of the star. Before the discovery of these exoplanets, simulations based on nebular theories produced a variety of planetary systems, but always with terrestrial-type planets within the ice line, and giant planets rich in hydrogen and helium *beyond* the ice line. Two explanations have been proposed for the exceptions. First, the giant planets could have formed where they are, but are massive because they are rich in iron rather than because they have huge quantities of hydrogen and helium. This explanation requires the stellar nebula to have been particularly rich in heavy elements. Unfortunately, the stars with exoplanets are not sufficiently enriched in such elements.

The second explanation is more likely — that the giant planets were formed further out, but have subsequently had their orbits greatly reduced in size. This is possible if the nebular disc is massive enough. If a large mass of nebular gas is added to the simulations then gravitational interactions between the disc and the planet cause the planet to spiral inwards in a time shorter than the lifetime of the disc. This inward migration can be halted either by tidal interactions between the protostar and the planet, or by the disc being excluded from the region near the star. Exclusion, out to about 0.1 AU, can result from magnetic interactions between the protostar and the disc, but regardless of how it is formed, when the planet enters the empty hole its inward migration ceases. Inward migration can also result from interactions with planetesimals, and from close encounters between planets. Close encounters can give rise to large orbital eccentricities. In the case of 16 Cygni B (Table 2.2), the high eccentricity could be the result of the gravity of the companion star — 16 Cygni is a binary star, with 16 Cygni A some way off.

The timing of the T Tauri phase with respect to the evolution of the nebula is clearly crucial to the final configuration of a planetary system.

❐ Why is this?

The T Tauri phase clears away the nebula, and thus halts its effect on planetary orbits.

The Solar System is different from the systems in Table 2.2 either because the solar nebula was always insufficiently massive or because the present planetary system formed from a nebular remnant. However, it is likely that a degree of migration of planets has also occurred in the Solar System, not only by the gravitational action of nebular gas but also by the gravitational scattering of planetesimals by a planet. It can be shown that in both cases Jupiter would move inwards, but the other three giants would move outwards. Migration is a possibility that needs further exploration in the models. Nevertheless, the great majority of astronomers believe that solar nebular theories are in fairly good shape, and offer by far the best type of theory that we have for the origin of the Solar System.

2.5 Summary

The origin of the Solar System cannot be *deduced* from its present state, though this state is an essential guide for the construction of theories, as are observations of other planetary systems, of star formation, and of circumstellar discs.

Most astronomers believe that solar nebular theories offer the correct explanation of the origin of the Solar System. In these theories the Solar System, including the Sun, forms from a contracting fragment of a dense interstellar cloud. As the fragment contracts it becomes disc-shaped, and at its centre the protoSun begins to form. Dust in the inner disc evaporates. As the temperatures in the disc decline, dust condenses and, along with pre-existing dust, settles towards the mid-plane of the disc, where it coagulates into planetesimals, and these undergo accretion to form planetary embryos. In the inner Solar System embryos come together and accrete smaller bodies, ultimately to form the terrestrial planets, consisting of rocky materials. In current theories the Moon is the result of a collision between an embryo and the Earth late in the Earth's formation.

In the outer Solar System the embryos are each several Earth masses, the result of fewer embryos forming and the condensation of water beyond the ice line. These embryos — called kernels — thus have an icy–rocky composition. They are too far apart to come together but are massive enough to capture nebular gas, mainly hydrogen and helium, a process that stops when the protoSun goes through its T Tauri phase and blows the gas out of the Solar System. Icy–rocky planetesimals are also captured, and this capture continues at a low rate today. The rate of growth of the giants decreases with increasing solar distance, so the T Tauri phase, if correctly timed, can explain the decrease in mass from Jupiter to Saturn to Uranus/Neptune. We can thus account not only for the existence of giants beyond the terrestrial planets but also for the broad differences between them.

An alternative (less favoured) means of forming the giant planets (though still within the context of solar nebular theories) is by a one-stage process in which each giant forms from a fragment of the nebula that contracts to become a protoplanet, and then contracts further to become the giant.

Estimates of the time it took for the Solar System to evolve from the formation of a nebular disc to the virtual completion of planetary formation are of order 100 Ma.

The distribution of angular momentum in the Solar System is thought to be the result of the transfer of angular momentum by the protoSun to the disc through turbulence in the disc. The Sun also lost some of its angular momentum via its T Tauri wind.

The disc of gas and dust that gave birth to the planets would have been rotating in the same direction as the solar rotation, giving rise to prograde planetary orbits roughly in the same plane. The axial inclinations are less well ordered because of off-centre acquisition of material as the planets grew.

The asteroids are the result of failed accretion due to the gravitational influence of Jupiter. Jupiter also stunted the growth of Mars, though it speeded the final stages of growth of the other terrestrial planets. The comets are thought to be icy–rocky planetesimals. There are two distinct populations, the far-flung Oort cloud, and the Edgeworth–Kuiper (E–K) belt extending from just beyond Neptune. Pluto and Chiron might be large members of the E–K belt, and so too might Triton.

The rings and most of the satellites of the giants are derived from discs of material in orbit around the giants.

Solar nebular theories are successful in that they account for most of the broad features of the Solar System in Table 2.1.

3 Small Bodies in the Solar System

We turn now from the Solar System in general to look in more detail at the smallest bodies that orbit the Sun — asteroids and comets. These are interesting in their own right, but they are also of importance in our attempts to understand the larger bodies, so it makes sense to consider the small bodies first. I shall start with the asteroids, then go on to comets, and conclude with meteorites — small interplanetary bodies that have reached the Earth's surface, where they can be collected and studied in much greater detail than we can study any body in space.

3.1 Asteroids

Until 1 January 1801, interplanetary space between Mars and Jupiter seemed empty, puzzlingly so, because the Titius–Bode rule (Section 1.4.3) had indicated the existence of a planet about 2.8 AU from the Sun. Therefore, a systematic search for the 'missing' planet was started in 1800 by twelve German astronomers. On 1 January 1801 the missing planet was discovered, though not by a member of the German team, but by the Italian astronomer Giuseppe Piazzi (1746–1826) during routine stellar observations at Palermo. The new body was called Ceres, and though it is close to 2.8 AU from the Sun (2.768 AU), it was regarded as disappointingly small. The modern value of its radius is 457 km — about a quarter that of the Moon. The German search therefore continued, and by 1807 had revealed three further asteroids, Pallas (at 2.774 AU), Juno (at 2.669 AU), and Vesta (at 2.361 AU). Each of these bodies is much smaller than Ceres, which is by far the largest asteroid.

Modern catalogues list nearly 8000 asteroids with accurately determined orbits. These tend to be the larger bodies, and Table 1.3 lists the orbital elements of the largest fifteen of them. Thousands more asteroids have been *seen*. Overall, we have probably seen all those greater than 100 km across (238), but only a tiny fraction of the small ones — it is estimated that there are of order 10^9 greater than 1 km across. The smaller the size, the greater the number of bodies, but the mass is dominated by the largest few bodies. If the estimate of 10^{22} kg for the total mass of the asteroids that exist today is correct then Ceres accounts for about 10% of this total. At a size of about 1 m across there is a somewhat arbitrary change in terminology, smaller bodies being called meteoroids. At sizes below a few millimetres we have micrometeoroids, and below about 0.01 mm we have dust.

Even the larger asteroids are so small that in nineteenth-century telescopes they looked like points, as do the stars — 'asteroid' means 'resembling a star'. Alternative names are 'minor planet', and 'planetoid'.

In Chapter 2 you saw that the asteroids are thought to be derived from the planetesimals and embryos that were in the space between Mars and Jupiter. Jupiter prevented the build-up of a major planet in this region, and scattered much of the material to other regions. Interactions between the asteroids and Jupiter have continued, and also between the asteroids themselves. This has resulted in considerable collisional fragmentation of the asteroid population over the 4600 Ma history of the Solar System, and also in a huge loss of material from this region. One estimate is that 4600 Ma ago there was about 10^3 times more mass than there is today in the space between Mars and Jupiter — 10^{25} kg, or about the same as the Earth's mass.

3.1.1 Asteroid Orbits in the Asteroid Belt

Figure 3.1 shows the distribution of the semimajor axes of the orbits of the asteroids. You can see that the great majority of values lie in the range 1.7–4.0 AU, with a particular concentration in the range 2.2–3.3 AU. The asteroids with semimajor axes in these two ranges constitute respectively the asteroid belt and the main belt. The orbital inclinations in these belts are fairly small, with few values above 20°, so the asteroids are part of the prograde swirl of motion in the Solar System, though on the whole the inclinations are larger than those of the major planets' orbits. The orbital eccentricities are also somewhat larger, with values of 0.1–0.2 being typical.

❐ If the semimajor axis a of an asteroid's orbit is 3.0 AU, and the eccentricity e is 0.30, what are its perihelion and aphelion distances?

From Figure 1.8 (Section 1.4.1), the perihelion distance is $(a-ae)$, which is 2.10 AU, and the aphelion distance is $(a+ae)$, which is 3.90 AU. This shows that a main belt asteroid even with an atypically large orbital eccentricity does not stray from the space between Mars and Jupiter.

Figure 3.1 also shows that the semimajor axes are not distributed smoothly. A prominent feature is the **Kirkwood gaps**, named after the US astronomer Daniel Kirkwood (1814–1895) who first detected them. These gaps are semimajor axis values around which there are few asteroids (because of orbital eccentricity it does *not* follow that these are depleted zones *in space*). They correspond to orbital periods that form simple ratios with the orbital period of Jupiter. For example, in the gap labelled 3:1, an asteroid would orbit the Sun three times while Jupiter orbited the Sun once. This means that if Jupiter and the asteroid are lined up, then three orbits later the line up is exactly repeated, as illustrated in Figure 3.2. This is an example of an **orbital resonance**, and because of the periodically repeated gravitational influence of Jupiter many resonant orbits are unstable, leading to a change in orbit that might be abrupt and large, characteristic of chaotic behaviour. This results in the asteroid

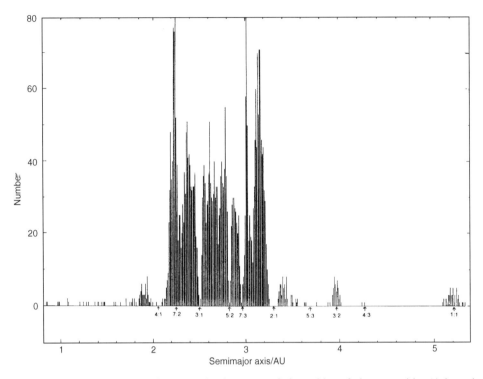

Fig. 3.1   The distribution of the semimajor axes of the orbits of the asteroids. (Adapted from R. P. Binzel, *Asteroids II*, 6, University of Arizona Press, 1989)

crossing the orbits of Mars and Jupiter, and even the Earth. The ultimate outcome is the removal of the asteroid from the belt.

In Figure 3.1 the Kirkwood gaps are particularly noticeable at the resonances 4:1, 3:1, 5:2, 7:3, and 2:1. Simulations show that at the 2:1 resonance the removal process is inefficient, and so the depletion of asteroids here might owe something to the original distribution of matter in the asteroid belt. The dearth of asteroids beyond 3.3 AU, which defines the outer edge of the main belt, can be explained by present-day resonances, plus an inward migration of Jupiter by about 0.2 AU (early in Solar System history — Section 2.4), which would have swept certain resonances through this region. Close to Jupiter asteroids have been removed by one-shot processes, capture or scattering, as can be performed by any planet. The location of the inner edge of the main belt near 2.2 AU seems also to be the result of Jupiter's gravity, though in this case orbital resonances are not involved.

Curiously, in a few cases, resonant orbits have an *excess* of asteroids. In Figure 3.1 the 3:2 resonance with Jupiter shows just this effect, the corresponding asteroids constituting the Hilda group. A factor that might help explain why they have not been removed by Jupiter is that when the Hilda group have Jupiter near opposition they are near perihelion, and so close approaches are avoided.

Another feature of asteroid orbits is grouping into families. In the early years of the twentieth century the Japanese astronomer Kiyotsugu Hirayama (1874–1943)

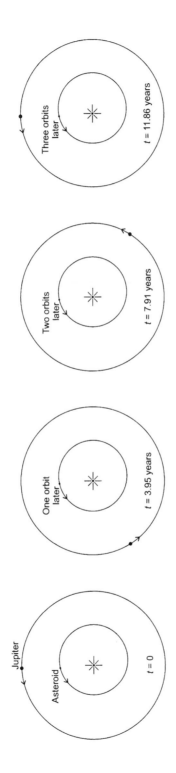

Fig. 3.2 The 3:1 orbital resonance of an asteroid with Jupiter

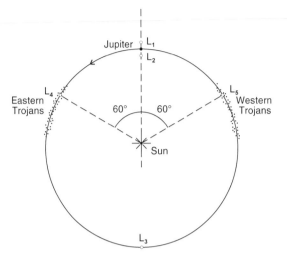

Fig. 3.3 The five Lagrangian points with respect to Jupiter and the Sun, and the Trojan asteroids

discovered several groups of asteroids, the members of a group having similar semimajor axes, orbital inclinations, and eccentricities. Each of these groups is called a Hirayama family, and the orbital similarities within each group indicate that the group members are the collisional fragments of a larger asteroid. This view is supported by observed compositional similarities within some families.

3.1.2 Asteroid Orbits Outside the Asteroid Belt

About 200 asteroids are known to have orbits with semimajor axes outside the range 1.7–4.0 AU. They thus lie outside the asteroid belt, and several interesting groups have been identified.

Figure 3.3 shows where the Trojan asteroids are located. Over a hundred are known, and there are probably hundreds of smaller ones yet to be discovered. They cluster around two of what are called the **Lagrangian points** of Jupiter and the Sun. Figure 3.3 shows all five of these points, labelled L_1–L_5. They are named after the Franco-Italian mathematician Joseph Louis Lagrange (1736–1813), who predicted their existence. They arise in a system of two bodies in orbit around their centre of mass plus a third body with a much smaller mass. The five points are where the third body can be located and remain in position relative to the other two. Thus, the whole configuration can be thought of as rotating like a rigid body about the centre of mass. Regardless of the ratio of the masses of the two main bodies, the points L_4 and L_5 are located as shown in Figure 3.3. By contrast, the locations of L_1 and L_2 do depend on the mass ratio, and lie closer to the less massive body. For the remaining point, if, as in the case of the Sun and Jupiter, the mass of one of the two bodies is much greater than that of the other, then L_3 lies very close to the orbit of the less massive body. The positional stability of a small mass placed at L_1,

L_2, or L_3 is poor, but at L_4 and L_5 it is much better. It is at L_4 and L_5 that the Trojans cluster.

In the case of the Sun and Jupiter, the stability of L_4 and L_5 is disturbed by the presence of the other planets. Nevertheless, the Trojans can reside at and even some way from these points for a long time. Each planet plus the Sun has Lagrangian points. In 1990 an asteroid was discovered near the trailing (L_5) point of the Sun and Mars.

Among the other asteroids outside the asteroid belt are some that can get very close to the Earth, and others that can collide with it. There are several groups, each one taking its name from a prominent group member. The Amor group have semimajor axes greater than 1 AU, but perihelion distances between 1.017 AU and 1.3 AU.

❑ What is the significance of 1.017 AU for the Earth's orbit?

This is the aphelion distance of the Earth. Twenty or so Amors are known, and it has been estimated that there are 1000–2000 with diameters exceeding 1 km. The Apollo group have perihelia less than 1.017 AU, so, if their orbital inclinations were zero, orbits with sufficiently large aphelia would intersect the Earth's. Even with non-zero inclination, intersection occurs if the ascending or descending node intersects the Earth's orbit (Section 1.4.2). A few hundred Apollo asteroids are known, and about 500–1000 are estimated to exist with diameters exceeding 1 km. The Aten group have semimajor axes less than 1 AU. With non-zero eccentricity intersection with the Earth's orbit can occur. About 20 are known. The Amor, Apollo, and Aten asteroids, plus others with as yet poorly known orbits, are collectively called the near-Earth asteroids.

Through the influence of the terrestrial planets, the orbital elements of the near-Earth asteroids vary, and therefore most of them will collide with the Sun sooner or later, and others will collide with the Earth or with another terrestrial planet. It is estimated that the average orbital lifetime is only about 1 Ma. This is very short compared to the 4600 Ma age of the Solar System, so replenishment must occur. The asteroid belt is undoubtedly a major source, the 3:1 orbital resonance being particularly copious. Comets are another source too, as you will see (Section 3.2.1). Currently (1998), the closest observed approach to the Earth was in December 1994, when a body passed the Earth at a range of only 10^5 km — about a quarter of the distance to the Moon. Though it was only 9 m across, its impact energy would have been several times the nuclear energy released by the Hiroshima atomic bomb. Earth has certainly experienced such hits from time to time, and occasionally far larger ones. A famous big hit was 65 Ma ago, when an asteroid 10–14 km across fell in Yucatan, and probably put enough debris into the atmosphere to cool the Earth sufficiently to hasten the death of many species, including the dinosaurs.

There is a handful of asteroids with orbits that lie among the giant planets. The first of these to be discovered, Hidalgo in 1920, is a body 10–20 km across, in an orbit with an inclination of 42.5°, and so eccentric that it extends from 2.01 AU to 9.68 AU from the Sun. The remainder are further out, and are known as the **Centaurs**. The first Centaur to be discovered was Chiron, in 1977. It is about 200 km across, and ranges from 8.46 to 18.82 AU from the Sun. However, it now seems that Chiron is not an asteroid, but probably a comet of low activity. In 1991 Pholus became the third asteroid to be discovered beyond Jupiter, in an orbit that takes it from 9 AU to 33 AU. It is about the same size as Chiron. About half a dozen or so have since been added to the list, ranging in size down to a few tens of kilometres. At such large distances from us, bodies smaller than this are difficult to discover, so the ones we have seen might only be the largest of a considerable number of asteroids in the outer Solar System.

Question 3.1

For the fifteen asteroids in Table 1.3
(a) Identify any unusual orbital elements, and state why each is unusual
(b) Discuss whether any of the fifteen could be at the Lagrangian points L_4 or L_5 of the Sun plus any planet.

3.1.3 Asteroid Sizes

With few exceptions the asteroids are too small and too far away from the Earth to be seen as anything other than points of light in the sky. The exceptions are Ceres, a handful of Apollos, plus a few that have been imaged from close range by spacecraft. Their sizes are thus known from direct observations. In addition, a dozen or so have passed between us and a star, the size then being obtained from the accurately known rate at which the asteroid moves across the sky, and the length of time for which the star is blocked. For the great majority of asteroids, sizes have to be obtained by indirect means.

An important indirect method depends on the measurement of the flux density of the reflected solar radiation that we receive from the body. **Flux density F** is a general term defined as the power of the electromagnetic radiation incident on unit area of a receiving surface. The receiving surface of our detector is perpendicular to the direction to the asteroid, and F spans the wavelength range of the whole solar spectrum.

Let's assume that the asteroid is in opposition, so that the Sun, Earth and asteroid are near enough in a straight line. In this case the asteroid is seen from the Earth at what is called zero phase angle, as in Figure 3.4(a), and the flux density received by reflection is labelled $F_r(0)$. It can be shown that

$$F_r(0) = kp\,A \tag{3.1}$$

where k is a combination of known factors involving the Sun and the distance to the asteroid, A is the projected area of the asteroid in our direction (Figure 3.4(a)), and p is a quantity called the **geometrical albedo**. This is the ratio $F_r(0)/F_L(0)$, where $F_L(0)$ is the flux density we *would* have received from a flat Lambertian surface perpendicular to the direction to the Sun and Earth, and with an area equal to the projected area of the asteroid (Figure 3.4(b)). A Lambertian surface is perfectly diffuse (the opposite of a mirror) and reflects 100% of the radiation incident upon it.

❑ Is the form of equation (3.1) reasonable?

It is reasonable that $F_r(0)$ increases with p and also with A.

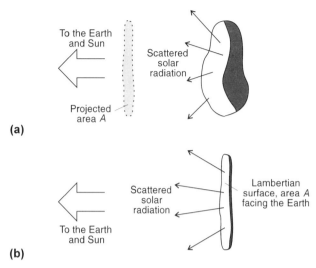

Fig. 3.4 (a) An asteroid in opposition. (b) A flat Lambertian surface with the same projected area as the asteroid

If we know p, and if we have measured $F_r(0)$, then A can be obtained from equation (3.1), and the mean radius can then be estimated. The value of p depends on the composition and roughness of the surface. Plausible surfaces having values of p ranging from about 2% to about 40%, and so, unless we can reduce this range, the size of the body can only be obtained to an order of magnitude. Comparisons of $F_r(0)$ with the flux density F_e of the infrared radiation *emitted* by the asteroid by virtue of its temperature have provided estimates of p, as have studies of the polarisation of the reflected solar radiation. The details will not concern us.

Indirect techniques have provided nearly all the size data in Figure 3.5. The data have not been extended to sizes much below 10 km radius because at smaller sizes there must be a significant proportion of undiscovered asteroids, the proportion increasing with decreasing size. Even with this cut-off, the general increase in the number of bodies with diminishing size is clear. The general shape of the curve is consistent with an initial population modified by mutual fragmentation as a result of collisions. You have seen that it is thought that during the formation of the Solar System there were several embryos between Mars and Jupiter, plus a host of smaller bodies, and that further growth was halted by the formation of Jupiter, whose gravitational influence 'stirred up' the asteroid orbits. This increased collision speeds to the point where net growth was replaced by net fragmentation. Note that this is *net* fragmentation — some reassembly of fragments is to be expected. And so it continues today.

Question 3.2

Even though the number of asteroids increases as size decreases (Figure 3.5), it does *not* follow that the host of tiny asteroids doubtless awaiting discovery will add significantly to the mass of those already discovered. Explain why this is so.

3.1.4 Other Properties of Asteroids

The term 'radius' in Figure 3.5 has to be interpreted with care. The self-gravitational forces inside an asteroid result in an increase of pressure with depth. If the pressure exceeds the strength of the solid materials in the interior, then the materials yield, and a roughly spherical body results, flattened by rotation at high rotation rates. The internal pressures decrease as the size of the body decreases, and there comes a point where the materials do not yield, and so the body need not be even approximately spherical. For an asteroid made of silicates and iron, the critical radius is about 300 km. Therefore, below this size 'radius' means an average distance from the surface to the centre of the asteroid — the body is not necessarily anywhere near spherical.

❐ Are any asteroids certain to be (very nearly) spherical?

From Table 1.3 you can see that only Ceres is significantly larger than the critical radius. Therefore, we can be confident of a near-spherical shape.

A few asteroids have had their shapes determined directly, either from images or from the way that the light we receive from them varies when a planet passes between the asteroid and us. Ceres, as expected, is fairly spherical, but significant departures from spherical form occur at smaller sizes, also as expected. Figure 3.6 and Plate 10 show images of asteroids that display their non-spherical shapes. Three main belt asteroids have been imaged — Vesta, Gaspra, and Ida. The image of Vesta in Figure 3.6(a) was obtained from Earth orbit by the Hubble Space Telescope. The tiny asteroid Gaspra (Plate 10) was imaged in October 1991 by the Galileo spacecraft en route to Jupiter, and in August 1993 the same spacecraft imaged the larger asteroid Ida (and a previously unknown tiny satellite, named Dactyl). Outside the main belt, but within the asteroid belt, Mathilde was imaged in

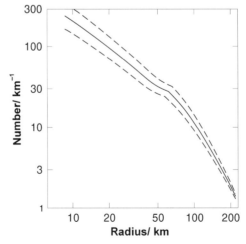

Fig. 3.5 The number of asteroids with mean radii per kilometre interval of radius. The dashed lines indicate uncertainties

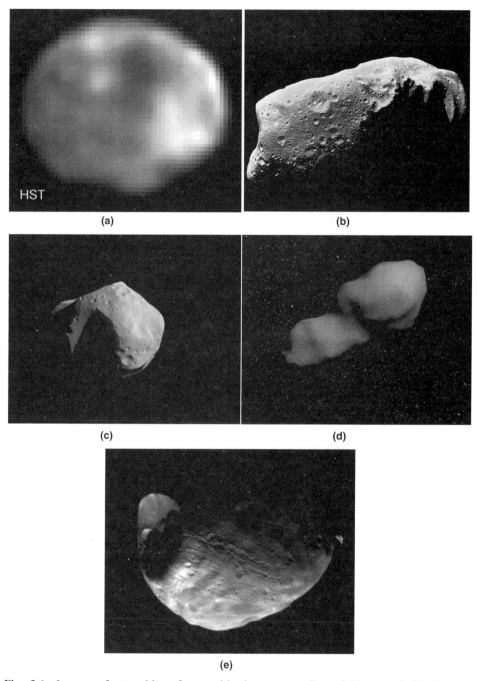

Fig. 3.6 Images of asteroid surfaces, with the mean radius of Vesta and the longest dimension of the others. (a) Vesta, 251 km. (AURA/STScI, NASA, PRC97-27, P. Thomas and B. Zellner.) (b) Ida, 53 km. (NASA/NSSDC P42964.) (c) Mathilde, 59 km. (The Johns Hopkins University, Applied Physics Laboratory.) (d) Toutatis, 4.6 km. The dots are noise. (NASA/JPL P46256, S. J. Ostro and S. Hudson.) (e) Phobos, 26 km. (NASA/NSSDC 357A64)

1997 by the spacecraft NEAR. Toutatis is a member of the Apollo group, and is one of a few near-Earth asteroids that have been imaged by Earth-based radar. The image of Toutatis in Figure 3.6(d) was obtained in December 1992 when it passed the Earth at a range of only 4×10^6 km. If, as is thought, the Martian satellites Phobos and Deimos are captured asteroids then these also belong to the list of those imaged. Figure 3.6 includes a Viking Orbiter image of Phobos; Deimos is broadly similar.

Many asteroids have had their shapes determined indirectly. All asteroids rotate, almost all with periods in the range 4–16 hours, and as they rotate, the flux densities F_e and $F_r(0)$ vary with time. There are two possible contributory factors — surface features, and changes in the projected area resulting from non-spherical form. From the variations in F_e and $F_r(0)$ these two factors can be separated. The outcome, as expected, is that large departures from spherical form are common among asteroids.

The rotational data also show a broad range of axial inclinations. This, and the range of rotational periods, is consistent with models in which small bodies frequently collide. Irregular shapes are also expected from collisional fragmentation, though in some cases a small asteroid might retain the irregular form it acquired at its origin. Irregular forms can also arise from the subsequent sublimation of volatile components. Conversely, small bodies can become more spherical through erosion by dust impact.

Very few asteroids have had their masses measured with useful precision. Close approaches of small asteroids to Ceres, Pallas, and Vesta have yielded mean densities of $(2700 \pm 140)\,\mathrm{kg\,m^{-3}}$, $(2600 \pm 900)\,\mathrm{kg\,m^{-3}}$, and $(3300 \pm 1500)\,\mathrm{kg\,m^{-3}}$ respectively. The orbit of Ida's tiny satellite Dactyl indicates a density for Ida of $(2600 \pm 500)\,\mathrm{kg\,m^{-3}}$. All these densities are consistent with high proportions of rocky materials. By contrast, the effect of Mathilde on the path of NEAR corresponds to a density of only $(1300 \pm 200)\,\mathrm{kg\,m^{-3}}$. Such a low density could be due to hydrated substances in abundance, but analysis of the light reflected from the surface indicates otherwise, and so the low density might indicate high porosity. High porosity helps to explain how Mathilde could have survived the large impacts that produced its heavily cratered surface — porosity cushions impacts, thus preventing disruption. The craters on all the surfaces in Figure 3.6 are typical of any solid surface long exposed to meteoroid bombardment. Such surfaces are also expected to acquire thin layers of dust, and observations of the surfaces of these and other asteroids indicate that a thin dust cover is indeed a common feature.

Question 3.3

(a) Why have gravitational forces failed to make the bodies in Figure 3.6 more spherical?
(b) If a small asteroid were spherical, what would this tell us about its possible histories?

3.1.5 Asteroid Classes

The surface composition of an asteroid can be inferred from a combination of various types of data. These include ratios of the reflectance in different wavelength

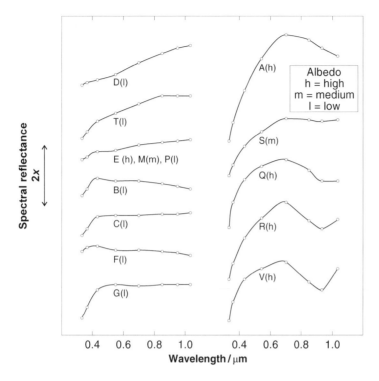

Fig. 3.7 Reflectance spectra and geometrical albedos of the fourteen Tholen asteroid classes. Note that the reflectances show spectral *shape* on a logarithmic scale, not absolute values, and that the spectra are offset vertically by arbitrary amounts. (Adapted from D. J. Tholen and M. A. Barucci, *Asteroids II*, 309, University of Arizona Press, 1989)

bands, derived from flux density measurements. If the bands are very narrow, numerous, and contiguous the measurement is called **spectrometry**, otherwise the measurement is called **photometry**. Further data include the geometrical albedo, and the polarisation induced in reflected solar radiation.

A first step towards compositional determination is to use the observational data to divide asteroids into different classes. This makes the problem more manageable — if we can obtain the (surface) composition of one member of a class, then this is probably similar to that of the other members of the class. Various classification schemes have been proposed. The one I shall describe was devised by the US astronomer David J. Tholen in 1983, and it is widely used. It is based on the narrow-band reflectances at eight wavelengths in the range 0.3–1.1 μm, plus the geometrical albedo. Fourteen classes are recognised, and Figure 3.7 shows the mean reflectance spectrum of each class, plus an indication of the albedo (high, medium, or low). Note that the classes E, M, and P are distinguished only by albedo. Some classes are represented by very few members. For example, until recently the V class was represented only by Vesta (hence the 'V'), and even now only a handful have been added, all of them very small, and in orbits that suggest they are collisional fragment of Vesta. Classes R and Q also contain only a handful of members.

About 80% of classified asteroids fall into the S class, and about 15% into the C class. Members of the C class have geometrical albedos in the approximate range 2–7%, so they are very dark, like soot.

❐ From Figure 3.7, how would you characterise the *colour* of S and C class asteroids?

At visible wavelengths the reflectances of C class asteroids do not vary much with wavelength, so they are rather grey. The slightly greater reflectances at longer wavelengths gives them a hint of red. S class asteroids are distinctly red, and they have higher albedos, about 7–20%. In Figure 3.6, Ida is S class (as is Gaspra, Plate 10), whereas Mathilde is C class.

By comparing reflectance spectra with laboratory spectra of various substances, and with the aid of albedo and other observational data on asteroids, likely surface materials can be identified. The outcome is that class M match alloys of iron with a few per cent nickel, mixed with little or no silicates. Class S match mixtures of similar iron–nickel alloys with appreciable proportions of silicates. Class C match a type of meteorite called the carbonaceous chondrite, of which more later, but which consist of silicates mixed with hydrated minerals, plus small quantities of iron-nickel alloy, carbon, and **organic compounds**. These are compounds of carbon and hydrogen, often with other elements. Carbon and organic compounds are collectively called **carbonaceous materials**, and they are mainly responsible for the low albedos of class C. Classes P and D are broadly like class C, but correspond to material that is richer in carbonaceous materials. Class V match some sub-classes of a type of meteorite called the achondrite, silicate meteorites of which, again, more later.

Note that these mineral matches are to the surface of the asteroid — it could be very different in its interior. Note also that there could be cases where the matches are merely coincidences, and that the surface composition of the asteroid is quite different from that of the corresponding type of meteorite.

3.1.6 Asteroid Classes across the Asteroid Belt, and Asteroid Differentiation

Figure 3.8 shows the distribution with heliocentric distance of the five most populous classes. Fractions are shown, such that at each distance the fractions of all asteroids (not just those shown) sum to 1. Some idea of the actual numbers can be obtained by comparing Figure 3.8 with Figure 3.1, though it must be noted that Figure 3.1 shows *observed* asteroids, whereas in Figure 3.8 an attempt has been made to correct for various observational biases, for example that a greater proportion of high-albedo asteroids must have been discovered than of low-albedo asteroids.

It is clear that the distributions differ from one class to another. *If* the mineralogical interpretations outlined above are correct, then the broadest trend is that mixtures of silicates and iron–nickel predominate in the inner belt (class S), and that carbonaceous materials and hydrated minerals become increasingly predominant as heliocentric distance increases (classes C, P, D).

An explanation of these trends is that the materials now in the outer belt formed there, where the cooler conditions allowed condensation of the more volatile substances, such as carbonaceous materials and hydrated minerals. In the warmer

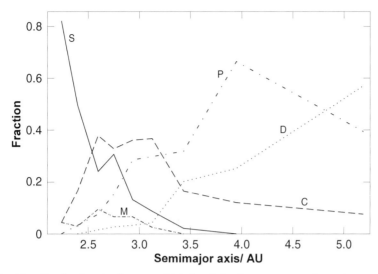

Fig. 3.8 The distribution in the asteroid belt of the five most populous classes of asteroid

inner belt this was not possible, so we get only silicates and iron–nickel alloy. This distinction could have been enhanced during the T Tauri phase of the protoSun, when the solar wind would have heated the asteroids by magnetic induction, i.e. by the heating from electric currents induced in the asteroids by the action of the magnetic fields entrained in the wind. The heating decreased with heliocentric distance, so asteroids in the inner belt would have been heated more than those in the outer belt. This explanation requires that there has been only limited migration of the different classes across the asteroid belt, and that differences across the belt never became obscured by migration into the belt of planetesimals that formed elsewhere in the Solar System. An alternative explanation places more weight on the loss of volatile materials from the inner belt throughout Solar System history, in which case substantial inward migration of C class with subsequent modification is a possibility.

Class M, which have surfaces that are largely or entirely iron–nickel, are presumably iron–nickel throughout — there is no plausible way of getting such an iron-rich surface and an iron-poor interior. Nor is there a plausible scheme of condensation and accretion in the solar nebula that would give such a silicate-free composition throughout, and it is therefore necessary to assume that internal temperatures in some asteroids rose to the point where the interiors became partially or wholly molten. This allowed a process called **differentiation** to occur, whereby denser substances settled towards the centre of a planet, and the less dense substances floated upwards to form a mantle, overlain in turn by a mineralogically distinct crust. The melting could have resulted from heat released by the decay of short-lived unstable isotopes, notably the isotope of aluminium ^{26}Al, nearly all of which decayed in a few million years. The asteroid interior would then have cooled, and became solid after a further interval of a few million years.

The temperature rise caused by isotope heating is greater, the larger the body. This is because the mass of the isotopes present is proportional to the volume of the body, whereas heat losses from the body are proportional to its surface area, and the ratio of volume to surface area is greater, the larger the body. In a body consisting of mixtures of silicates and iron–nickel alloy, differentiation would have occurred at sizes larger than about 200 km across, and would have resulted in a predominantly iron–nickel core overlain by a mantle and crust largely composed of silicates. There will also be a core–mantle interface consisting of a mixture of iron–nickel alloy and silicates. Collisions can break up these bodies, and fragments of the cores give us chunks of iron–nickel alloy, i.e. class M. Fragments of the mantle and the mantle–core interface could be an important source of S class. The surface properties of Vesta are consistent with the sort of silicates that would form the crust of a fully differentiated body.

The scarcity of M and S classes in the outer belt indicates that differentiation was uncommon there. One explanation is that supplementary heating by T Tauri magnetic induction was too weak at these greater heliocentric distances. Further discussion of the composition of asteroids is in Section 3.4.

Question 3.4

The reflectance spectrum and albedo of the asteroid Eros are shown in Figure 3.9. Allocate it to one of the five most populous asteroid classes, justifying your choice. Hence deduce its likely surface composition, and where it is likely to be found in the asteroid belt.

3.2 Comets

A comet differs from an asteroid in that a typical comet develops a large thin atmosphere when it is within 10 AU or so of the Sun. This is called the coma, and from it develops a very tenuous hydrogen cloud and two tails, as in Plate 22. One of these tails consists of dust, the other of ionised gas. The coma can become as large as Jupiter, the hydrogen cloud larger than the Sun, and the tails as long as 1–2 AU. As seen from the Earth, a comet can be a spectacular sight in the sky for several months. Figure 3.10 illustrates the growth and shrinkage of the tails versus the heliocentric distance as the comet goes around its orbit. The source of the coma, cloud, and tails is a solid nucleus, typically a few kilometres across, and this is all that exists of the

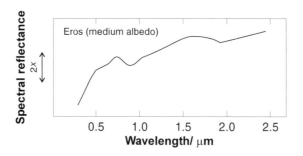

Fig. 3.9   The reflectance spectrum (log scale) and albedo of the asteroid Eros

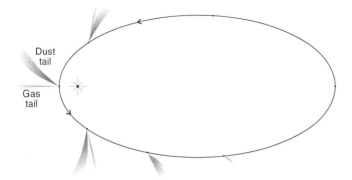

Fig. 3.10 The growth and shrinkage of the tails of a comet versus its heliocentric distance

comet when it is in the outer Solar System. The nucleus differs from an asteroid in that it contains sufficient quantities of icy materials to generate the coma, hydrogen cloud, and tails. This indicates that comets formed further out in the solar nebula than the asteroids, sufficiently far that solid icy materials were present.

3.2.1 The Orbits of Comets

Cometary orbits fall into two main categories, long-period, and short-period. As the names indicate, the categorisation depends on the orbital period. There is no sharp division, but the defining orbital period is exact.

Long-period comets have orbital periods in excess of 200 years. In most cases the periods are *greatly* in excess of 200 years, with values extending up to about 10 Ma.

❐ Use Kepler's third law to calculate the semimajor axis a of the orbit of a comet with an orbital period $P = 1$ Ma. Express your answer in AU.

From equation (1.3) (Section 1.4.1)

$$a = (P/k)^{2/3}$$

where $k = 1$ year AU$^{-3/2}$. Thus, with $P = 1$ Ma (10^6 years), $a = 10^4$ AU. Such comets are observed only because they have highly eccentric orbits that bring them within a few AU of the Sun. A small number of long-period comets are in parabolic or hyperbolic orbits, though this could be due to the perturbation of an eccentric elliptical orbit as a comet is on its way towards us. This perturbation can be caused by a close approach to a planet, or by an eruption of gas from the comet. Thus, there is no incontrovertible evidence for any comet having come from interstellar space, though this can't be ruled out in some cases. If a comet is leaving the inner Solar System on a parabolic or hyperbolic orbit, and if this orbit is not perturbed into an ellipse, the comet will certainly escape.

The huge orbital periods of most long-period comets means that most of them have been observed only once in recorded history. About 1000 different long-period comets have been recorded, and about 600 of these have well-known orbits. On average, about half a dozen long-period comets are observed per year, and about one per decade becomes noticeable to the unaided eye. A recent spectacular example was comet Hale–Bopp in 1997 (Plate 22), and its orbit in the inner Solar System is shown in Figure 3.11(a). One of the factors that determines how spectacular a comet becomes is its proximity to the Earth. Another is its perihelion distance. If this is less than the radius of the Sun (poor comet!), or not much greater (Sun-grazing comets), then magnificent tails develop.

The orbital inclinations of the long-period comets are randomly distributed over the full range, as are the longitudes of the ascending node and of perihelion (Section 1.4.2). The long-period comets thus bombard the inner Solar System from all directions.

The **short-period comets** are defined as having periods of less than 200 years, and therefore they must have semimajor axes less than 34 AU — comparable to the outermost planets Neptune and Pluto. However, unlike the planets, most of the short-period comets are in eccentric orbits, in some cases with $e > 0.9$ (Table 1.4). About 180 short-period comets are known. Of these, roughly half a dozen per year grow bright enough to be visible in a modest telescope.

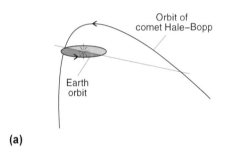

(a)

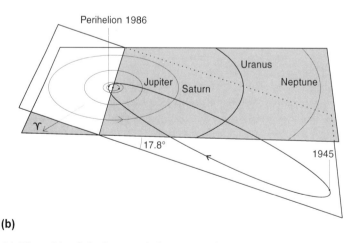

(b)

Fig. 3.11 (a) The orbit of the long-period comet Hale–Bopp in the inner Solar System. (b) The orbit of the short-period comet Halley

Most of the short-period comets have periods less than 20 years. They move in moderately eccentric orbits with perihelion distances in the range 1 AU to a few AU, aphelion distances 4–7 AU, and in most cases the inclination is less than 35°. Therefore, their aphelia are broadly in the region of Jupiter and so they are often called the Jupiter family comets. Though all comets are prone to orbital changes through gravitational interaction with a planet, this is particularly so for the Jupiter family. This is because their low inclination orbits, traversing the planetary zone, give a comparatively high probability of a close encounter with a planet. Such an encounter will result in a drastic alteration of the orbit, with an outcome that can be anything from solar capture to ejection from the Solar System.

The remaining short-period comets are typified by the most famous short-period comet of all, Halley's comet, and so are called Halley-type. The members of this small group have periods in the range 15–200 years, and their inclinations are typically larger than for the Jupiter family, with a few in retrograde orbits. Halley's comet itself has an orbital period of 76.1 years, and an orbital inclination of 162°, i.e. retrograde. It has been observed every 76 or so years as far back as reliable records go — at least as far back as 240 BC. One of its more famous apparitions was in the year of the Battle of Hastings, AD 1066. It is named after the British astronomer Edmond Halley (1656–1742), who noticed that the orbits of the bright comets of 1531, 1607, and 1682 were very similar. He deduced that this was the same comet each time, and he predicted its reappearance in 1758. Halley's comet duly appeared, but alas! Halley had died 16 years earlier. It was last at perihelion 9 February 1986, and in March 1986 it became the first (and so far only) comet to be imaged at close range by a spacecraft. Figure 3.11(b) shows the orbit of Halley's comet.

Question 3.5

The comet Temple–Tuttle has a perihelion distance $q = 0.9766$ AU, an orbital eccentricity $e = 0.9055$, and an orbital inclination $i = 162.49°$. Show that this is a Halley-type comet.

3.2.2 The Coma, Hydrogen Cloud, and Tails of a Comet

The coma that grows when a comet comes typically within 10 AU or so of the Sun is a result of the heating of the nucleus by solar radiation. The coma is a large, tenuous atmosphere, consisting of gases derived from the more volatile constituents of the nucleus, mixed with dust carried aloft by the outgassing. Spectroscopic studies, and measurements made by spacecraft, have shown that the dust in the coma consists of rocky materials like silicates, and carbonaceous materials. Such studies also show that except in the innermost part of the coma the gases are predominantly *fragments* of molecules, rather than intact molecules. Such fragmentation is to be expected as a result of the disruptive effect of solar radiation. This is called **photodissociation**, and UV photons are particularly effective. The fragments, and intact molecules, can also be ionised in a process called **photoionisation**, where a solar UV photon ejects an electron.

From the molecular fragments the parent molecules can be identified. Hydrogen atoms (H), hydroxyl (OH), and oxygen atoms (O) are particularly common, and must have been derived from water molecules (H_2O). Other molecular fragments have been derived from carbon dioxide (CO_2) and carbon monoxide (CO). H_2O, CO_2, and CO have also been detected as intact molecules in the coma. From the relative abundances of molecules and fragments in the coma, it is inferred that the predominant volatile constituent of the nucleus is water, typically accounting for over 80% of the mass of the volatile substances. Carbon monoxide and carbon dioxide are the next most abundant volatiles. Note that these substances would be present in the nucleus as *solids*. They are evaporated from the nucleus to form gases, and are then photodissociated and photoionised. The more volatile the material, the greater the heliocentric distance at which it can evaporate.

The hydrogen cloud is derived from the coma through the photodissociation of the molecular fragment OH. Though the cloud can greatly exceed the size of the Sun there is very little mass involved.

The tails can likewise be huge, extending as far as about 2 AU from the coma, but they are also very tenuous, so again little mass is present. There are two sorts of tail — a tail stretching almost along the line from the Sun to the coma, and a curved tail that points away from the Sun only in the immediate vicinity of the coma (Plate 22 and Figure 3.10). The spectrum of the radiation received from the curved tail shows it to be a solar spectrum modified in a way consistent with scattering from micron-sized dust particles. Therefore, this is a dust tail, and it is seen by the solar radiation that it scatters. By contrast, the radiation from the straight tail shows it to consist of electrons and ionised atoms, plus a trace of very fine dust with particle sizes less than a micron. The evidence is a very weak spectrum like that from the dust tail, and strong spectral lines emitted by the ions, consequent upon absorption of solar radiation. This is the ion tail. One common ion is OH^+, produced from OH in the coma by photoionisation. All the ions that have been identified in the ion tail could have been produced by the ionisation of atoms and molecular fragments in the coma.

It is the strikingly different composition of the two tails that underlies their separation in space as they stream away from the coma. The ion tail is swept from the coma by the force exerted on the coma ions by the magnetic field in the solar wind as the field moves across the ions. The details are complicated, but the outcome is that the ions are swept in the direction of the solar wind, i.e. radially from the Sun. The wind speed is much higher than the orbital speed of the comet, so in the time it takes an ion to travel from the coma to where the tail is indistinguishable, the comet does not move very far. Consequently, the tail is fairly straight. Because of the intricate structure of the solar wind, the ion tail too is highly structured, with filaments and knots, and it can temporarily break away from the coma. The trace of submicron dust in the ion tail is carried by the ion flow.

The dust tail is driven from the coma through bombardment by the photons that constitute solar electromagnetic radiation. These act like an outflow of tiny bullets, and the resulting force is called **radiation pressure**. The dust is driven away from the Sun, but only reaches speeds comparable to the orbital speed of the comet.

❐ So, why is the dust tail curved?

This tail is curved because the comet moves appreciably around its orbit in the transit time of the dust to the end of the visible tail.

The smaller the particle, the greater the acceleration caused by radiation pressure. This is because the area to mass ratio is greater for smaller particles, as discussed in relation to the Poynting–Robertson effect (Section 2.3.2). As a result, particles in the coma much greater than a few tens of micron in size are retained. The greater effect of radiation pressure on smaller particles raises the question of why it is not an important force on the ions in the tail. This is because when the particle size is less than the predominant wavelength of the photons the interaction is enfeebled. For solar radiation the predominant wavelength is about 0.5 micron, which is very much greater than ion radii.

Often, more than two tails are seen. These extra tails are often ion or dust tails with slightly different properties, such as the thin tail of neutral sodium atoms seen streaming away from comet Hale–Bopp.

3.2.3 The Cometary Nucleus

From their starlike images it has long been known that cometary nuclei are small. Size can be estimated in a similar indirect manner to asteroids (Section 3.1.3), but the uncertainties are great. Sizes of the order of a few kilometres are typical, extending up to a few tens of kilometres for the largest comets. The composition of the nucleus can be deduced from the composition of the coma, as outlined in Section 3.2.2. The typical nucleus is deduced to consist of ices, predominantly water, mixed with some carbon dioxide and carbon monoxide ices, plus rocky and carbonaceous materials.

A great variety of other molecules and molecular fragments have been identified in comas and ion tails, implying the existence of small quantities of other icy substances in the nuclei of some comets. Methanol (CH_3OH), methanal (HCHO), and nitrogen (N_2) are usually the most abundant traces inferred to exist. Traces in some comets indicate that the icy dust grains that formed their nuclei are interstellar material that has not been heated above about 100 K. On the other hand, the long-period comet Hale–Bopp has isotope ratios for C, N, and S that are the same as in the Solar System in general, indicating that Hale–Bopp's icy grains came from a well-mixed solar nebula.

❐ Why does this suggest that Hale–Bopp comes from the Oort cloud?

Solar nebular theories derive the Oort cloud from icy–rocky planetesimals that formed in the giant region, where the nebula was well mixed (Section 2.2.5). Therefore, the composition of Oort cloud comets is expected to resemble that of the nebula. The low abundance of neon in Hale–Bopp supports a giant region origin. Beyond Neptune, in the E–K belt, it would have been cold enough for neon to condense and thus neon would now be a more significant component.

Our knowledge of cometary nuclei received a huge boost in 1986 when five spacecraft made close observations of Halley's comet. The European Space Agency's Giotto flew closest, sweeping past at a range of only about 600 km from the nucleus, and it obtained the image in Figure 3.12. The peanut-shaped nucleus is 16 km long, and 8 km by 7 km in typical cross-section. It rotates around its long axis with a period of 170 hours, and this axis precesses with a period of 89 hours around an axis inclined at 66° with respect to the long axis.

Fig. 3.12 The nucleus of Halley's comet. The long dimension is 16 km, and the Sun is to the left. (ESA 3416 etc. composite. Reproduced by permission of the European Space Agency)

The mass of the nucleus was estimated from the effect on Halley's orbit of the forces exerted by gas jets erupting from surface vents. The estimated mass is 10^{14} kg, give or take 50%, and therefore the density of the nucleus is only 100–250 kg m^{-3}, considerably less than the 920 kg m^{-3} of water ice at the Earth's surface. Therefore, the nucleus is not so much a block of dirty ice as a fluffy aggregation of small grains. The effect of jets on the orbits of other comets has yielded similar densities, though with much greater uncertainty. The fragility of cometary nuclei is indicated by several that have broken up through their degassing as they approached the Sun, or in the case of Shoemaker–Levy 9 by a close approach to Jupiter.

The spacecraft observations of the coma of Halley's comet, supplemented by ground-based observations, added many details to our knowledge of the composition of the nucleus, but did not change the broad picture very much. It is deduced that Halley's nucleus consists of 80% water, 10% carbon monoxide, 3.5% carbon dioxide, by numbers of molecules. Definite evidence for methane ice was not obtained, even though this was expected to be relatively abundant. An important detail is evidence that some of the water is probably present in chemical combination with rocky and carbonaceous materials, as water of hydration. Moreover, it seems likely that a proportion of the different icy materials are present in what are called **clathrates**, where one material is enclosed in the crystal structure of another. In particular, the rather open crystal structure of water ice can readily enclose molecules of other icy substances, such as carbon dioxide.

The Giotto flyby also confirmed that cometary nuclei can be very dark. Halley has a geometrical albedo of only 3–4%, the result of carbonaceous materials at the surface. Low albedos have since been established for the nuclei of other comets. It seems that, as the icy materials evaporate near the Sun to give the coma and tails, a residue of dust depleted in icy materials concentrates at the surface, where it forms an insulating protective fluff over the ice-rich grains beneath. This fluff is broken by the vents that spew forth the coma and tail material. Vents tend to switch on when they face the Sun, and switch off when they turn away from the Sun. For Halley's comet this phenomenon is apparent in Figure 3.12.

Vents can explain the transient brightening that some comets exhibit when they are *more* than a few AU from the Sun. Slow evaporation beneath the protective fluff would build up the gas pressure to the point where the fluff ruptures, and a vent forms. An interesting example is Schwassmann–Wachmann 1, which has a nearly circular orbit between Jupiter and Saturn (Table 1.4). Though it is fairly large for a comet — about 40 km across — it would have gone unnoticed but for its outbursts, which occur every year or so. Halley's comet suddenly brightened in 1991, when it was 14 AU from the Sun. This might have been due to a large solar flare which caused shock waves that ruptured the fluff.

Question 3.6

In 150–200 words, describe the visual appearance of a comet from when it is about 30 AU from the Sun on its way in, to when it is outgoing at the same distance. Relate the visible changes to events at the nucleus.

3.2.4 The Death of Comets

When beggars die, there are no comets seen;
The heavens themselves blaze forth the death of princes.
                    William Shakespeare (*Julius Caesar*)

Comets die too, because the loss of volatiles is acute within a few AU of the Sun. If perihelion is at 1 AU then the order of 10^2 perihelion passages will suffice to evaporate all the ices from a nucleus of typical size, leaving a loose aggregate of silicate and carbonaceous dust particles, no longer with any capability to develop a coma and tails. Comets are seen in such last throes of activity — some of the short-period comets have very small comas and tails. The Infrared Astronomical Satellite (IRAS) that gathered data for nearly the whole of 1983 discovered many Solar System objects with small dusty envelopes. Though some of these might be asteroids that never had ices, but are surrounded by fine collisional debris, others might be devolatilised comets.

Elsewhere in the Solar System some members of the low-albedo classes of asteroid, such as the C and D classes, might also be devolatilised comet nuclei. The reflectance spectra and albedos of cometary nuclei resemble those of these classes. For example, Hidalgo (Section 3.1.2) is a D class asteroid with an unusually distant aphelion for an asteroid, and so it is a good candidate for being a comet remnant. Some of the small satellites of the giant planets, particularly those in unusual orbits, might also be cometary remnants, captured by the planet.

A devolatilised comet nucleus has lost not only ices but also a proportion of its dust. The inner Solar System is pervaded by dust, much of it cometary. The average density of the dusty medium is about 10^{-17} kg m^{-3}, but greater along the orbits of comets. (It is greater still in the asteroid belt where dust from asteroid collisions makes a large additional contribution.)

As well as suffering slow attrition, comets can also have spectacular ends. Devolatilised or not, comets have been seen to collide with the Sun, and in July 1994 one was seen to collide with Jupiter, namely Shoemaker–Levy 9, or rather its fragments. There must have been many other collisions with the planets, including the Earth.

Some short-period comets have orbits that resemble the Amor and Apollo asteroids, and it is thought that some of these are devolatilized cometary nuclei. The collision of one such nucleus with the Earth might account for the huge explosion in 1908 in the Tunguska river area of central Siberia, though a small asteroid proper is another possibility. There is archaeological evidence for earlier impacts, and in the future the Earth must surely collect further comets.

Question 3.7

As well as perihelion distance, what other orbital property influences the mass of volatile material lost by a comet per orbit? Justify your answer.

3.2.5 The Sources of Comets

The Oort cloud

You have seen that the orbits of the long-period comets have aphelia far beyond the planets, and that the aphelia lie in all directions from the Sun. This was the sort of data that led the Estonian astronomer Ernst Julius Öpik (1893–1985) to suggest in 1932 that there was a huge cloud of comets surrounding the Solar System, but so far away that only those members with perihelia less than a few AU became visible, through the growth of coma and tails. In 1950 this idea was developed by the Dutch astronomer Jan Hendrick Oort (1900–1992). The cloud is usually known as the Oort cloud, sometimes as the Öpik–Oort cloud. Direct observations of the Oort cloud are still beyond our instrumental capabilities. Models suggest that today the cloud needs to consist of about 10^{13} comets (perhaps 10^{12}), in orbits confined to a spherical shell centred on the Sun and extending from about 10^3 AU (perhaps 10^4 AU) to about 10^5 AU (Figure 3.13). In spite of the huge number of comets in the cloud, their mass is estimated to be only of order 10^{25} kg, about the same as the Earth's mass.

The outer reaches of the Oort cloud are at a significant fraction of the distances between neighbouring stars in the solar neighbourhood — currently the nearest star (Proxima Centauri) is 2.7×10^5 AU away. The stars are in motion with respect to each other, so it is to be expected that from time to time a passing star will perturb these outer comets. As a result, some comets are drawn out of the Solar System, while others have their perihelion distances greatly reduced, so that near perihelion the comet develops a coma and tails, and a new long-period comet is observed. Giant molecular clouds in the interstellar medium can have similar effects to stars, as can the Galactic tide (Section 2.2.5). These perturbations on the comets in the outer Oort cloud explain the highly eccentric, large, randomly oriented orbits of the long-period comets, and the frequency with which these comets are observed. Comets reaching us from beyond the Solar System might constitute a small proportion.

One way in which the Oort cloud might itself have been created was outlined in Section 2.2.5.

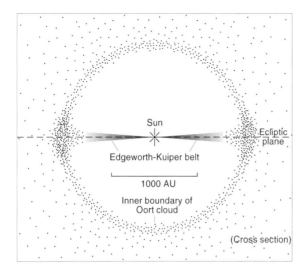

Fig. 3.13 The inner part of the Oort cloud of comets, and the Edgeworth–Kuiper belt of comets

❐ Which region was a particularly copious source of Oort cloud comets?

The Uranus–Neptune region would have been a particularly copious source of icy–rocky planetesimals that were ejected to form the Oort cloud.

The Edgeworth–Kuiper belt

It used to be thought that the short-period comets were long-period ones that had had their orbits perturbed by the giant planets. However, detailed simulations have failed to produce an essential feature of the orbits of the short-period comets — orbital inclinations predominantly less than 35°. By contrast, it is easy to produce this feature from a source population already in low-inclination orbits. Therefore, it has been proposed that the main source of the short-period comets is a low-inclination belt of comets extending from about 34 AU (somewhat beyond the orbit of Neptune) to about 45 AU.

Because the short-period comets have unstable orbits, and because of devolatilisa-tion, the population of active short-period comets needs to be resupplied. A reservoir of about 10^{10} comets is needed to meet the required rate, and the resupply would occur in two stages. First, an orbit is modified by gravitational perturbations, partly by the outer planets, but particularly by the larger members of the belt itself, perhaps of order 10^3 km across. Second, if the new orbit is such that the comet can approach a giant planet, a possible outcome is that the orbit is further modified into one typical of a short-period comet.

Such a belt is also expected for a different reason. As outlined in Section 2.2.5, it is predicted that beyond Neptune, icy–rocky planetesimals were left over after the Solar System formed, and that accretion was so ineffective out there that they still remain. Such a belt of left-overs is expected to extend well beyond 45 AU, perhaps to 10^3 AU, and to contain about as many comets as the Oort cloud. The short-period comets would be derived from the inner belt; further out, the original population would have changed little. In 1943 the Anglo–Irish astronomer Kenneth Essex Edgeworth (1880–1972) proposed the existence of such a belt, and eight years later so did the Dutch–American astronomer Gerard Peter Kuiper (1905–1973). It is therefore called the Edgeworth–Kuiper (E–K) belt (sometimes, the Kuiper belt).

The great majority of E–K belt comets would be too small to be seen from Earth, though discs of dust observed around some other stars might be associated with comparable belts there. The inner part of the E–K belt is, however, sufficiently close that its larger members should be visible, and in Section 2.2.5 you learned that about 60 members have been detected. The first was discovered in 1992, and has the name 1992 QB_1 (QB_1 identifies when in 1992 it was discovered). It is about 200 km across and occupies an orbit with a semimajor axis of 43.8 AU, an eccentricity of 0.088, and an inclination of 2.2°. By mid 1998 over 60 objects in this trans-Neptunian region had been discovered, mostly in low-eccentricity, low-inclination orbits, and with comparable sizes to 1992 QB_1. In addition, a few 100 km E–K objects have been seen around 45 AU, though this is fewer than expected in this size range. The Hubble Space Telescope has possibly detected about 30 E–K objects a few km across, in a patch of sky a few arc minutes wide. With this new perspective, Pluto and its satellite Charon could be the largest members of the belt, Neptune's large satellite Triton, which has a peculiar orbit (Section 2.3.1) and resembles Pluto, could have been captured from the belt, and also some of the small icy–rocky satellites.

Though the inner part of the Edgeworth–Kuiper belt is thought to be the main source of the short-period comets, models show that at least some of the Halley-type comets could have the inner Oort cloud as their source. Rather more Halley-type comets are then expected to have been detected, but observational selection effects might be leading to under-representation.

If we regard asteroids as bodies that have always lacked sufficient icy materials to have the potential to generate a coma, then the Centaurs (Section 3.1.2) should not be regarded as asteroids — they are undoubtedly rich in icy materials. For example, in 1988 when it was several years past aphelion, Chiron developed a coma, and it has had one ever since. Chiron is still beyond Saturn, and at such a large heliocentric distance the ices involved must be very volatile, such as CO, CH_4, and N_2. The Centaurs are in unstable orbits, and could be the larger members of a population in transition from E–K belt orbits to short-period comet orbits. Some other asteroids in cometary orbits could have feeble comas and tails that have so far escaped detection, though some such asteroids could be rocky planetesimals flung far from the asteroid belt during the formation of the Solar System (Section 2.2.4).

The icy composition of comets depends on where they were formed. Water ice always dominates, but the proportions of the minor components depend on the temperature of formation. The Jupiter–Saturn region was the warmest place of formation, and the icy–rocky planetesimals there would have contained negligible proportions of the very volatile ices. Those in the cooler Uranus–Neptune region, and that now dominate the Oort cloud, would have contained significant highly volatile proportions, and those that formed in the E–K belt even larger proportions.

Question 3.8

Discuss which feature(s) of the orbits of Halley-type comets indicate an inner Oort cloud origin for some of them, rather than an origin in the Edgeworth–Kuiper belt.

3.3 Meteorites

3.3.1 Meteors, Meteorites, and Micrometeorites

You have probably seen a 'shooting star', a bright streak of light that flashed across the sky for a second or so before disappearing. You might even have been lucky enough to see a spectacularly bright example, called a fireball, or a bolide if it explodes. These phenomena are caused by **meteors**, small bodies that have entered the Earth's atmosphere at great speed, mostly in the range 10–70 km s^{-1}. Sometimes the sonic boom produced by the supersonic speed of the body can be heard. They ionise the atmosphere as they travel, and become very hot themselves. The streak of light is the glow from the ionisation. In space, the parent body of a meteor is typically less than a few millimetres across.

❐ What are such bodies called?

Such bodies are called micrometeoroids, or dust if smaller than about 0.01 mm (Section 3.1).

Most meteors vaporise completely at altitudes above 60 km. Some of the larger ones reach the ground, perhaps fragmenting before they do so, or on impact. A fragment, or the whole object from which a set of fragments came, is called a **meteorite**. Meteorites that are seen to fall are unsurprisingly called falls, and there can be no doubt that a fall came from the sky. For four falls there are sufficient observations of the path through the atmosphere for accurate orbits to have been obtained. These orbits resemble those of the near-Earth asteroids, suggesting an ultimate origin in the asteroid belt.

The majority of meteorites are not seen to fall, but are found on the Earth's surface some time later. Naturally, these are called finds. You might wonder why a rock on the ground should be thought to have fallen there from the sky. One indicator is a fusion crust on its surface (Plate 25a). This is evidence of high-speed travel through the atmosphere. Some of the meteorite burns off in a process called ablation, and the fusion crust is the millimetre or so layer of heat-modified material overlying almost pristine material underneath. However, though this indicates that a rock has arrived at a location via a rapid passage through the atmosphere, it does not establish that it came from interplanetary space. This can be established from detailed study of its structure and composition, a topic for Section 3.3.2.

Deserts and the Antarctic ice sheets are particularly good places to find meteorites, because small rocky bodies on the surface stand out. Also, in the Antarctic, ice flows concentrate meteorites into glaciers, where subsequent sublimation of ice exposes meteorites long buried in the ice.

A typical unfragmented meteorite is of the order of 10 centimetres across, and has a mass of a few kilogrammes. Bigger parent bodies tend to fragment. For example, the known fragments of the Murchison meteorite that fell near the town of Murchison, Australia in 1969 amount to about 500 kg. The larger the meteorites, the rarer they are. A meteorite of the mass of Murchison, or larger, will arrive at the Earth's surface roughly once a month, but most of these land in the oceans, or in remote areas where they go undiscovered.

Conversely, the smaller the meteorites, the more common they are. The really small ones, a few millimetres or less across, are placed in a separate category called micrometeorites. One type is found in abundance in ocean sediments, where their nature is recognised through their spherical form. They are resolidified small bodies that melted in the atmosphere, or resolidified droplets from larger bodies. At sizes below about 0.01 mm, the most common type is a fluffy aggregate of tiny particles, also found in sediments, but also collected by high-flying aircraft. These have traversed the Earth's atmosphere without melting because they are slowed down before they reach their melting temperatures. In many cases those collected might be fragments of larger fluffy aggregates. These dust particles float gently to Earth, and are so common that if you spend a few hours out of doors, even as small a target as you is likely to collect one. Alas! you do not recognise this extraterrestrial mote among all the dust of terrestrial origin that you collect.

Overall, extraterrestrial material is currently entering the Earth's atmosphere at a rate of about 10^8 kg per year, mainly in the form of meteors that completely vaporise.

❐ What is this as a fraction of the Earth's mass?

This is only just over 1 part in 10^{17} of the Earth's mass. This infall has a variety of sources, as you will see in Section 3.4.

Question 3.9

Why are most meteorites never found? (The answer lists four short reasons.)

3.3.2 The Structure and Composition of Meteorites

Three main classes of meteorite are defined: stones, stony-irons, and irons. Figure 3.14 shows these classes in the relative numbers in which they occur in *falls*. Finds are excluded because of a strong observational bias that favours irons. As their name suggests, irons are composed almost entirely of iron, and resemble more or less rusty lumps of metal. Stones, as their name suggests, look superficially much like any other stone. Irons thus look much odder than stones, with the result that a much larger fraction of irons are found than of stones. Furthermore, some stones suffer more rapid degradation than irons.

Irons, as has just been mentioned, consist almost entirely of iron. This is alloyed with a few per cent by mass of the metal nickel, and small quantities of other materials. Naturally occurring terrestrial iron is almost always combined in compounds with non-metals, and so an extraterrestrial origin for irons is at once suspected, particularly given the variety of geological environments in which irons are found. This suspicion can be reinforced by cutting an iron, polishing the fresh surface, and etching it with a mild acid. A pattern emerges like that in Plate 25(b). This is called a Widmanstätten pattern, after the Austrian director of the Imperial Porcelain works in Vienna, Alois von Widmanstätten (1754–1849), who discovered the pattern in 1808. The pattern arises from adjacent large crystals that differ slightly in nickel content. The large size of the crystals is the result of very slow cooling, 0.5–500 K per Ma, indicating that solidification took place deep inside a body at least a few tens of kilometres across. Such slow cooling in the rare bodies of metallic iron in the Earth's crust is extremely unusual.

Stones are constituted mostly of various sorts of silicate, though small quantities of iron and nickel are usually present, plus other substances.

❐ What do you think are the main constituents of a stony-iron meteorite?

A **stony-iron** is a mixtures of iron–nickel alloy and silicates, with small quantities of other materials (Plate 25(c)).

Stones comprise about 95% of all falls (Figure 3.14) and presumably of all meteorites. The two subclasses are chondrites and achondrites. **Achondrites** are defined on the basis of something that they (and the irons and stony-irons) *haven't* got, namely chondrules. But the great majority of stones do have chondrules, and they are accordingly called **chondrites** (Plate 25(d)). A chondrule is a globule of silicates, up to a few millimetres diameter, thought to have been formed by the rapid cooling of liquid droplets. Such droplets could have been produced by electrical discharges in the sheet of dust in the solar nebula, or by impacts between

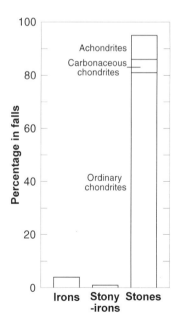

Fig. 3.14 The proportions of the three main classes of meteorite in falls, with stones divided into their main subclasses

planetesimals. By contrast, the silicates outside the chondrules would have been formed by condensation of nebular gas directly to the solid phase. Chondrules are not found in terrestrial rocks.

Ordinary chondrites are the most abundant sort of chondrite. In the matrix in which the chondrules are embedded there are more silicates and 5–15% by mass iron–nickel alloy. The alloy further distinguishes the ordinary chondrites from terrestrial rocks. The **carbonaceous chondrites** are distinguished by a few per cent by mass of carbonaceous material, and up to about 20% water bound in hydrated minerals. The presence of these volatile components suggest that the carbonaceous chondrites have suffered little heating since they formed. Moreover, they are not fully compacted, indicating that they have never been greatly compressed. These are two of the indicators that carbonaceous chondrites have never been in the interiors of bodies more than a hundred or so kilometres across. They are therefore primitive, in that they have been little altered since their formation.

Further evidence that the carbonaceous chondrites are primitive bodies comes from the relative abundances of the chemical elements in them. Apart from the depletion of hydrogen, helium, and other elements that would have been concentrated in the gas phase of the nebula, the abundances in carbonaceous chondrites are similar to those in the Sun, which is an unbiased sample of the solar nebula. This indicates that these meteorites are not from differentiated bodies, because on fragmentation this would lead to non-solar ratios in each fragment. Therefore, in carbonaceous chondrites it seems we have the least altered samples of the materials that condensed from the solar nebula when the Solar System was forming.

As well as volatile compounds, carbonaceous chondrites also contain white inclusions that are rich in non-volatile calcium and aluminium minerals such as corundum (Al_2O_3) and perovskite ($CaTiO_3$). Unsurprisingly these are called calcium–aluminium inclusions, and are thought to be refractory condensates from the solar nebula. Radiometric dating (Section 3.3.3) shows that some chondrules solidified a few Ma after some inclusions, so the melting of inclusions might be a further source of chondrules.

More evidence that meteorites of all classes are of non-terrestrial origin comes from the isotope ratios of certain elements, such as oxygen. The isotope ratios are strikingly different from those found in the Earth's crust, oceans, atmosphere, and Antarctic ice. In most cases the non-terrestrial ratios are consistent with general Solar System values. However, in many meteorites there are tiny refractory grains with very different ratios, indicating that these grains have survived from before the birth of the Solar System. It is thought that they were condensed in the winds from various stars. Isotope ratios in a small proportion of organic molecules and hydrated minerals in certain meteorites indicate that these molecules formed in the interstellar medium, and therefore have also survived the formation of the Solar System.

Question 3.10

In what sort of meteorite would you expect the ratio of carbon to iron to be much the same as that of the Sun? Why are the helium to carbon ratios far smaller in such meteorites than in the Sun?

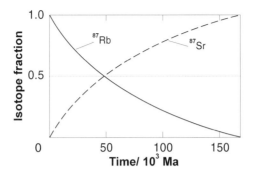

Fig. 3.15 The principle of radiometric dating using the decay of ^{87}Rb into ^{87}Sr

3.3.3 Dating Meteorites

There are various events in the life of a meteorite that can be dated, but we shall concentrate on two important ones, namely, the time that has lapsed since a meteorite, or a component within it, last solidified, and the time for which a meteorite was exposed to space rather than protected by some overlying material.

Imagine that a component in a meteorite solidifies from a gas or a liquid, and that this component contains minerals that include the chemical element rubidium. A small proportion of the rubidium atoms will be of the unstable isotope ^{87}Rb that radioactively decays to form the stable strontium isotope ^{87}Sr. Assume that initially there is no ^{87}Sr in the component, but that it builds up as the ^{87}Rb decays, and that neither of these isotopes escape from the component. The relative quantities of ^{87}Sr and ^{87}Rb in the component thus change with time as in Figure 3.15. If, at some time, we measure the ratio N_{Sr}/N_{Rb}, where N_{Sr} and N_{Rb} are the numbers of atoms of ^{87}Sr and ^{87}Rb respectively, then this measurement will tell us how long ago the mineral last solidified, *provided* that we know the rate at which ^{87}Rb disappears. Such rates are known, and are usually expressed as the **half-life** — the time for half the atoms to disappear. For ^{87}Rb the half-life is 48 800 Ma.

❏ What would be the value of N_{Sr}/N_{Rb} after 48 800 Ma, and after twice this time?

After 48 800 Ma, there would be an equal number of the two isotopes, so the ratio would be 1.0. After a further half-life N_{Rb} will have again halved and N_{Sr} would have increased by half, so the ratio would be 1.5/0.5, i.e. 3. This general method of dating is called **radiometric dating**.

Thus, by measuring an isotope ratio, and knowing the half-life of the unstable isotope, we can calculate the solidification age. The method is complicated by the possibility that both isotopes were present in the component on solidification, which could leave us to overestimate the age. The extent to which each was present can be determined from the relative abundances of various other isotopes in the meteorite component, and so allowance can be made.

The solidification ages of meteorite components concentrate in the range 4450–4530 Ma. The Solar System is thus taken to be only a bit older than 4500 Ma. Furthermore, small components of chondrites, such as the calcium–aluminium inclusions, could not have survived in isolation for more that a few Ma, and so the formation of meteorite parents must have been fairly rapid. This is consistent with the time scale of the formation of planetesimals in Chapter 2. Some solidification ages are younger, but very few are less that 1600 Ma. These younger ages are the result of some later melting or vaporisation that reset the radiometric clock.

The time for which a meteorite has been exposed to space is obtained from the action of cosmic rays on the parent meteoroid. **Cosmic rays** are atomic particles that pervade interstellar space, moving at speeds close to the speed of light. They are primarily nuclei of the lighter elements, notably hydrogen. When a cosmic ray strikes a solid body it will penetrate up to a metre before it stops, leaving a track, and creating unstable and stable isotopes via nuclear reactions. The quantities of these isotopes increase with the duration of the exposure, and so, by measuring the quantities among the tracks, and knowing the cosmic ray flux in interplanetary space, the cosmic ray exposure age can be calculated.

Many meteorites have exposure ages considerably shorter than their solidification ages — strong evidence that solid bodies larger than the metre or so cosmic ray penetration depth have been disrupted in space long after they solidified. Some stones have particularly short exposure ages, as little as 0.1 Ma. This is presumably because stony materials are less strong than iron, and are thus more readily broken in collisions and eroded by dust. This gives a constant supply of unexposed material in which cosmic rays can lay their tracks.

Question 3.11

If a certain meteorite is a piece of a larger body, how could it nevertheless have an exposure age far greater than the time ago that it was liberated from the larger body? Why could its calculated exposure age never exceed its solidification age?

3.4 The Sources of Meteorites and Micrometeorites

3.4.1 The Sources of Meteorites

A clue to the sources of the meteorites is in the few known orbits of the parent meteoroid, which resemble the orbits of the near-Earth asteroids (Section 3.3.1).

❐ What does this suggest is the source region of these meteorites?

This suggests an ultimate origin in the asteroid belt. The short cosmic ray exposure ages of the stones supports this conclusion, the ages being consistent with the high rate of collisional disruption expected in the asteroid belt, and the relatively short times before many of the meteoroids so generated will collide with the Earth. Many meteorites show evidence of collisional disruption, notably in minerals that have been shocked, and in a structure indicating broken fragments that have been cemented together. Sometimes the fragments seem to have come from different bodies, or to have been subject to different processes. To get a meteoroid from an orbit within the asteroid belt into a near-Earth orbit, it is usually necessary for its orbit to be perturbed by Jupiter or perhaps by Mars.

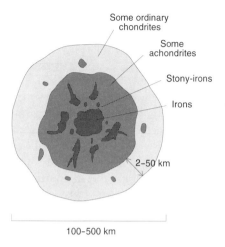

Fig. 3.16 A partially differentiated asteroid, showing regions from where various sorts of meteorite could originate

Further support for the view that meteorites are derived from the asteroids comes from comparing the reflectance spectra of the various classes of asteroid with those of the various classes of meteorite. As noted in Section 3.1.5, a clear correspondence exists between the carbonaceous chondrites and class C asteroids. The carbonaceous chondrites are presumably collisional fragments of an asteroid that never became sufficiently heated to lose its carbonaceous materials and hydrated minerals, and was far too cool to differentiate. The parent asteroid might itself have been a fragment of a larger unheated body.

There is also a clear correspondence between irons and class M asteroids. As pointed out in Section 3.1.6, early in Solar System history the larger asteroids (a few hundred km across, or larger) could have become warm enough to fully or partially differentiate. Figure 3.16 shows the resulting layered structure in the partially differentiated case. The Widmanstätten pattern in irons is indicative of the slow cooling that would occur in the iron core of a large asteroid. Fragmentation of the asteroid can expose the core, which itself could subsequently be fragmented. The core or its fragments are the class M asteroids, and the smaller fragments are the parent meteoroids of the iron meteorites. A complication is that the magnesium-rich silicate called enstatite could be mixed with iron–nickel without betraying its presence. Therefore, some class M asteroids might be a mixture of iron–nickel with this type of silicate.

Stony-irons show some correspondence with S class asteroids.

❐ What is a possible origin of such asteroids?

These asteroids could come from the interface between the iron core and the silicate mantle of differentiated asteroids.

There are only a few asteroids that match the achondrites. For example, two sub-classes of achondrite, the eucrites and the diogenites, show a good match with the rare class V asteroids, which includes Vesta and a handful of small asteroids. These achondrites consist of the sort of silicates (basalts) that would form the crust of a fully (not partially) differentiated asteroid. Therefore, this parent must have been more than a few hundred km across, and so the small V class asteroids are probably chips off Vesta, a view supported by their orbits being similar to that of Vesta, and by HST images of Vesta that show a big impact crater (Figure 3.6(a)). Radiometric dating of these achondrites shows that the asteroid parent(s) solidified 4500 Ma ago.

The most common class of meteorite is the ordinary chondrite (Figure 3.14), and yet it was only in 1993 that an asteroid was discovered that provided a good spectral match. This is Boznemcova, and it is only 7 km across. Other candidate asteroids are the Q class, though these are few in number. A particularly promising candidate is the asteroid Hebe. This is at least 200 km across, and in 1996 its surface spectrum was shown to match that of a subclass of ordinary chondrites (H-type) that accounts for about 40% of them. Moreover, Hebe orbits near to the 3:1 resonance (Figure 3.1), and so chips off its surface would readily find their way to the Earth. Some other ordinary chondrites could originate from the outermost zone of a partially differentiated asteroid (Figure 3.16).

A more copious source of ordinary chondrites could be a large proportion of the S class asteroids. Though the spectral correspondence is not impressive, it is possible that exposure of the asteroid surfaces to space has modified their spectral reflectances such that many S class asteroids are ordinary chondritic material in disguise.

A further explanation of the (apparent) scarcity of ordinary chondrite asteroids starts with the origin of the asteroids, as outlined in Section 3.1.6. When the asteroids formed they would have been undifferentiated and with a volatile component that increased as the heliocentric distance increased. In the inner belt the composition would have been that of the ordinary chondrites, and in the outer belt that of the carbonaceous chondrites. Subsequently, their interior temperatures were raised through the decay of ^{26}Al, and by T Tauri magnetic induction heating. The larger the asteroid, the greater the temperature rise through ^{26}Al decay, and the smaller the heliocentric distance, the greater the effect of magnetic induction heating. It is thus plausible that in the inner belt only the small asteroids escaped differentiation. Ordinary chondritic material is rather weak and so collisions would have readily broken these small asteroids into pieces that are now too small to have been included in reflectance spectra surveys. Likewise, an outer undifferentiated shell as in Figure 3.16 could have been eroded away from most of the partially differentiated asteroids, leaving non-chondritic surfaces. So, even without S class, the ordinary chondrite asteroids could be out there in quantity, but predominantly less than 20 km across.

In the outer belt we see class C asteroids in abundance, indicating that large asteroids avoided differentiation, perhaps because of less magnetic induction heating. From their position in the outer belt there could well be a low probability of transfer to a near-Earth orbit, which would explain why class C asteroids are common, but the corresponding meteorite, the carbonaceous chondrite, is rare.

Question 3.12

Asteroids that are the exposed silicate mantles of *fully* differentiated asteroids are rare. Give plausible reasons for this, and hence explain why achondrites that could have come from these mantles are also rare.

3.4.2 Martian and Lunar Meteorites

Eleven meteorite finds contain minerals that are of volcanic origin, with solidification ages in the range 180–1300 Ma. The oxygen isotope ratios are similar throughout the group and are distinctly non-terrestrial. We thus seek an extraterrestrial parent body that could have produced molten rock at its surface by volcanic processes 180–1300 Ma ago. It also has to be relatively nearby, and with at most a thin atmosphere so that a huge meteorite impact could throw surface materials into space. Among our neighbours, only Venus and Mars could have had volcanic processes so comparatively recently.

❐ Venus has to be ruled out. Why?

Venus has a very thick atmosphere, inhibiting the escape of rocks. Also, the impact on Venus would have to be so violent that either the rocks would be vaporised completely or they would bear tell-tale signs of extreme violence, and these are not seen.

Mars is thus the best candidate. That Mars is indeed the parent body is strongly indicated by gases trapped within one of the eleven meteorites — these have a similar composition to the Martian atmosphere. A twelfth and much older Martian meteorite has now been identified, ALH84001. Though it has a solidification age of 4500 Ma, its Martian origin is indicated by its oxygen isotope ratios, which are similar to the other Martian meteorites. Recently (1998), a thirteenth has been added to the list. These meteorites provide us with important information about Mars, as you will see in later chapters.

Twelve meteorites from the Moon have also been found, a largely undisputed origin because of the compositional similarities with the lunar surface samples that have been returned to Earth by many lunar expeditions. Further material from the Moon might be some of the many tektites found on Earth. These are rounded glassy objects, typically 10 mm across, with a presumed volcanic or impact melt origin.

3.4.3 The Sources of Micrometeorites

Most micrometeorites are derived from bodies that in space must have been less than a few millimetres across. The great majority of bodies of this size vaporise completely in the Earth's atmosphere and account for most of the meteors. Therefore, if we can find the source(s) of the meteors we will have found the source(s) of the micrometeorites.

If you were to go out on a clear dark night, then on most days in the year you would see, on average, about 10 meteors per hour. On or around a few dates, the same each year, the hourly rates are considerably greater. These enhanced rates are called meteor showers. Just how much greater the hourly rate becomes in a shower varies from year to year, but in exceptional years rates of the order of 10^5 meteors per hour are observed; these are called meteor storms. Observations show that the meteors in a shower very nearly share a common orbit, and for many showers this orbit is the same as the orbit of a known comet. Some of the comet-related showers are included in Table 1.4.

Figure 3.17 shows how a comet gives rise to a meteor shower. Rocky particles are lost by the comet and initially do not get far away. They are estimated to vary in size from submicron dust particles to loose aggregates up to several millimetres across, and sometimes far larger. With each perihelion passage of the comet the debris accumulates, and various perturbations gradually spread it along and to each side of the orbit. The debris moves around the orbit, and when the Earth is at or near the comet's orbit at the same time as the debris, a shower results. The year-to-year variations are the result of a non-uniform distribution of debris along the orbit. The cometary origin of shower meteors is further supported by estimates of the particle densities, obtained from the rate at which the Earth's atmosphere slows them down. Values in the range $10–1000 \, \text{kg m}^{-3}$ are obtained, suggesting loose aggregates of dust particles of the sort that comets could yield.

Micrometeorites are also loose aggregates, indicating that they are comet debris that has survived atmospheric entry. This possibility is strongly supported by the composition of the micrometeorites, which is in accord with remote observations of comets and with *in situ* measurements made by Giotto on dust lost by Halley's comet. Micrometeorite composition is something like that of the carbonaceous chondrites, though sufficiently different to indicate a source other than class C asteroids. It therefore seems that most meteors, and hence most micrometeorites, originate from the rocky component of comets.

Of the meteors that do not belong to showers, most are thought to be comet debris no longer concentrated along the comet's orbit. Some meteors are, of course, asteroidal material. A few meteors have entry speeds that are so high ($>72 \, \text{km s}^{-1}$) that they might have come from beyond the Solar System. This interpretation is supported by the greater fluxes of fast meteors when the Earth is at points in its orbit when it is either travelling in the same direction through the Galaxy as the Solar System as a whole, or travelling towards nearby massive stars.

Question 3.13

Give two plausible reasons for why some meteor showers are in orbits in which no comet has been seen.

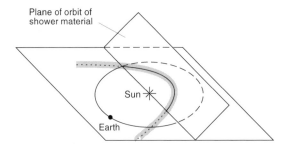

Fig. 3.17   Meteor showers and the orbit of a comet

3.5 Summary

The asteroids are small bodies, the great majority being confined to the space between Mars and Jupiter in orbits that are prograde but, on average, somewhat more eccentric and more inclined than the orbits of the major planets. The total mass of the asteroids is of order 10^{22} kg, and there are the order of 10^9 bodies greater than 1 km across. Ceres, with a radius of 457 km, is by far the largest asteroid, containing the order of 10% of the total mass.

The asteroids are thought to be the remnants of material in the space between Mars and Jupiter that the gravitational field of Jupiter prevented from forming into a major planet. Throughout Solar System history collisions in the asteroid belt have been common, and so the population has evolved considerably. There has also been a net loss of material, the present mass being only about 0.1% of the original mass.

An asteroid is categorised according to its reflectance spectrum and geometrical albedo. There are fourteen Tholen classes (Figure 3.7), with about 80% of classified asteroids falling into the S class, and about 15% into the C class. Class C dominates the outer asteroid belt and Class S the inner belt. A class C asteroid is thought to consist of an undifferentiated mixture of silicates, iron–nickel alloy, hydrated minerals, and carbonaceous materials. A class S asteroid has a surface that consists of mixtures of iron–nickel alloy and silicates. The population variations across the belt, and those seen in other classes of asteroid could be the result of lower temperatures and weaker heating at greater heliocentric distances in the solar nebula.

Comets are small bodies that are distinct from asteroids in that they develop a coma, a hydrogen cloud, and tails when within 10 AU or so of the Sun. Studies of these huge structures indicate that the solid nucleus of a comet — typically only a few kilometres across — contains a significant proportion of icy materials, particularly water, in a loose aggregation with rocky and carbonaceous materials. The volatile materials are liberated by solar radiation to form the coma and hydrogen cloud, and then driven off by the solar wind and by solar radiation to form the tails.

The comets have a wide variety of orbits. Long-period comets have orbital periods greater than 200 years and enter the inner Solar System from all directions. About 1000 have been recorded. It is inferred that they are a very small sample of a cloud of 10^{12}–10^{13} comets 10^3–10^5 AU from the Sun, called the Oort cloud. This cloud, with a present-day mass of about 10^{25} kg, is thought to consist of icy–rocky planetesimals ejected into large orbits from the giant planet region during the formation of the giants.

Short-period comets have orbital periods less than 200 years, and most have orbital inclinations less than 35°. About 180 are known. It is inferred that most of them (the Jupiter family) are a sample of a belt of 10^{10} comets in low inclination orbits at 34–45 AU. This is the inner edge of a presumed much larger belt, the Edgeworth–Kuiper (E–K) belt, planetesimals left over because of ineffective accretion in the solar nebula beyond Neptune. The remaining short-period comets, in higher-inclination orbits (the Halley-type), might in some cases be samples of the inner Oort cloud. The Centaurs could be the larger members of a population in transition from the E–K belt to the family of short-period comets.

Meteorites are small rocky bodies that survive passage through the Earth's atmosphere to reach the Earth's surface. The three classes are stones, stony-irons and irons. Stones account for about 95% of the meteorites observed to fall to Earth. Most of them contain silicate chondrules that define the subclass called chondrites, the remainder being achondrites. About 6% of the chondrites are carbonaceous chondrites, primitive bodies that contain not only silicates but also hydrated minerals and carbonaceous material. Radiometric dating shows that the components of most meteorites solidified about 4500 Ma ago — interpreted as a bit less than the age of the Solar System. A small number of meteorites have come from Mars and the Moon. Micrometeorites are largely derived from the rocky component of comets.

Spectral reflectances of asteroids and meteorites show matches between

- The carbonaceous chondrites and class C asteroids
- The iron meteorites and class M asteroids, which are the iron–nickel cores of differentiated asteroids
- The stony-irons and some S class asteroids, which are from the core–silicate mantle interfaces of differentiated asteroids
- Most achondrites and the rare V class asteroids, which are from the silicate crust of fully differentiated asteroids, of which Vesta is the only known parent.

The ordinary chondrites have close matches to very few asteroids, though many S class asteroids with space-weathered surfaces might also have an ordinary chondritic composition, as might asteroids too small to have been included in spectral reflectance surveys.

4 Interiors of Planets and Satellites: The Observational and Theoretical Basis

Our understanding of planetary and satellite interiors is considerable, but will always be limited by their inaccessibility — we have to rely on *external* observations. These are made by telescopes on the Earth or in Earth orbit, and by instruments on spacecraft in the vicinity of the planetary body. For Mars, Venus, the Moon, and, of course, the Earth, we also have observations made at the surface of the body. For the Earth and the Moon we have also sampled materials from below the surface, though for the Moon only the upper metre or so has been sampled, and even in the case of the Earth, the deepest samples are from only about 100 km — less than 2% of the distance to the centre, and brought to us by volcanoes. Nevertheless, models of planetary and satellite interiors have been developed.

Such a model has as its basic feature a specification of the composition, temperature, pressure, and density, at all points within the interior. Planets and large satellites are close to being spherically symmetrical, so, at least as a first step, this reduces to a specification of properties versus radius from the centre, or equivalently, versus depth from the surface. Not all the features of a model are independent. For example, the density of a substance depends on the pressure and temperature. However, models usually retain more than the minimum number of features, partly because some of the relationships between them are poorly known, and partly because some observational data are very powerful indicators of a specific feature, such as gravitational data for density.

A model will embody certain physical principles. For example, if the material at some depth is neither rising nor falling then the net force on it in the radial direction must be zero. The model is then used, with initial depth profiles of the various features, to derive properties that are observed externally, and the depth profiles are varied until an acceptable level of agreement with the actual observations is obtained.

We will now have a look at the main types of external observations that are available for modelling planetary and satellite interiors. Table 4.1 lists some

Table 4.1 Some pioneering missions of planetary exploration by spacecraft

Object	Spacecraft Name	Country of origin	Encounter date(s)[a]	Some mission features
Mercury	Mariner 10	USA	March 74, Sept. 74, March 75	The only mission — flybys
Venus	Venera 4	USSR	Oct. 67	First penetration of Venusian atmosphere
	Venera 7	USSR	Dec. 70	First lander
	Pioneer Venus Orbiter	USA	Dec. 78–Oct. 92	First global radar mapping
	Magellan	USA	Aug. 89–Oct. 94	Detailed radar mapping; most recent mission
Earth	Explorer 1	USA	Jan. 58	First artificial satellite to yield scientific data
Moon	Luna 3	USSR	Oct. 59	First images of lunar farside
	Apollo 11	USA	July 69	First manned landing
	Clementine	USA	March–Apr. 94	First polar orbiter
	Lunar Prospector	USA	Jan. 98–	Polar orbiter; most recent mission
Mars	Mariner 9	USA	Nov. 71–Oct. 72	Orbiter; first global survey
	Viking 1 Orbiter	USA	June 76–Aug. 80	Orbiter; second global survey
	and Lander	USA	July 76–Nov. 82	First soft lander
	Viking 2 Orbiter	USA	Aug. 76–Jul. 78	Orbiter; third global survey
	and Lander	USA	Sep. 76–Apr. 80	Second soft lander
	Mars Global Surveyor	USA	Sep. 97–	First polar orbiter; most recent orbiter
	Mars Pathfinder	USA	July–Sep. 1997	Lander, and first rover; most recent lander
Jovian system	Pioneer 10	USA	Dec. 73	First flyby
	Pioneer 11	USA	Dec. 74	Second flyby
	Voyager 1	USA	March 79	Third flyby
	Voyager 2	USA	July 79	Fourth flyby
	Galileo Orbiter	USA	Dec. 95–	First orbiter and probe; most recent mission
Saturnian system	Pioneer 11	USA	Sept. 79	First flyby
	Voyager 1	USA	Nov. 80	Second flyby
	Voyager 2	USA	Aug. 81	Third flyby; most recent mission
Uranian system	Voyager 2	USA	Jan. 86	First flyby; the only mission
Neptunian system	Voyager 2	USA	Aug. 89	First flyby; the only mission
Asteroids	Galileo	USA	Oct. 91	Images of Gaspra
	Galileo	USA	Aug. 93	Images of Ida
Comets	Giotto	ESA[b]	March 86	First image of comet nucleus (Halley)
			July 92	Close approach to comet Grigg–Skjellerup

[a]This is the encounter date. For orbiters and landers a range of dates is given if the encounter spanned more than a month.
[b]European Space Agency.

pioneering spacecraft missions that have made a particularly important contribution to the observations that will be described.

4.1 Gravitational Field Data

4.1.1 Mean Density

The mean density of a body is an important indicator of bulk composition. It is said to be an important *constraint* because it places useful limits on the range of possibilities. If the volume of a body is V and its mass is M, then the mean density is given by

$$\rho_m = M/V \tag{4.1}$$

For a sphere, $V = (4/3)\pi R^3$ where R is the radius. Planets and satellites are not perfectly spherical; Saturn, for example, is clearly flattened by its rotation (Plate 16). We can, however, measure the shapes of planetary bodies sufficiently accurately for the actual volumes to be obtained.

There are various ways of measuring the mass M. You have seen in Section 1.4.4 how the mass of the Sun can be obtained from the semimajor axis and period of the orbit of a planet. The same procedure can be applied to get the mass of a planet itself, from the orbit of a satellite or spacecraft around it. If the mass of the satellite or spacecraft is m, then from Newton's laws of motion and gravity we get an equation like equation (1.6) (Section 1.4.4), which can be rearranged as

$$M + m = \frac{4\pi^2 a^3}{GP^2} \tag{4.2}$$

where a is the semimajor axis of the orbit of m with respect to M, P is the orbital period, and G is the gravitational constant. If m is much less than M, it can be omitted from equation (4.2), and we get the value of M without knowing the value of m.

If m is not negligible, for example the Moon, then we can still get M by finding the centre of mass of the system. It is the centre of mass that moves in an elliptical orbit around the Sun, and M and m are each in orbit around the centre of mass, as you saw in Section 1.4.5. The position of the centre of mass can be determined from these orbital motions, and the ratio of M and m is then given by an equation like equation (1.7) (Section 1.4.5):

$$M/m = r_m/r_M \tag{4.3}$$

where r_M and r_m are the distances of M and m from the centre of mass at any instant. From equations (4.2) and (4.3) we can obtain both masses, and we can then use equation (4.1) to get the mean density of each body in turn.

The two bodies do not have to be in orbit around each other. Though the equations are different, the masses can be obtained from the change in their trajectories as they pass each other. If the one body has a much smaller mass than the other, then only the smaller body's trajectory will change appreciably. This is the case, for example, when a spacecraft passes near a planet.

Table 4.2 lists the mean densities, and other data, for the planets and larger satellites. For a wider range of bodies Figure 4.1 shows the mean density along with the radius. Radius and density are more indicative of composition than density alone, as you can see if we consider two bodies with the same mean density but with greatly different radii. You might think that, with equal mean densities, the two bodies could have the same composition. This is not so. The internal pressures in the larger and thus more massive body will be much greater than in the less massive body.

❑ What effect does this have on the mean densities?

This results in greater compression in the more massive body resulting in a greater density. The hypothetical **uncompressed mean density** of the more massive body must be lower than that of the other body, and therefore the more massive body must contain a higher proportion of intrinsically lower-density substances.

In Figure 4.1 some distinct groups can be recognised. On the basis of size alone the planets can be divided into four giants, four terrestrial planets, and Pluto. With the mean densities included, this broad division is reinforced by a strong indication of compositional differences between the groups. Consider the giants. It is clear that these bodies are so large that the compression is considerable. If the compression were slight their mean densities would be lower than those in Figure 4.1 by a large factor. The mean densities of the other bodies would be much less reduced, and so the already clear distinction between the giants and the rest would sharpen. The giant planets thus form a quite distinct group.

For the remaining bodies in Figure 4.1, the uncompressed mean densities of Venus and the Earth would be about 20% less than the values shown, with smaller reductions for the rest. Therefore, the groupings among the non-giants are not sharply defined. The bodies in the centre-right region, namely the four terrestrial planets plus the Moon, Io, and Europa, constitute the **terrestrial bodies**. The remaining large satellites, namely Ganymede, Callisto, Titan, and Triton, group with Pluto to constitute the **icy–rocky bodies**, a term that reflects their composition. The other bodies in Figure 4.1 are the two largest asteroids, namely Ceres and Pallas, and nearly all of the intermediate sized satellites. These satellites have mean densities comparable to those of the icy–rocky bodies, while those of the asteroids are more comparable to the terrestrial bodies, indicating broad compositional affinities in each case.

In Section 3.1.4 you saw that there is a threshold size below which a body could have an irregular shape. Bodies large enough to be approximately spherical are often called **planetary bodies**. The threshold radius is very roughly 300 km, and therefore among the bodies in Figure 4.1, Enceladus, Mimas, Miranda, and Pallas could be irregular.

Question 4.1

Use Figure 4.1 to argue that the giant planets divide into two subgroups on the basis of size and composition.

4.1.2 Radial Variations of Density: Gravitational Coefficients

You have seen that to obtain the mean density of a body we must first measure its mass. This measurement, however it is made, relies on the gravitational force that the

Table 4.2 Some physical properties of the planets and larger satellites[a]

Object	Equatorial radius[b]/ km	Flattening	Mass/10^20 kg	Mean density/ kg m^{-3}	Sidereal rotation period[c]/ days	J_2[d]	C/MR_e^2	Mag dipole moment[e]/ A m^2	Energy outflow[f]/ 10^{-12} W kg^{-1}
Mercury	2 440	~0	3 302	5430	58.646	~0.00006	?	4×10^{19}	?
Venus	6 052	~0	48 690	5240	243.019	~10^{-7}	?	$<8.4 \times 10^{17}$	6
Earth	6 378	0.00336	59 740	5520	0.9973	0.001901	0.3308	7.9×10^{22}	5.9
Mars	3 397	0.00649	6 419	3940	1.0260	0.001959	0.365	$<2 \times 10^{18}$	3.3
Moon	1 738	~0	735	3340	27.3217	0.0002037	0.394	$<1 \times 10^{17}$	15
Io	1 821	~0	893	3530	1.769	0.00186	0.378	$\leqslant 4 \times 10^{20}$	~1000
Europa	1 565	~0	487	3020	3.551	0.000389	0.347	~0	?
Ganymede	2 635	~0	1 490	1940	7.155	0.000127	0.311	1.4×10^{20}	?
Callisto	2 405	~0	1 075	1850	16.689	0.000034	0.358	$<2 \times 10^{18}$	?
Titan	2 575	~0	1 346	1880	15.9?	?	?	?	?
Triton	1 350	~0	214	2070	5.877	?	?	?	?
Pluto	1 150	?	150	2100	6.3872	?	?	?	?
Jupiter	71 490	0.0649	18 987 000	1330	0.4135	0.014750	0.264	1.54×10^{27}	94
Saturn	60 270	0.098	5 685 000	700	0.4440	0.016479	0.21	4.6×10^{25}	90
Uranus	25 560	0.0229	866 200	1300	0.7183	0.003516	0.23	3.8×10^{24}	0–10
Neptune	24 765	0.0171	1 027 800	1760	0.6712	0.00354	0.29	2.0×10^{24}	18

[a] A '?' denotes that no useful value for the quantity has been obtained, and '~0' indicates an (unknown) extremely small value.
[b] For the giants the equatorial radius is to the atmospheric altitude where the pressure is 1 bar (10^5 Pa).
[c] For the giants, the equatorial period is given.
[d] For the giants J_2, J_4, and J_6 are used in preference to C/MR_e^2 to constrain radial variations of density.
[e] To convert to T m^3 divide these values by 10^7.
[f] This assumes that Venus has much the same energy outflow as the Earth.

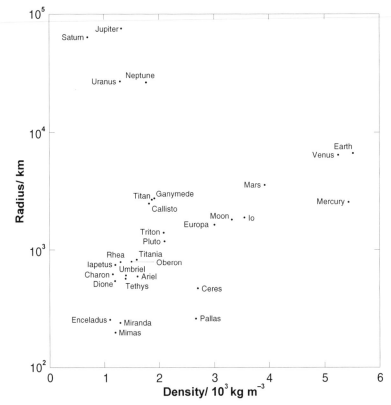

Fig. 4.1   Radii and mean densities of Solar System bodies

body exerts on some *other* body. From more detailed measurement of this force we can go further, and obtain information on the variation of the density from point to point in the body.

Newton's law of gravity (Section 1.4.4) tells us that the magnitude of the gravitational force that a body of mass M exerts on a body of mass m at a distance r from M is given by

$$F = GMm/r^2 \tag{4.4}$$

This equation underpins equation (4.2), and requires either that M and m are point masses or that they are spherically symmetrical, i.e. the density varies only with radius. In this latter case r is measured from the centres of the bodies, and we have to be outside them for the equation to apply. The direction of the force on m is towards the centre of M.

❑ By measuring F and using equation (4.4) can we learn anything about how the density inside a spherically symmetrical body varies with radius?

For a spherically symmetrical body we can learn nothing from external gravity measurements about the variation of density with radius: all possible variations of density with radius for a given total mass M would give the same external force. Fortunately, each Solar System body exerts a gravitational force that is not quite that given by equation (4.4), and therefore we *can* learn something about the variation of density with radius.

It is useful to introduce the concept of **gravitational field** — this will allow us to ignore the mass m, which is of no interest in itself. For *any* gravitational force F on a body of mass m at a point in space (not just the force in equation (4.4)) the magnitude g of the gravitational field at the point is defined as the force per unit mass, i.e.

$$g = F/m \qquad (4.5)$$

The direction of the field is the same as the direction of the force F. We can now refer to the gravitational field of M without specifying the mass of the body used to measure it. For example, from equations (4.4) and (4.5) it follows that

$$g = GM/r^2 \qquad (4.6)$$

and so g is independent of m. Like equation (4.4), equation (4.6) applies if M is a point mass or if it is spherically symmetrical. More generally, note that any gravitational force F at a point in space can be thought of as the combination of the forces of a set of point masses, and so g is independent of m for any mass distribution. From Newton's second law we see that g is the gravitational acceleration of m.

Because the gravitational forces exerted by planetary bodies do not conform to equation (4.4), it follows that their gravitational fields do not conform to equation (4.6). Instead, the magnitude of the field is given by

$$g = GM/r^2 + \text{ extra terms} \qquad (4.7)$$

The extra terms at some point of measurement P depend on the position of P. Figure 4.2(a) shows P at some distance r from the centre of mass of a planetary body, with the direction to P specified in terms of the angles θ and ϕ. For planetary bodies the extra terms are small, and the largest is approximately equal to $J_2 \, (GMR^2/r^4) f(\theta)$ where $f(\theta)$ varies with θ, though not with ϕ or r. J_2 ('jay two') is the **gravitational coefficient** of this term. It measures the strength of the extra term, and a particular planetary body will have a particular value of J_2. Note the rapid decrease with r — as $1/r^4$. This means that at large distances this extra term becomes negligible, but that at sufficiently small distances it is appreciable. The other extra terms decrease no less rapidly with r.

For most planetary bodies the main source of the J_2 term is the rotation of the body. This flattens the sphere as in Figure 4.2(b), and it is the rotational flattening of the body that generates this extra term. For a small degree of flattening the planetary body is an oblate spheroid (Figure 1.13(a), Section 1.4.5) — note that it is symmetrical around the rotation axis. The rotational flattening (also called the oblateness) is quantified by $f = (R_e - R_p)/R_e$, where R_e is the equatorial radius and R_p is the polar radius. Figure 4.2(b) also shows how, at some fixed value of r, the magnitude of the total field g varies with direction when J_2 is the only extra term. The field is symmetrical around $\theta = 0$ (the rotation axis) and it is a maximum at $\theta = 90°$. The larger the value of J_2, the greater the variation of g with θ. With $J_2 > 0$ the field

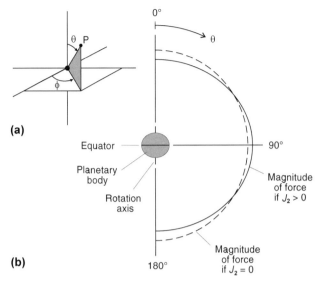

Fig. 4.2 (a) Specifying the position of a point P. (b) The variation of gravitational field with θ at a fixed distance r from the centre of a planetary body with $J_2 = 0$ and $J_2 > 0$

now points towards the centre of the body only at $\theta = 0$, 90°, and 180° — this is not apparent in Figure 4.2(b) because magnitudes alone are shown, not directions.

Regardless of its origin, the J_2 term can be deduced from the gravitational field as mapped by small bodies in the vicinity of the planetary body. J_2 provides a constraint on the variation of density with radius, in that only certain variations are consistent with the measured value of J_2. However, a more powerful constraint is obtained if J_2 is combined with other data to obtain a property of the body called its polar moment of inertia.

Question 4.2

In doubling the distance r from a planetary body, show that the gravitational field becomes more closely like that of a spherically symmetrical body.

4.1.3 Radial Variations of Density: the Polar Moment of Inertia

A moment of inertia is a quantity involving a body and some axis. Any axis can be used, and the corresponding moment of inertia calculated. If we imagine the body divided up into small equal volumes δV, as in Figure 4.3, then the moment of inertia for the chosen axis is the sum of $(\rho \delta V)x^2$ over the whole body, where x is the perpendicular distance from δV to the axis, and ρ is the density in the volume δV. ($\rho \, \delta V$ is the mass δM in δV.) The moment of inertia therefore depends on the choice of axis. For a rotating planetary body the rotation axis is a natural choice, in which case we have the **polar moment of inertia** C. C depends on the variation of density from place to place in the body, which is why it is a powerful constraint.

C can be obtained from the precession of the rotation axis (Section 1.5.1). Precession is caused by the gravitational torque on the non-spherical mass

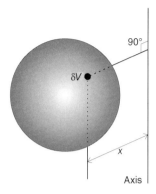

Fig. 4.3 The basis of calculating a moment of inertia

distribution of the rotating body. For the planets the torque is provided by any large satellites and by the Sun. In order to calculate C, the precession period has to be known, and also the mass M of the rotating body, its equatorial radius R_e, its sidereal rotation period T, and J_2. So far it has been possible to apply this precession method with useful accuracy only to the Earth, the Moon, and Mars.

To obtain C for the other planetary bodies we have to assume that the interiors are in **hydrostatic equilibrium**, i.e. that the interior has responded to rotation as if it had no shear strength — as if it were a fluid. Solid materials in planetary interiors can achieve such equilibrium in times *very* short compared to the age of the Solar System because of the huge internal pressures that overwhelm the shear strength of the solids. For a planetary body in hydrostatic equilibrium, the value of C can be calculated from M, R_e, J_2, and T — the details will not concern us. If J_2 is unknown, then the rotational flattening f can be used instead. In fact, J_2 is preferred to f because any departures from hydrostatic equilibrium are likely to be confined to the outer layer of the body, which influences f more than J_2. Furthermore, terms additional to the J_2 term can reveal departures from hydrostatic equilibrium, and these can be used to adjust J_2 so that C can be calculated more accurately.

Table 4.2 lists the values of $C/(MR_e^2)$ for those planetary bodies for which C is known. The division of C by MR_e^2 is useful, because it gives an at-a-glance indication of the degree of concentration of mass towards the centre. For a hollow spherical shell $C/(MR_e^2)$ is 0.667 (2/3 exactly), for a sphere with the same density everywhere it is 0.4 exactly, and for a sphere with its mass concentrated entirely at its centre it is zero.

The giant planets are so rotationally flattened that J_2 is comparatively large, and further terms in equation (4.7) are also significant. Their strength is measured by the gravitational coefficients J_4 and J_6 (the odd terms like J_3 and J_5 are zero because of the northern–southern hemisphere symmetry of the planets). In this case J_2, J_4, and J_6 are in themselves a useful constraint on radial variations of density, and for the giant planets are often used in preference to C.

Question 4.3

Show that $C/(MR^2) = 1$ exactly for a hollow, cylindrical shell, with a radius R and a mass M, for rotation around its longitudinal axis.

4.1.4 Local Mass Distribution, and Isostasy

Close to the surface of a planetary body the gravitational field is not only sensitive to radial variations of density, but also to *local* mass distributions. For the modelling of planetary interiors one of the most important things we can learn from such data is whether the planet is in **isostatic equilibrium**. A familiar example of isostatic equilibrium is a block of ice floating on water, as in Figure 4.4(a),(b). The ice is less dense than the water, and in equilibrium floats with its base at a certain depth such that in any vertical column of given cross-sectional area extending from this depth to the surface, there is the same amount of mass. This is where the term 'isostatic' comes from — 'equal standing'. If the ice is raised and released there is a net downward force on it, and if it is pushed down and released there is a net upward force on it. Because liquid water is very fluid the ice quickly regains its equilibrium level, and isostatic equilibrium is restored rapidly. The minimum depth above which there is equal mass in each column is called the depth of compensation.

Imagine measuring the gravitational field at a fixed altitude above the water surface. If the ice is in isostatic equilibrium then the field changes very little over the block, because of the 'equal standing'. Thus, in spite of the mountain of ice, gravity hardly varies across it. By contrast, if there is no isostatic equilibrium the amount of mass in a vertical column changes as we cross the ice, and so the variations in the field are larger. By measuring the field as we cross the ice we can determine whether the ice is in isostatic equilibrium.

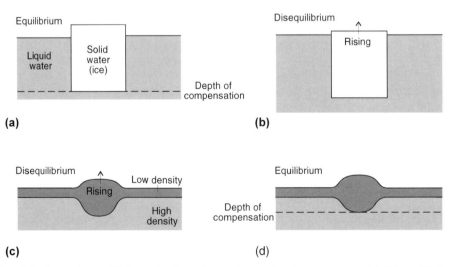

Fig. 4.4   Isostatic equilibrium. (a) Equilibrium in ice floating on water. (b) Disequilibrium with ice and water. (c) Disequilibrium at a planetary surface. (d) Equilibrium at a planetary surface

❐ How does the field vary if the ice is below its equilibrium level?

If the ice is below its equilibrium level there is a deficit of mass in its vicinity and so the field is slightly weaker than to each side. If it is above its equilibrium level there is an excess of mass and the field is slightly stronger.

A planetary body typically has a crust of less dense solids of non-uniform thickness on top of a mantle of more dense solids. For the crust to be in isostatic equilibrium the underlying mantle has to be able to flow in response to departures from isostasy, and the upper layer must be able to deform to take up the equilibrium shape. This is illustrated in Figure 4.4(c), (d). Given sufficient time isostasy will be achieved. The greater the plasticity of the interior and the greater the flexibility of the surface layer, the shorter the time required. The plasticity of the interior increases with temperature and pressure, and depends on composition. The flexibility of the surface layer also increases with temperature and depends on composition, and it also increases as its thickness *decreases*. It is quite possible for the adjustment time to be many thousands of years, or longer, and so departures from isostasy are to be expected.

4.2 Magnetic Field Data

Magnetic fields are caused by electric currents. Therefore, if a planetary body has a strong magnetic field there must be large electric currents within it, and this can tell us a lot about the interior.

Whereas mass is the source of gravitational field, electric charge in motion — electric current — is the source of **magnetic field**. For magnetic field, the basic source is an electric current loop, and a circular one is shown in Figure 4.5(a). At distances that are large compared to the distance across the loop the magnetic field is independent of the shape of the loop and the field is then called a dipole field. It has the form shown schematically in Figure 4.5(b), where the loop is presented edge-on. The lines show the direction of the magnetic field, and their spacing is a qualitative indication of the magnitude of the field. Note that the field is rotationally symmetrical around the magnetic axis, which is perpendicular to the loop and passes through its centre. Near to the loop the spatial form depends on the shape of the loop.

❐ The basic source of gravitational field is the point mass. Describe the equivalent drawing to Figure 4.5 in this case.

The gravitational field lines of a point mass would be radial, and would point inwards. The increase in the spacing of radial lines as distance from the mass increases is a qualitative representation of the decrease in the magnitude of the field with distance.

At any point of measurement at a given distance and in a given direction, the magnitude of the magnetic field is proportional to $i \times A$, where i is the electric current circulating around the loop, and A is the area of the loop. The product $i \times A$ is the magnitude of the magnetic dipole moment, μ. The *direction* of the magnetic dipole moment is the same as the direction of the magnetic axis.

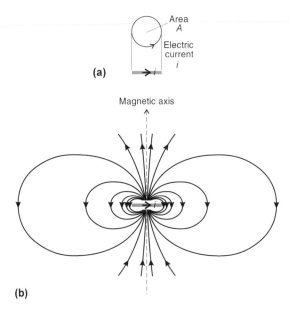

(a)

(b)

Fig. 4.5 (a) A circular electric current loop. (b) Dipole field due to a current loop

The mechanism for sustaining magnetic fields in planetary bodies is not fully understood; indeed, some astronomers regard it as very poorly understood. The general idea is that the process starts with an electrically conducting fluid in motion because of thermal convection. The presence of a small external magnetic field will set up electric currents in the fluid that generate further magnetic fields. This influences the fluid motion in such a way that the electric currents are enhanced and modified so that the field strength is increased until there is as much energy in the magnetic field as in the fluid motion. It is plausible that if the rotation of the planetary body is sufficiently rapid then the fluid motions are coordinated such that a strong magnetic field is established. Once the field is established, it is probably unnecessary to have a small external magnetic field to sustain it. The crucial role of rotation leads to a fundamental theorem, the Cowling theorem, which states that the field and the rotation cannot have the same symmetry. This means that the rotation and magnetic axes cannot coincide — there must be an angle between them.

The energy in the planetary magnetic field comes from the electric currents and this in turn comes from the kinetic energy of the fluid. Conversion of kinetic to magnetic energy also occurs in an electrical dynamo, so this mechanism is called the **self-exciting dynamo**. It is also thought to sustain the Sun's magnetic field. The fluid motions tends to decline through viscous dissipation, and the currents themselves tend to diminish through the electrical heating they cause. These losses need to be offset by internal energy sources to prevent the field decaying.

The currents inside a planetary body are not a single simple loop as in Figure 4.5, and consequently the near field is very different from that of a simple loop. The departures from dipole form near the body can be used to constrain the distribution of the electric currents in the interior, just as the extra terms in the gravitational field can be used to constrain point-to-point variations in density. Nevertheless, the existence of a dipole field at sufficiently great distances means that we can specify the strength of the source of the field by the corresponding magnitude of a magnetic dipole moment. The known values are given in Table 4.2. From a large dipole moment we infer the presence of a considerable body of electrically conducting fluid in motion within the planetary body.

Figure 4.6 shows the magnetic axes of the five planetary bodies that have large magnetic dipole moments—the Earth, and the giant planets. Where the magnetic axis passes through the surface of the body we have the magnetic poles. The magnetic equatorial plane (Figure 4.6) is perpendicular to the magnetic axis, and they intersect at a point that is the centre of the far field. You can see that this point is displaced from the centre of the planetary body, and so the currents are not symmetrically distributed around its centre. The magnetic axis does not coincide with the rotation axis—except for Saturn. Saturn thus violates the Cowling theorem, and this creates unease about the self-exciting dynamo as the mechanism sustaining planetary fields. It is, however, the best theory we have.

Magnetic fields are subject to small variations in magnitude and in the direction of the magnetic axis—those in Figure 4.6 are for the present time. These variations can be traced through rocks that retain the imprint of past magnetisation. One way in which this can happen is through the solidification of a magnetised molten rock. This remanent magnetism can be used to trace the history of the dipole field, particularly if the time of solidification has been radiometrically dated. For the Earth it is known that the field direction has varied erratically, though keeping roughly the same inclination with respect to the rotation axis. In addition, the dipole component of the field has reversed in direction many times. Each reversal takes about 0.01 Ma, and during it the dipole field declines in strength to zero, and then grows in the opposite direction. During the last 80 Ma these magnetic reversals have occurred at intervals of 0.1–1 Ma; the last time this happened was about 0.7 Ma ago. Numerical simulations of the dynamo theory as applied to the Earth's interior do, in some cases, exhibit magnetic reversals.

Other sources of magnetic fields, from electric currents in the upper atmosphere to permanently magnetised rocks at and below the surface, are weak, and are sufficiently disorganised on a planetary scale that their net contribution to the dipole field is zero.

Question 4.4

Making use of what you learned about the planets in Chapter 1, outline reasonable hypotheses based on the self-exciting dynamo mechanism to explain why the Earth and the giant planets have large magnetic dipole moments, but the other planets do not.

4.3 Seismic Wave Data

4.3.1 Seismic Waves

A **seismic wave** travels through a planetary body, and bears information about the conditions along its path. Seismic waves are mechanical waves in that the wave

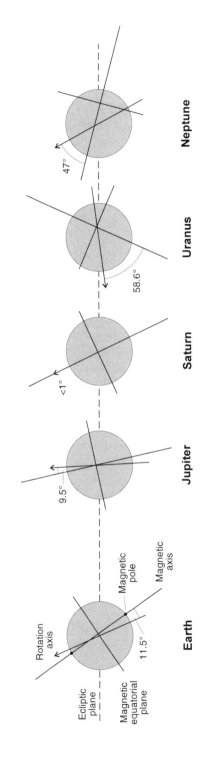

Fig. 4.6 The magnetic axes of the planetary bodies that have large dipole moments. North is at the top

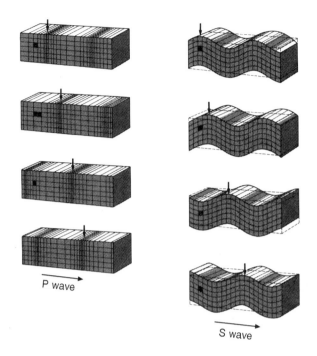

Fig. 4.7   P and S waves

proceeds via the direct push of an atom or molecule on its neighbours. Other mechanical waves include sound waves in the air and the ripples on the surface of a pond. There are several types of seismic wave. Rayleigh and Love waves, rather like water ripples, can exist only at an abrupt change of density, such as at the surface of a planet. But it is two other types that have been of particular value in elucidating the structure of planetary interiors. These types are illustrated in Figure 4.7. In a P wave the oscillatory motion of the substance is longitudinal — to and fro in the direction in which the wave progresses, as in a sound wave in the atmosphere. In an S wave the oscillatory motion is transverse — perpendicular to the progression. S waves require the medium to have shear strength, so they do not travel in liquids or in gases, and they are strongly attenuated in soft, plastic materials.

Seismic waves can be generated by surface impacts, by surface explosions, and by the sudden yielding of interior rocks in which strains (distortions) have built up as a result of some applied stresses. These stresses can result from the tidal action of external bodies, or from internal heating. Considerable surface motion occurs in the vicinity of the source, and in the case of the Earth this constitutes an earthquake. Source sizes are typically only a few kilometres, and seismic waves travel from the source in all directions. For a planet with a liquid core and a solid mantle some typical paths are shown in Figure 4.8. The curvature of the paths is a consequence of the gradual increase in wave speed with depth. You can see that the path bends away from the direction in which the speed is increasing.

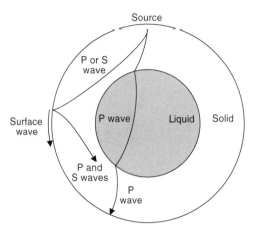

Fig. 4.8 Seismic wave paths in a hypothetical planet. (In reality, the P and S waves would follow different paths in the solid material)

The importance of P and S waves is that they can reach a considerable depth in the planetary body. For example, in Figure 4.8 the P wave that enters the liquid core passes near the centre of the body and reaches the surface on the far side. If it is detected there, and if we know the source position and the time that the wave set off, we can calculate the average speed of P waves along the path.

❐ If P waves sets off at time t_1 from a shallow depth, pass through the centre of a planet with a radius R, and are detected at t_2 on the opposite surface of the planet, what is their average speed?

The average speed is the total distance divided by the total time: $2R/(t_2-t_1)$. If average speeds are determined for many different paths, then we can deduce the P wave speed at all points in the interior, and not just as an average along a path. A similar feat can be performed for S waves. What can we learn from these data?

The P wave speed in a material depends on just two of its properties — its density ρ and its axial modulus a. The axial modulus is one of several elastic moduli of a material. In general, an elastic modulus is the stress to which a material is exposed, divided by the resulting strain. For P waves the relevant stress is the rapid application of a force along the direction of wave motion, and the resulting strain is the compression or rarefaction along the wave direction with no expansion or contraction of the material perpendicular to this direction (Figure 4.7). In this case the relevant modulus is called the axial modulus, and the P wave speed is given by

$$v_P = \sqrt{\frac{a}{\rho}} \qquad (4.8)$$

It is beyond our scope to derive this equation, but you can see that it depends on a and ρ in a reasonable way. Thus, if the material is less deformable then a is larger and it passes the wave along more rapidly. On the other hand, if it is denser then it is more sluggish in its response, and the wave travels less rapidly.

For S waves the speed is given by

$$v_S = \sqrt{\frac{r}{\rho}} \qquad (4.9)$$

where r is a different modulus, called the rigidity (or shear) modulus. This is a rapidly applied shear stress divided by the resulting shear strain (Figure 4.7).

Equations (4.8) and (4.9) show that if we know the P and S wave speeds at a point in a material then we know something about the density and the elastic moduli there. Materials differ in these properties so we seem to have valuable information on composition. Unfortunately, because we have just *two* pieces of information — S wave speed and P wave speed — but *three* unknowns — density and two elastic moduli, we cannot deduce the density and the elastic moduli. Therefore, we have only a weak constraint on composition. Tighter constraints are obtained by combining seismic wave speeds with gravitational and other data. The outcome is that the density versus depth is strongly constrained, even deducible with sufficient data, and this provides a strong constraint on composition.

Seismic data alone do, however, provide other types of information on the interior. Depth ranges that are liquid, and therefore inaccessible to S waves, can be identified through the absence of S waves over that range. Also, if the S and P wave speeds change suddenly at a certain depth, this is a strong indication of a change in composition of some sort.

For a body that is fluid throughout, seismic wave speeds can be obtained from the Doppler effect of the wave motions on visible spectral lines in the radiation from its outer regions (Section 2.1.2). In the case of the Sun such studies constitute helioseismology, a well-established technique that has yielded much information about the solar interior. In the near future it might be possible to apply a similar technique to spectral lines in the outer atmospheres of the giant planets.

4.3.2 Planetary Data

To obtain the P and S wave speeds versus depth requires a large network of seismic observatories well spread over the surface of a planetary body, with data from a large number of seismic sources. This has only been achieved for the Earth, and Figure 4.9(a) shows the P and S wave speeds versus depth in our planet.

❒ What is the reason for the absence of S waves from 1215 km to 3470 km from the centre?

A liquid shell is clearly indicated by this absence of S waves. At all depths above and below the shell S waves do exist, and so there are no other *extensive* regions that are liquid. (Note that S waves are generated beneath the liquid shell by P waves that traverse the shell.)

Layering of the Earth is also indicated by large changes in the wave speeds over very short depth ranges, and these generally indicate changes in composition at these boundaries. Between these boundaries there is a general increase of speed with depth. Given that the density in planetary interiors increases with depth this would seem to

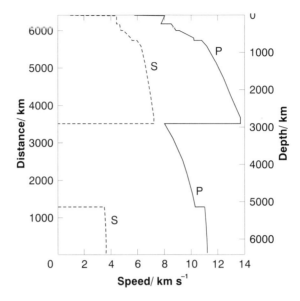

(a)

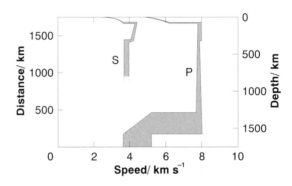

(b)

Fig. 4.9 Seismic wave speeds in (a) the Earth (b) the Moon

be contrary to equations (4.7) and (4.8) which show a *decrease* in speed for an increase in density. However, it is also the case that the elastic moduli of most substances increase with an increase in density (though decrease with an increase in temperature), so the outcome is not obvious in advance. The seismic data are so extensive for the Earth that small changes have been detected in the wave speed profiles, and some of these correspond to slow convective motions within the various layers.

Figure 4.9(b) shows the P and S wave speeds versus depth in the Moon, the only other body for which there are data at present. These are from seismometers set up at the landing sites of Apollos 11, 12, and 14–16 (Table 4.1). The large uncertainties in the speeds, indicated by the width of the graphs, are the combined result of only having these five seismic stations, their location all on the near side of the Moon, some not working very well, and the Moon being seismically less active than the Earth, so the rate of 'moonquakes' is lower and, on average, they are weaker. Nevertheless, it is clear that at depths greater than about 800 km S waves are strongly attenuated, indicating that the material deeper down has little shear strength. At shallower depths there are indications of composition changes. The data in Figure 4.9 will be further interpreted in Chapter 5.

Question 4.5

Sketch the variation of seismic wave speeds versus depth in an imaginary planet that has both of the following properties:

(1) An entirely fluid interior
(2) A change in composition half-way to the centre, with the material just below the boundary having twice the axial modulus of the material just above it, and a density 20% greater.

4.4 Composition and Properties of Accessible Materials

4.4.1 Surface Materials

Even though the only planetary materials available for direct compositional analysis are from the surface or near surface, the composition of such materials can give us a useful indication of the composition, temperature, and density deeper down. For example, the Earth's surface consists of rocks with a density of about 2600 $kg\,m^{-3}$ on the continents, and about 2900 $kg\,m^{-3}$ in the ocean basins. (In many regions these rocks are covered by a very thin veneer of soil or sediment, but that is of no relevance here.) By contrast, the Earth has a mean density of 5520 $kg\,m^{-3}$, and it is not possible for the internal pressures to be high enough to squeeze the surface rocks to such a high mean value. Therefore, the interior *must* consist largely of intrinsically denser materials.

As well as densities, we can also constrain the range of minerals that could comprise the interior. This is done by choosing a mineral mix that when subjected to various geological processes that produce surface rocks, would produce those that are actually in place. We will look at this more closely in Chapters 5–8.

For the giant planets there is no solid surface, and they are all covered in atmospheres that blend into deep outer envelopes. Therefore, determination of the composition of these atmospheres is of particular importance in building models of their interiors.

4.4.2 Elements, Compounds, Affinities

A further powerful constraint on composition is provided not by observations of the body in question but of other bodies, notably the Sun and meteorites. The relative abundances of the elements in the Solar System are obtained mainly from the

Table 4.3   The densities of some important substances

Substance	Chemical formula	Density^a/kg m^{-3}
Silicon carbide	SiC	3217
Corundum	Al_2O_3	3965
Iron–nickel	Fe plus $\sim$6% Ni by mass	7925
Diopside^b	$CaMgSi_2O_6$	3200
Forsterite^c	Mg_2SiO_4	3270
Alkali feldspars	$(Na, K)AlSi_3O_8$	2700
Troilite	FeS	4740
Hydrated minerals^d	$X(H_2O)_n$ or $X(OH)_n$	$\lesssim$2000
Water (liquid)	H_2O	998
Carbon dioxide^e	CO_2	1.98
Carbon monoxide^e	CO	1.25
Hydrogen^e	H_2	0.08987
Helium^e	He	0.179

^{a}At 293 K and 1.01×10^5 Pa unless otherwise indicated.
^{b}A particular form of pyroxene, $(Ca, Fe, Mg)_2Si_2O_6$.
^{c}A particular form of olivine, $(Mg, Fe)_2SiO_4$.
^{d}X can be a molecule of a variety of rocky minerals, and n is greater or equal to 1.
^{e}As a gas, at 273 K.

composition of the observable part of the Sun (Section 1.3), and from the composition of the least altered meteorites, the carbonaceous chondrites (Section 3.3.2). Table 1.5 shows the relative abundances of the fifteen most abundant chemical elements. It is to these elements, and to compounds containing them, that we must look to account for most of the mass of a planetary body. In doing so, we must take account of the density of the substance. Table 4.3 lists the densities of some important substances, and distinguishes the more dense from the less dense.

For example, suppose that we are building a model of the Earth on the basis of its mean density. In spite of their abundance we can rule out the two lightest elements hydrogen and helium as important constituents — the Earth is *far* too dense to contain much of these intrinsically low-density substances. But when it comes to choosing between iron–nickel (7925 kg m^{-3}) and zinc (7140 kg m^{-3}) for the dominant constituent of a dense central core, density alone is a poor guide — both substances are denser than the mean density of the Earth. There are many grounds for choosing iron (and nickel), but among them is the relative abundances of the elements in the Solar System — iron is about 1000 times more abundant than zinc.

Of course, most elements will be present as compounds, and of particular importance are abundant rocky materials, notably silicates, and abundant icy materials, notably water.

❏ Do silicates and water include abundant elements?

Silicates are based on the abundant elements oxygen and silicon (Section 2.2.2), and water (H_2O) is a compound of oxygen with the most abundant element of all, hydrogen. Rocky materials are intrinsically denser than icy materials, and icy materials are intrinsically denser than hydrogen and helium.

Chemical affinities are also important. These express the tendency for certain elements to occur together. Elements can be grouped on the basis of these affinities, and three groups of particular relevance to planetary interiors are the siderophiles, chalcophiles, and lithophiles. The **siderophiles** ('iron-lovers') comprise iron, and various metallic elements that tend to be present with metallic iron, such as nickel. The **chalcophiles** ('sulphur-lovers') tend to form compounds with sulphur. Zinc is one example. The **lithophiles** ('rock-lovers') tend to be found in silicates and oxides— magnesium and aluminium are important examples. Iron is so abundant that it occurs in sulphides, oxides and silicates, as well as in elemental form.

To illustrate how chemical affinities can be used, consider a planetary body with a surface that is heavily depleted in iron. It can then be argued that iron has been concentrated into the interior, perhaps into a central core of metallic iron. Support for this possibility would be a surface depletion in the siderophilic element nickel and an enrichment in the lithophilic element aluminium.

4.4.3 Equations of State, and Phase Diagrams

In deciding whether a substance could be a significant component at some depth in the interior of a planetary body, we need to know its density at the pressure and temperature at that depth. The equilibrium relationship between the density of a substance and its pressure and temperature is known as the **equation of state** of the substance. This equation not only allows us to calculate the density at some depth, it also enables our models to achieve consistency. Thus, if we specify a substance at a certain depth, we are only free to specify two of density, pressure, and temperature—the equation of state determines the third.

For the great majority of plausible substances for the interiors of most planetary bodies, pressure is by far the more important determinant of density. Though the temperatures in the deep interior are far higher than at the surface, the thermal expansion is slight, resulting in only a slight decrease in density. By contrast, the increases in density arising from the great pressures deep in the interior are far larger. So, to a first approximation we can ignore temperature and concentrate on pressure.

Pressures in the interior

To find the pressures in the interior we use the fact that if a body is neither contracting nor expanding then the pressure at any radius must be just right to support the overlying weight. Consider a thin spherical shell of material of inner radius r and thickness δr, where δr is small compared to r, as in Figure 4.10. To a first approximation we can assume that the planetary body is close enough to spherical symmetry so that we can regard its density as varying only with radius. In this case it is remarkable but true that the mass above the shell exerts zero net gravitational force at all points in the shell. The net gravitational force on the shell due to the rest of the body is thus the gravitational force exerted on it by the inner sphere of radius r; the net force is thus downwards, towards the centre. There is also a downward force on the shell due to the pressure at $r + \delta r$, and an upward force due to the pressure at r. In equilibrium the total downwards force must equal the upwards force, and so the pressure at r must exceed that at $r + \delta r$. It follows that *pressure increases with depth at all values of r*. This is an important general result.

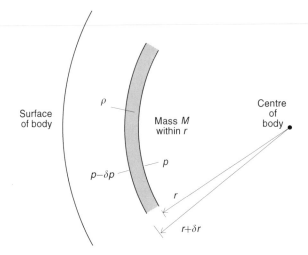

Fig. 4.10   The balance of forces on a spherical shell in a planetary interior

If p is the magnitude of the pressure at r, and $p - \delta p$ that at $r + \delta r$, then the net upward force F_{up} on the material in the shell has a magnitude $A \times \delta p$, where A is the surface area of a sphere of radius r. Thus

$$F_{up} = 4\pi r^2 \times \delta p$$

The magnitude of the gravitational force on the shell is obtained from Newton's law of gravity, equation (1.5) (Section 1.4.4).

❐ For a spherically symmetrical body of radius r, where can all of its mass be regarded as concentrated for calculating the gravitational force it exerts?

It exerts a gravitational force as if all its mass were concentrated at its centre. We can thus use equation (1.5) as it is, with M as the mass of the inner sphere and m the shell mass. Therefore, the gravitational force has a magnitude GMm/r^2 where G is the gravitational constant. The mass of the shell is its density times its volume, i.e. $\rho(4\pi r^2 \times \delta r)$, and so

$$F_{down} = GM \times \rho(4\pi r^2 \times \delta r)/r^2$$

In equilibrium F_{up} must equal F_{down}. Thus, from the equations for these forces, and after tidying up the algebra, we obtain

$$\delta p = -\frac{GM}{r^2}\rho \times \delta r \qquad (4.10)$$

where the minus sign expresses the decrease in p as r increases. Equation (4.10) is called the hydrostatic equation. Planetary rotation adds a small term to the right-hand side of the equation, but we can ignore it for our purposes.

The central pressure is obtained by applying equation (4.10) step by step from the surface to the centre, calculating the increase in pressure as we proceed. To obtain an exact value we clearly need to know the density versus depth, yet this is one of the things that we *don't* know until our modelling is complete. However, even without this knowledge the useful rough-and-ready result can be established from equation (4.10) that the central pressure p_c is approximately given by

$$p_c \approx \frac{2\pi G}{3} \rho_m^2 R^2 \tag{4.11}$$

where ρ_m is the mean density and R is the radius of the surface of the body.

Equation (4.11) indicates how huge the pressures are deep in the interior of a planetary body. In the case of the Earth the equation yields a value of 1.7×10^{11} Pa — over a million times greater than atmospheric pressure at the Earth's surface. For most substances the equation of state is not very well known at such high pressures. But we can still deduce whether a substance is a plausible ingredient. For example, molecular hydrogen (H_2) and atomic helium require pressures in the Earth *far* in excess of 1.7×10^{11} Pa to become even as dense as liquid water at the Earth's surface (1000 kg m^{-3}). The same is true of atomic hydrogen. Therefore, these potentially abundant substances can account for no more than a tiny fraction of the Earth's mass. Water at comparable pressures is much denser than hydrogen and helium, but still well short of the Earth's mean density. Thus, neither can water, H_2O, a compound of the first and third most abundant elements, account for much of the Earth's mass.

Equation (4.11) also shows how much greater are the interior pressures in larger, denser bodies. Thus, at the centre of the Moon, equation (4.11) gives 4.7×10^9 Pa, nearly 40 times less than at the centre of the Earth. Moreover, the higher the pressure given by equation (4.11), the more this approximate equation underestimates the true pressure — by about a factor of two in the Earth's case.

The general conclusion is that a given substance will have a markedly different density in the interiors of different planets. For comparing compositions, it is therefore useful to obtain the uncompressed mean density. Unlike the mean density it is not directly measurable, but must be calculated from equations of state, assuming a composition. Equations of state are measured in the laboratory up to pressures that fall short of those in the centres of the larger planetary bodies. For higher pressures it is necessary to do the best one can with theoretical equations or reasonable estimates.

Question 4.6

Calculate an approximate value for the pressure at the centre of Jupiter. What can you say about the pressure at shallower depths?

Phase diagrams

The pressure and temperature in the interior determine whether a substance there is liquid or solid, and this is of importance to the behaviour of the interior. You have already seen this to be the case with regard to magnetic fields and the passage of

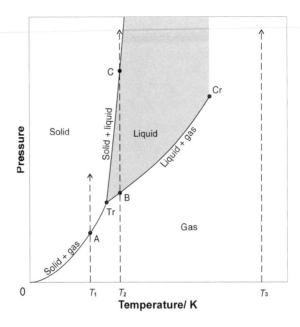

Fig. 4.11 The typical form of a phase diagram. For *very* few substances, notably water, the solid + liquid phase boundary slopes to the left, and not to the right as here

seismic waves, and more examples will arise shortly. If a substance were to be a gas in the interior it is unlikely to be retained, but would escape to the surface. We therefore need to know the ranges of pressure and temperature over which a substance will be liquid, solid, or gas. This information is provided by a **phase diagram**, 'phase' being a generic term for solids, liquids, and gases.

Figure 4.11 shows the typical form of a phase diagram. It consists of the boundaries between the phases. If the substance is in equilibrium, and its pressure and temperature are *anywhere* within the area to the left of the boundaries between solid and gas, and solid and liquid, it will be solid. Unsurprisingly, the solid phase is confined to the lower temperatures, particularly at low pressure. The liquid phase requires a minimum pressure and temperature, marked by point Tr in Figure 4.11. At each temperature up to Cr the gas phase can exist only up to a certain pressure. At temperatures above Cr the gas phase becomes difficult to distinguish from a liquid, hence the absence of a distinct boundary there. Tr and Cr are, respectively, the triple point and the critical point of the substance.

❐ What do you think is the origin of the term 'triple point'?

The triple point is so named because only at this point do all three phases co-exist. By contrast, on each phase boundary just two phases co-exist — the phases on each side of the boundary.

Numerical values have not been attached to the axes in Figure 4.11 because the pressures and temperatures of the phase boundaries vary from substance to substance. Note that the density varies across the phase diagram. The density at a specified temperature and pressure is provided by the equation of state of the substance, and so a phase diagram contains only a subset of the information encapsulated in the equation of state.

It is instructive to take a substance on three imaginary journeys across the phase diagram, each one at constant temperature and starting in the gas phase. If we start at T_1 and increase the pressure, the substance gets denser, and it then meets the phase boundary at A, where a further increase in pressure causes a huge increase in density as it solidifies. Within the solid phase there might be structural changes from one crystal form to another, which might give small density changes. If we now reverse the track the substance sublimes at A. If we start at T_2, and increase the pressure, then at B the gas condenses, with a huge increase in density, and it becomes a liquid. At C there is a much smaller increase in density as the substance solidifies. Reversing the track, the substance melts at C and then vaporizes at B. Finally, if we start at T_3, higher than the temperature of the critical point, and increase the pressure, then the density goes on increasing to liquid-like values with no sudden changes in density such as occur across phase boundaries.

The pressure and temperature of the triple point is a better measure of whether a substance is volatile or refractory than the sublimation temperature used in Section 2.2.2, which clearly depends on pressure. The lower the triple point values, the more volatile the substance; the higher the triple point values the more refractory the substance. For water the triple point values are 6.1 Pa and 273.16 K, whereas for iron the values are $5–8 \times 10^{10}$ Pa and 2000 K. Water is volatile, as are all icy materials, whereas rocky materials, which include iron and other metals, are refractory. Carbonaceous materials are intermediate.

Question 4.7

Water is unusual in that the solid–liquid phase boundary slopes to the *left* from the triple point, and not to the right as in Figure 4.11. Describe the phase changes that occur when water in the gas phase at 273.10 K is gradually compressed at this temperature, to pressures greatly in excess of 6.1 Pa.

4.5 Energy Sources, Energy Losses, and Interior Temperatures

Interior temperatures are an important property of a planetary body. For example, along with composition, they largely determine whether there are any liquids in the interior, and consequently whether a powerful magnetic field is possible. Temperatures in the past can determine whether core formation has occurred, and whether other compositional divisions have become established. Temperatures also determine the dynamics of the interior, such as the rate of any convection. And the thermal history of a planetary body helps determine the nature of the surface and of any atmosphere, as you will see in later chapters. The temperatures inside a body at any particular moment in its history depend on the accumulated effects of the energy sources and the energy losses at all earlier times.

It is important to distinguish between the internal energy of a body, its temperature, and heat. The **internal energy** consists of kinetic and potential energy. The kinetic energy is almost entirely in the random energy of motion of the particles (atoms and molecules) that constitute the body (the motion of the body as a whole is excluded). The potential energy is the energy of interaction between the particles, be this gravitational or electrical. It generally increases as the mean separation between the particles increases. The temperature of the body is proportional to the random kinetic energy per particle. **Heat** is a particular form of energy, namely, energy that is transferred *randomly* at a *microscopic* level. For example, when a body at a certain temperature is in contact with a body at a lower temperature there is a net transfer of random energy from the higher to the lower temperature region through the random collisions of the microscopic particles at the interface. 'Heat' is often used interchangeably with 'energy'. This is avoided here unless the language would otherwise sound peculiar. **Power** is the rate of energy transfer, by heat, or by any other means.

If a region is a net receiver of heat its temperature might rise, and if it is a net loser its temperature might fall. Temperature changes can also be caused without heating, by any other process that causes a rise in the random kinetic energy part of the internal energy. One example is the temperature rise when two bodies collide. However, heat transfer does not always give a temperature change. For example, heat fed into ice at its melting point does not cause a rise in temperature but a change in phase from solid to liquid. When water solidifies at its melting point it gives out heat with no change in temperature. Heat that produces no temperature change is called **latent heat**. It is the energy that has to be supplied either to increase the average separation between the particles, or to break chemical bonds between them. In either case there is an increase in the potential energy part of the internal energy.

It is important to stress that the present-day interior temperatures do not depend on the *present* energy sources and losses but on their accumulated effect right back to the formation of the body. A familiar example of this dependence on history is the heating of water in an electric kettle. The temperature of the water at any instant does not depend on the flow of electric current nor on the energy losses at that instant, but on their earlier values. We cannot therefore ignore the past when we consider energy sources and losses in a planetary body.

4.5.1 Energy Sources

Energy sources no longer active; primordial energy sources

You saw in Chapter 2 that planetary bodies are thought to have been built up from planetesimals and other bodies in a process called accretion. This led to temperature rises as the kinetic energy of the infalling material was partially converted to internal energy of the surface materials at the point of impact. For the giant planets there was also infall of nebular gas. The kinetic energy of the infall was derived from gravitational accelerations, and so this accretional energy, as it is called, came from the conversion of gravitational energy to internal energy.

If the rate of accretion was low, much of the internal energy would have been lost by radiation from the surface to space, so the interior temperatures would not have risen much. But if the rate of accretion was high, the hot surface would have been buried, leading to raised temperatures in the whole interior. For a given mass of impactor, the energy transfer is greater the more speed the impactor picks up as it is accelerated by the gravitational field of the body to its point of impact on the surface. This leads to accretional energy per unit mass being roughly proportional to $\rho_m R^2$, where ρ_m and R are the mean density and radius respectively of the planetary body. Therefore, in large bodies, rapid accretion will lead to extensive interior melting.

Accretion is a primordial source of energy, in that accretion is not continuing today on an important scale. But the *effect* of accretional energy lives on, as you will see later. Another primordial source is any magnetic induction during the T Tauri phase of the protoSun (Section 3.1.6). This could have given considerable temperature rises in bodies of asteroid dimensions.

The other primordial source of importance was short-lived radioactive isotopes. Of particular importance was ^{26}Al, which decays to ^{26}Mg. Studies of isotope ratios in meteorites indicate that this isotope accounted for a significant proportion of the aluminium present in the Solar System at its birth. ^{26}Al has a half-life of only 0.72 Ma, and so there are 64 half-lives in just the first 1% of Solar System history.

❏ By what factor did the initial quantity of ^{26}Al decrease during this time?

With a halving of the quantity every half-life the decrease in 46 Ma was by a factor of 2^{-64} of its initial quantity, i.e. 5.4×10^{-20}. ^{26}Al is thus long gone. But the shortness of its half-life and the estimates of a modest initial abundance indicate that the initial rate of energy release was high. If the isotope was distributed throughout the body, or concentrated at its centre, most of this energy would have been contained, and this **radiogenic heating** would have given considerable temperature rises in bodies of all but the smallest size. If the isotope was concentrated in the surface, then a greater proportion of its energy would have been lost to space. The loss of a greater proportion of energy when an energy source is concentrated at the surface occurs with other energy sources too.

The source of ^{26}Al, and of other short-lived isotopes, is thought to have been supernovae — massive stars ending their lives in huge explosions that generated the isotopes and implanted them in the interstellar medium from which the Solar System was born.

Differentiation

Primordial sources could have raised the temperatures of the interior of a planetary body to the point where an important process occurred that is itself an energy source, and in some bodies it might still be active today. This is differentiation (Section 3.1.6) — the separation of layers of different density. For differentiation to occur, the chemical forces binding different substances together must be weaker that the gravitational forces that tend to separate the denser substances downwards. For example, if the Earth formed with a homogeneous composition throughout (homogeneous accretion) and subsequently underwent partial or total melting, then most of the iron would drain downwards to form a core, and most of the other materials would float upwards to form a predominantly silicate mantle. This downward separation of denser materials would have converted gravitational energy into internal energy, with a consequent rise in temperature. There would have been no such energy source if the iron core formed first and the silicate mantle was acquired later (heterogeneous accretion).

Differentiation can also result from the *cooling* of the interior. For example, if two liquids are already mixed and the temperature falls, the liquids can unmix, the denser settling towards the centre of the planetary body.

Meteorites provide observational evidence that differentiation has occurred. For example, irons and achondrites are readily interpreted as fragments of an asteroid that had differentiated to form an iron core and a silicate mantle, irons being fragments of the core, and achondrites fragments of the mantle (Section 3.4.1). The Widmanstätten pattern in irons (Section 3.3.2) is just what is expected from differentiation in which a body of iron melted and then formed a core, which then cooled slowly under the insulating mantle of the asteroid. Stony-irons could come from the core–mantle interface. By contrast, the chondrites seem to represent the sort of material that was around in the asteroid belt before any differentiation occurred.

Energy sources that can still be active

Ongoing differentiation is one possible 'live' source of energy in planetary interiors. Another is latent heat released when a liquid solidifies.

A further type of active source is radiogenic heating. Of particular and widespread importance are the four long-lived radioactive isotopes, ^{235}U, ^{238}U, ^{40}K, ^{232}Th. Table 4.4 lists their half-lives and an estimate of the power they would have been releasing 4600 Ma ago in a typical rocky material. The same data are also included for ^{26}Al. You can see that, unlike ^{26}Al, their half-lives are comparable to or exceed the 4600 Ma age of the Solar System, and so these isotopes can provide energy throughout the life of the Solar System, though at a declining rate as they decay. They are distributed non-uniformly in the interiors of many planetary bodies. For example, in the Earth there is a significant concentration into the outer 30 km or so, a result of earlier partial melting at greater depths.

The other 'live' source of importance in some bodies is tidal energy.

❒ Recall what a tide is, and how a tide is caused.

In Section 1.4.5 you saw that a tide is a distortion produced by a differential gravitational force. Now that you have met the concept of gravitational field (equation (4.6)) it is better to think of this as a differential field, which we can call

Table 4.4 Radioactive isotopes that are important energy sources

Isotope	^{26}Al	^{235}U	^{40}K	^{238}U	^{232}Th
Half-life/Ma	0.72	710	1300	4500	13 900
Power[a]/10^{-12} W kg^{-1}	10^4	3	30	2	1

[a]This is per kg of 'typical' rocky material 4600 Ma ago. The values are only illustrative.

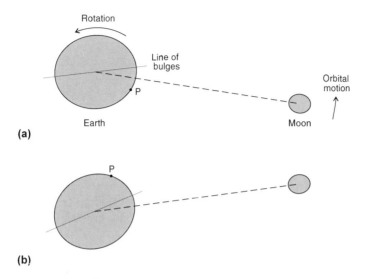

Fig. 4.12 Tides (greatly exaggerated) in the Earth due to the Moon (a) at some instant and (b) about 6 hours later, with the orbital motion of the Moon exaggerated five times. P is fixed to the Earth's surface

a tidal field. Figure 4.12 shows the elongation of the whole Earth caused by the tidal field of the Moon. The rotation of the Earth carries the elongation slightly ahead of the line connecting the centres of the Earth and Moon, but of crucial importance to tidal heating is that this rotation also moves material into and out of the elongation, as illustrated by the point P in Figure 4.12, which is fixed on the Earth's surface. From the viewpoint of the Earth's interior, the elongation sweeps like a wave through it — literally, a tidal wave. This flexing increases the interior temperatures in the Earth.

A second, smaller tidal effect is due to the variation in distance between the Earth and the Moon. As the Moon moves around the Earth in its elliptical orbit, the size of the tidal elongation varies, being greatest when the Moon is nearest to the Earth (at perigee), and least when it is furthest (at apogee). This is another way that the Earth's interior is flexed.

The Sun produces similar types of effect, but even though the Sun is far more massive than the Moon it is much further away. Therefore, the solar tidal field across the Earth is about 46% of that due to the Moon. Overall, the rate at which heat is being generated by tides in the Earth today is about 30 times less than the rate at which heat is being released from long lived radioactive isotopes. By contrast, in Io, the innermost large satellite of Jupiter, tidal energy is by far the largest internal energy source.

The ultimate source of tidal energy is the rotational energy of the bodies concerned, and the energy in their orbital motions.

❏ Is it then possible for tidal heating to be associated with changes in rotation rates, and changes in orbital motion?

This is not only possible, it *must* be the case — the internal energy gained through tides must equal a loss of energy in rotation or orbital motion.

Solar radiation

A small planetary body might for a long time have had no significant energy source except for solar radiation. In this case its interior will be at a uniform temperature determined by a steady state in which the rate at which the body absorbs solar radiation is equal to the rate at which it emits radiation to space. The mean surface temperature will equal the interior value. At a given distance from the Sun the exact value of the temperature will depend on the fraction of the incident solar radiation that the body absorbs, and on the efficiency with which it emits its own radiation. For typical planetary materials at 1 AU from the Sun, the value will be about 270 K. Most planetary bodies have far higher interior temperatures, so solar radiation is not an important energy source in such cases.

High interior temperatures do not necessarily mean that the rate of energy input from sources other than solar radiation greatly exceeds the solar input. A modest source confined to a small core would give high core temperatures and a temperature gradient to the surface. If the surface layer is a good insulator the interior can be hot throughout much of its volume. Moreover, temperature gradients are themselves important, because they can drive processes such as convection. Temperature gradients will occur whenever the energy source is not concentrated at the surface. Solar energy is concentrated in this way and therefore can never give rise to gradients in the interior.

Question 4.8

Draw graphs showing qualitatively how the rate of energy input to a planetary body from the various energy sources outlined above might have varied during the 4600 Ma of Solar System history.

Question 4.9

(a) Use equation (4.6) to show that the difference in gravitational field across a body of radius R due to a body with a mass M a distance r away, is approximately $4GMR/r^3$ if r is much greater than R.

(b) Hence show that, for the Earth, the solar tidal field is 46% of the lunar tidal field.

4.5.2 Energy Losses and Transfers

As well as gaining internal energy from various energy sources, planetary bodies also lose internal energy. In the near vacuum of interplanetary space energy is not conducted or convected away from the body, and so it is by emitting radiation that the internal energy is lost. For an atmosphereless body this radiation is from the surface. If there is a substantial atmosphere then a proportion, perhaps all, of the radiation is from the atmosphere. Overall, the radiation is somewhat like that from an ideal thermal source (Section 1.1.1), and so the wavelengths depend on the

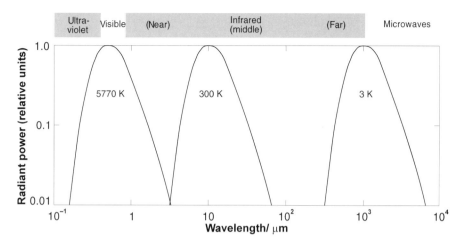

Fig. 4.13 The radiation spectra of ideal thermal sources at various temperatures

temperature of the source, as indicated in Figure 4.13. For planetary bodies the radiation is predominantly at infrared wavelengths.

In order to be radiated away, the internal energy must first reach the surface or atmosphere, and there are several ways in which this happens, as follows.

Radiative transfer

This process was outlined in Section 1.1.3, and can be thought of as an outward diffusion of photons. The rate of energy transfer increases very rapidly as temperature increases, and more sedately as the opacity of the interior decreases. Planetary interiors are too cool and too opaque for radiative transfer to be important, except perhaps in certain zones deep in the interiors of Jupiter and Saturn.

Thermal conduction

Thermal conduction is a process by which heat is transferred through the direct contact of two regions that are at different temperatures. In colliding with their neighbours, the more energetic particles in the higher temperature region lose some of their random energy of motion to the less energetic particles in the lower temperature region, just as a fast-moving ball transfers some of its energy of motion to a slower-moving ball with which it collides. There is thus a flow of random motion — of heat — from the higher- to the lower-temperature region. This flow will cool the region of higher temperature unless there is energy generation within it sufficient to offset the flow.

Clearly conduction must play a role in transporting internal energy from the interior of a planetary body to its exterior. Whether it is significant depends on the effectiveness of the two remaining mechanisms.

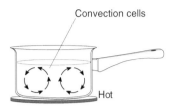

Fig. 4.14 Convection in a pan of liquid, showing two convection cells

Convection

Convection differs from conduction in that energy is transported by bulk flows, rather than on an atomic scale. It is a process of energy transfer in a substance heated from below, and requires that the substance can flow. The temperature of the heated substance increases, and it therefore expands, becomes buoyant, and ascends, displacing the overlying cooler material downwards, where it in turn is heated. The rising material loses energy to its surroundings, loses its buoyancy, and descends. This sets up the cycle exemplified in Figure 4.14 by the everyday case of heating a saucepan of liquid, where two so-called convection cells are shown in cross-section.

To obtain a closer look at convection in general, regardless of whether it is in a saucepan or in the atmosphere or interior of a planetary body, imagine a small parcel of material being swapped with a parcel immediately above it. The hydrostatic equation (equation (4.10)) shows that the raised parcel is now in a lower pressure environment. It quickly responds by expanding until its pressure is the same as the pressure at its new level. This expansion alone causes a decrease in the temperature of the parcel — no heat has had time to flow out of it, and the decrease is solely a consequence of the expansion. The lower parcel will be compressed, and this will raise its temperature, again with no flow of heat into it. If no heat has flowed into or out of a parcel, then it has undergone what is called an **adiabatic process**. Note that an adiabatic process does not rule out the transfer of energy, only of heat. Thus, energy is transferred through the expansions and compressions that takes place. Moreover, real processes are not strictly adiabatic — there will be some heat transfer by conduction and by radiation. Nevertheless, an adiabatic process is a very useful idealisation.

The crucial question is whether the new temperatures are greater or less than those of the new surroundings. Let's focus on the parcel that moves upwards. If its new temperature is greater than that of its surroundings then, the pressures being the same, its density must be lower than its surroundings. It is therefore buoyant, so it will continue to move upwards, and convection will start.

❐ What will happen if its temperature is lower than its surroundings?

Its density will then be higher than its surroundings, and it will sink back to its point of origin; there is no convection. Corresponding conclusions apply to the parcel that moves downwards.

Whether the temperature of the raised parcel is less than that of its surroundings depends on how rapidly the temperature of the surroundings decreases with altitude — this is the temperature gradient. The critical value is the adiabatic

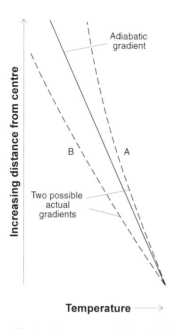

Fig. 4.15   The adiabatic gradient and some possible actual temperature gradients

gradient. If the actual temperature gradient is greater than the adiabatic value, as in case B in Figure 4.15, then a rising parcel is always hotter than its surroundings and it will continue to rise, and convection occurs. If the actual gradient is less than the adiabatic value, as in case A in Figure 4.15, then a rising parcel is always cooler than its surroundings and it will sink back, and convection will not occur. The value of the adiabatic gradient depends on the gravitational field, the greater the field, the greater the gradient. It also depends on various thermodynamic properties of the material of the parcel, though the details will not concern us.

If a substance can flow there is a strong tendency for the actual temperature gradient to become equal to the adiabatic value. If the actual gradient is greater (B in Figure 4.15), energy is transported at such a high rate that the deeper levels cool, thus reducing the gradient towards the adiabatic value. If the gradient is less than the adiabatic value then energy is transported by conduction, and this is at such a low rate that the deeper levels get hotter, thus increasing the gradient, until convection starts.

The existence of convective currents in planetary bodies does *not* make the hydrostatic equation inapplicable, even though this equation requires that the gravitational force on a shell is equal to the pressure gradient force, in which case there can be no convection. The convection currents are quite slow, and therefore the imbalance between the two forces is relatively slight. Moreover, averaged over large areas the rising and sinking motions tend to cancel out, and so the hydrostatic equation applies as a global average.

It is natural to think of convection as being confined to fluids, and indeed convection does occur in many of the fluid regions of planetary bodies. But it also occurs in some regions that are solid! Solid-state convection can occur if a solid behaves like a fluid, and this will be the case if the pressure exceeds the yield strength of the solid.

❑ In what sort of planetary bodies are the internal pressures high?

Equation (4.11) shows that high internal pressures will be found in large planetary bodies. Fluid behaviour is thus expected in a large planetary body, particularly if the yield strength is reduced by raised temperatures. The rate of convection will depend on the composition and on the buoyancy. Though the convective motions are extremely slow, for example of order 0.1 m per year in the Earth, the rate of outward heat transport can exceed that of conduction. In the case of the Earth convection is particularly well developed and is the dominant process of energy loss from the interior to the crust. Note that solid-state convection is far too slow for it to give rise to magnetic field generation.

Advection

If the uppermost part of the interior of a planetary body is solid it will be at too low a pressure and temperature to convect. But this does not leave conduction as the only mechanism of outward energy transfer. **Advection** is the process of energy transfer carried by local regions of moving liquids, such as molten rock or water. If these erupt onto the surface we have volcanic effusions of various sorts. But the liquids need not reach the surface. For example, if molten rock solidifies at some depth it has still contributed to the outward flow of energy. Table 4.5 indicates the relative importance of advection, convection and conduction for a few planetary bodies.

Heat transfer coefficient

Regardless of the energy transfer process, the intrinsic property of the material that determines the rate is encapsulated in a heat transfer coefficient. This is the rate at which heat crosses unit area of a material, per unit thickness, per unit temperature difference across the thickness. For many bodies the heat transfer coefficients are comparatively low at the surface and in any atmosphere, and so these regions can have a significant effect on the overall rate of energy loss.

Table 4.5   Mechanisms of heat reaching the surface regions of some planetary bodies today

	Conduction	Convection[a]	Advection[b]
Venus	Probably unimportant	Perhaps dominant	Perhaps dominant
Earth	Minor contribution	Dominant	Minor contribution
Mars	Dominant	Negligible	Minor contribution?
Moon	Dominant	Negligible	Negligible
Io	Minor contribution	Negligible	Dominant
Jupiter	Negligible	Dominant	Negligible
Neptune	Negligible	Dominant	Negligible

[a]Including any solid-state convection, and any large-scale lithospheric recycling.
[b]Includes any delamination (Section 8.1.2).

Question 4.10

Suppose that a rocky planetary body has a weak energy source concentrated at its centre. The interior temperatures throughout are modest, decreasing with radius, almost the entire decrease being across a thin surface layer.

(a) Explain why it is likely that the only way in which energy is being transferred up to the surface layer is by conduction.
(b) What can you deduce about the heat transfer coefficient of the thin surface layer compared to that of the rest of the body?

4.5.3 Observational Indicators of Interior Temperatures

If planetary modellers only had the various types of energy gains and losses that *might* be present, then models of the interior temperatures of planetary bodies would be very poorly constrained. Fortunately, there are observational data that are direct indicators of the thermal state. These include

- Energy flow from the interior
- Seismic waves and the level of seismic activity
- The degree of departure from isostatic equilibrium
- The level and nature of geological activity at the surface, today and in the past
- The magnetic field today, and evidence for its nature in the past.

The energy flow from the interior of a planetary body gives an indication of present-day interior temperatures. In principle the flow can be measured via the infrared radiation to space to which it gives rise, this being the way that the energy is ultimately lost by the body. If there is flow from the interior then the rate at which infrared radiation is emitted to space would exceed the rate at which radiation is absorbed from the Sun. In practice only for Jupiter, Saturn, and Neptune is this infrared excess large enough to have been measured accurately. For some other bodies the present energy flow from the interior has been obtained by estimating the outward energy flow at the surface by conduction, convection, and advection. Values for energy outflows from the interiors of some planetary bodies are given in Table 4.2.

Seismic wave speeds can provide an indirect indication of interior temperatures.

❐ If there is a zone in which S waves do not travel, what can you conclude about the physical conditions in the zone?

A zone free of S waves must have negligible shear strength, so it could be a liquid or be highly plastic. Departures from isostatic equilibrium also indicate plasticity, and changes in isostasy can give an indication of how dynamic the interior is, as can the level of seismic activity.

Geological activity is another indicator of interior temperatures, as you will see in later chapters. By examining the geological activity on surfaces of different ages it is possible to gain insight into interior temperatures not only today but in the distant past. The present-day magnetic field indicates whether there are liquids in

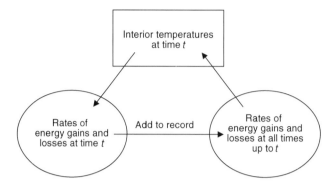

Fig. 4.16 The interplay between interior temperatures, and the rates of energy gains and losses in a planetary body

convection, and past magnetic fields leave an imprint in the rocks, and so we can infer the nature of the field in the past, and hence gain further insight into interior temperatures throughout the history of the body.

4.5.4 Interior Temperatures

It has been noted that the temperatures inside a planetary body at any time t depend on the energy gains and losses at all earlier times. But the gains and losses themselves depend on the temperatures. For example, at time t the temperature profile $T(t, r)$ will help determine whether there is any convection, and if there is any convection then this will increase the energy loss rate at t, which will influence subsequent temperatures. We thus have a complicated interplay, represented in the simplest manner by Figure 4.16. Planetary modellers have to grapple with this interplay until there is self-consistency.

Though the modelling process required to obtain the interior temperatures of a planetary body is intricate, there are three broad features that are readily appreciated. First, energy losses tend to win in the end. Energy sources can slow down the rate of decline of interior temperatures, and can hold interior temperatures steady, even for long periods, by producing energy at the same rate at which energy is lost to space. But most sources of energy have either died already, or are implanting less power as time passes. Sooner or later, interior temperatures will fall, and ultimately all bodies will come to have internal temperatures determined by the balance between solar radiation absorbed, and infrared radiation emitted to space.

Second, if sufficient time has passed since all energy sources at various depths became active, the temperature decreases monotonically as we move outwards from the centre to the surface. This is because the ultimate energy loss from the body is from its surface. The details of the temperature profile depend on the radial distribution of energy sources, and on the heat transfer coefficients at each radius. Only if all energy sources have long been concentrated at the surface will the interior temperature be the same everywhere.

❏ What can give this outcome?

This is the outcome if, for a long time, the only active source has been solar energy. Another way is through the concentration of all radioactive isotopes into the surface.

Finally, a planetary body will lose internal energy faster per unit mass than a larger planetary body, provided that the two bodies have similar heat transfer coefficients, similar interior energy sources per unit mass, and similar spatial distributions of energy sources. This is because the global rate of energy loss increases with the surface area A of the body, whereas the global rate at which energy is released in the interior increases with the mass M of a body. A/M is thus a measure of the rate of loss of internal energy per unit mass, and A/M increases as the size of the body decreases. For a spherical body of radius R, A/M is given by

$$\frac{4\pi R^2}{4\pi R^3 \rho_m /3}$$

where ρ_m is the mean density. This simplifies to

$$\frac{A}{M} = \frac{3}{R\rho_m} \tag{4.12}$$

showing indeed that as R decreases the ratio increases. Thus, in a sufficiently small body, from early in its history, any active energy sources will have been unable to compensate the energy losses, so the interior temperatures will have declined, and the interior will now be cool.

Question 4.11

Describe qualitatively the present-day temperature versus depth in a hypothetical planetary body if its *sole* energy source (apart from solar radiation) has always been each one of the following:

(a) Accretional energy confined to its surface
(b) Long-lived radioactive isotopes confined to a small central core
(c) As for (b), but in a body that is a scaled-up version of that in (b).

4.6 Summary

In order to develop our knowledge about the interior of a planetary body, a model has to be constructed and adjusted until its predictions of external observations match the actual observations as closely as possible.

A model has as its basic feature a specification of the composition, temperature, pressure, and density, at all points within the interior. For planetary bodies, variations versus radial distance from the centre are dominant.

Details of the gravitational field of the body, plus other data such as mass, radius, and rotation period, constrain the density at all points in the interior.

If the body has a large magnetic dipole moment then this is thought to be due to electrically conducting liquid layers that are convecting. If this explanation is correct, then the location of these layers can be deduced.

Seismic waves reveal the presence of liquid layers, the presence of regions that are plastic, and changes in composition. Seismic waves also constrain the composition and density versus depth, particularly if combined with gravitational and other data.

The composition is also constrained by the composition and properties of accessible materials, and by the relative abundances of the elements in the Solar System.

The temperatures in the interior of a body at any one time depend on the rates of energy gains and the rates of energy losses at all earlier times. The interior temperatures are a factor in determining whether differentiation has occurred, and they also determine the dynamics of the interior and of the surface.

Energy sources no longer active include accretional energy and heat from short-lived radioactive isotopes. Differentiation can be a primordial source of energy, or it can still be an active source. Other active sources include heat from long-lived radioactive isotopes, tidal energy, and solar radiation.

Energy is ultimately lost in the form of infrared radiation emitted to space by the surface or atmosphere of the body. To reach these outer regions energy is conveyed by radiation, conduction, convection (in liquids or solids), and advection. Radiative transfer is negligible (except perhaps in certain zones in Jupiter and Saturn), and the proportions of the overall flow carried by the other three means varies considerably from body to body.

Some insight into the thermal state of the interior of a body is provided by energy flow measurements, by gravitational and seismic data, by its history of geological activity, and by the magnetic field now and in the past.

There are three broad features of the thermal history of a planetary body:

(1) Except for very early on in Solar System history, interior temperatures have either been roughly constant, or have been declining.
(2) The temperatures today increase with depth.
(3) If all other things are equal, the smaller the planetary body, the faster it loses internal energy per unit mass.

Table 4.2 summarizes some of the observational constraints relevant to interior models of various planetary bodies. Other relevant data are given in Tables 1.5, 4.3–4.5, and in Figures 4.6 and 4.9.

5 Interiors of Planets and Satellites: Models of Individual Bodies

The models outlined here are of the interiors of the planetary bodies as scientists believe them to be today. The evolution of planetary interiors, from the past to the present, and into the future, is largely deferred to later chapters because of the consequences for surfaces and atmospheres.

Recall from the previous chapter that a model of the interior of a planetary body has, as its basic feature, a specification of the composition, temperature, pressure, and density at all radii from the centre. From this basic specification, other things follow, such as internal motions, magnetic fields, and so on. The model is arrived at through applying physical principles to calculate the observed properties of the body, and the model is varied until an acceptable level of agreement with the observations is obtained. It is invariably the case that a *range* of models can be made to fit the observational data. The range will be wide if the data are uncertain, or if some data are absent, such as seismic data. Therefore, a model is not unique, though for most bodies certain features are beyond reasonable doubt.

In considering individual planetary bodies, it would be a very lengthy task to relate all the various features of a model to the observational and experimental data that underpin it. Therefore, I will highlight just a few examples where specific data are strongly suggestive of specific model features.

5.1 The Terrestrial Planets

Figure 5.1 shows models of the interiors of the terrestrial planets that highlight the main compositional divisions within each of them. A common feature is an outer **crust** overlying a **mantle**, which itself overlies a central core. In considering the various layers in Figure 5.1, it is important to realise that it is cross-sections through

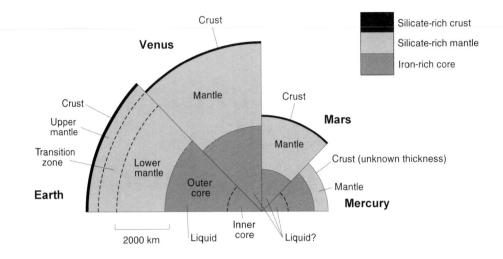

Fig. 5.1   Cross-sections through the terrestrial planets

spherical volumes that are shown. Therefore, it is easy to get the wrong impression of the volume ratios of the different layers. For example, from a superficial look it might seem that the volume of the Earth's core (inner plus outer) is a bit over half the total volume of the Earth—it extends a little over half-way to the Earth's surface. But the volume ratio is actually about one sixth. A much better impression is obtained from the cutaway drawing in Figure 5.2.

For all the terrestrial planets there is sufficient observational data to indicate that the increase of density with depth is too great for the body to consist solely of the material at the surface, getting denser as the pressure increases with depth. Gravitational data, as outlined in Section 4.1, have been of particular importance

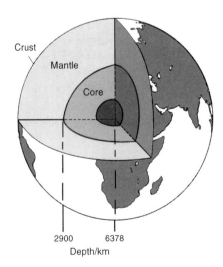

Fig. 5.2   Cutaway section through the Earth

in establishing density profiles. In accord with the observational data, the bodies are shown differentiated, with the intrinsically denser substances lying deeper. Thus, silicates dominate the outer layers, and iron, or iron-rich compounds such as iron sulphide (FeS), dominate the cores. We therefore have rocky materials throughout.

Regardless of the form in which a chemical element is present, over 90% of the mass of each of the terrestrial planets consists of oxygen, iron, silicon, magnesium, and sulphur, though the proportions within this group vary from body to body. These are among the fifteen chemically most abundant elements in the Solar System as a whole. Other abundant elements are less well represented, notably hydrogen, helium, carbon, nitrogen, and neon. This is because the pure element or the common compounds, including icy materials, have densities that are too low even at high pressures to fit the density data. It is their volatility that has led to their scarcity in the terrestrial planets.

Table 5.1 shows temperatures, densities, and pressures at a few depths in the Earth, and Table 5.2 gives the values at the centres of all the terrestrial planets, and the Moon. Note that these are model values, and therefore depend on the particular model adopted, and so the values are indicative and not definitive. Note also that for a given substance, its equation of state links the three quantities in Table 5.2, and so in principle if we know any two we can calculate the third. However, the equations are not well known, partly because the terrestrial planets consist of mixtures of minerals in somewhat uncertain proportions, and partly because the extreme conditions at great depths are beyond laboratory reach, though it helps that to a fair approximation in terrestrial interiors we can ignore the effect of temperature on density.

You can see in Table 5.2 that the central pressures are greater, the larger and denser the body, in accord with equation (4.11) (Section 4.4.3). In Table 5.1 the pressures and temperatures increase with depth.

❐ Could this be otherwise?

Table 5.1 Model temperatures, densities[a], and pressures in the Earth

Crust[b]	
T (surface)/K	288
ρ (ave)/kg m^{-3}	2 800
Mantle	
ρ (top)/kg m^{-3}	3 300
T (base)/K	4 000
p (base)/Pa	1.5×10^{11}
ρ (base)/kg m^{-3}	5400
Core	
ρ (top)/kg m^{-3}	9 900
ρ (centre)/kg m^{-3}	13 000
T (centre)/K	4 300
p (centre)/Pa	3.3×10^{11}

[a]The densities are the estimated values *in situ*, i.e. they are compressed densities.
[b]The crustal values are observed rather than modelled.

Table 5.2 Model temperatures, densities[a] and pressures at the centres of the terrestrial planets and the Moon

	Mercury	Venus	Earth	Mars	Moon
ρ /kg m^{-3}	9300	12 500	13 000	7000	8000
T/K	1370	3 700	4 300	2000	1600
$p/10^9$ Pa	44	290	330	41	4.9

[a]The densities are the estimated values *in situ*, i.e. they are compressed densities. In all cases a predominantly iron core is assumed, with Mars having an appreciable proportion of FeS.

In Section 4.4.3 it was shown that pressure must increase with depth, and in Section 4.5.4 it was argued that, given sufficient time, temperature will decrease outward from the centre.

For a given temperature and pressure, whether a region is liquid depends on the substances present. The mantle silicates have melting points at a particular pressure that are slightly lower than that of pure iron. However, even a small admixture of other substances into the iron reverses the situation, and so in several of the bodies in Figure 5.1 the mantle–core interface is also shown as a solid–liquid interface.

Only at a level of greater detail would there be much dispute about any of the features of Figure 5.1 and the data in Tables 5.1 and 5.2.

5.1.1 The Earth

Observational data pertaining to the Earth's interior are particularly copious, and the model is consequently tightly constrained. As well as an abundance of gravitational and magnetic field data, the composition of the Earth's surface is very well known, as are the details of its high level of geological activity. It is the only planetary body for which we have copious seismic data, far more than for the only other body for which we have any seismic data at all — the Moon. Figure 5.3 shows the P and S wave speeds versus depth in the Earth.

❐ Do the boundaries in the model of the Earth in Figure 5.1 match those in Figure 5.3?

The division of the Earth in the model in Figure 5.1 into a crust, mantle, outer and inner cores, is apparent in the seismic data. There is plenty of evidence that at all these boundaries there is a change in composition. The disappearance of seismic S waves in the outer core indicates that it is liquid (Section 4.3.1). Their reappearance at greater depths indicates a solid inner core.

Iron is highly favoured for the core, because of the density indicated at this depth by the seismic and gravitational data.

❐ Among the metals, what is another reason for favouring iron?

Iron is also favoured because of its high relative abundance in the Solar System. Though iron is dominant throughout the core, models that fit the observational data need up to about 4% nickel in the inner core (similar to iron meteorites), though the

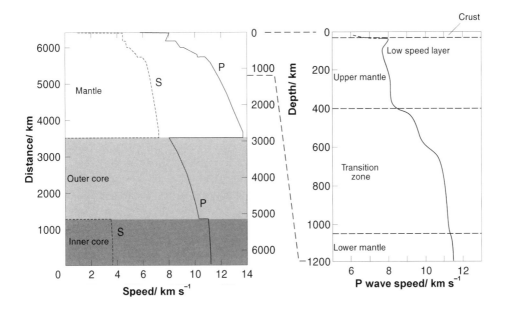

Fig. 5.3 P and S wave speeds versus depth in the Earth

exact composition of the inner core is uncertain. The density of the outer core needs to be about 10% lower than the inner core, and its melting point needs to be lower too. Various minor constituents in addition to more nickel can achieve this, such as a few per cent iron sulphide, or even iron hydrides (FeH_x). The inner core is solid because of its slightly different composition, and because it is at a higher pressure.

A liquid iron outer core is consistent with the strong magnetic field of the Earth, for which the dynamo theory requires an electrically conducting fluid in convective motion (Section 4.2). The crystallisation of iron from the outer core on to the inner core might be the main source of energy required at the base of the convecting outer core to maintain the convection. This crystallisation releases latent heat and energy of differentiation (Section 4.5). It also releases lighter elements that float upwards and this also promotes convection.

In the model the mantle consists of silicates, and from samples of the mantle it is known that the upper mantle consists almost entirely of a rock called **peridotite**. This is largely a mixture of two silicates—pyroxene (($Ca, Fe, Mg)_2Si_2O_6$, where rarely the metals can be Na, Al, or Ti), and olivine (($Mg, Fe)_2SiO_4$). From a depth of about 400 km to about 1050 km there is a gradual transition zone apparent in the seismic speeds (Figure 5.3) that can be explained by the gradual conversion of the minerals in peridotite into higher-density crystal structures that are more stable at high pressures, and that dominate down to the core, though the overall chemical composition remains much the same. By contrast, the sharp changes in the speeds at an average depth of about 30 km marks the boundary between the crust and the mantle, and this is a chemical change, though the crust, like the mantle, is dominated by silicates.

The overall model composition of the Earth follows fairly closely the Solar System's relative abundances of the less volatile elements. Iron is under-represented in the mantle and crust because of its concentration in the core. The composition of the crust, which is readily accessible, will be further discussed in Section 8.1.1.

Just below the crust–mantle division the seismic data in Figure 5.3 reveal a low-speed layer which is present in much of the upper mantle and extends over the approximate depth range 50–200 km. This indicates that around this depth the material is particularly plastic, and volcanic activity indicates that this is because some of its constituents are molten — this is an important process called **partial melting**. This plasticity is consistent with the isostasy of most of the crust.

The thermal structure of the model suggests that solid-state convection is occurring in the mantle. There is shallow convection around the low-speed layer, and larger-scale convection perhaps extending right down to the core. These two types of convective cell are shown schematically in Figure 5.4. The more plastic region of the mantle constitutes the **asthenosphere** (from the Greek 'asthenes', meaning weak). It extends from near the top of the low-speed layer, to considerable depth, right down to the liquid core if that's as far as convection extends. Above the asthenosphere, the uppermost part of the mantle and the crust are much more rigid. They constitute the **lithosphere** (from the Greek 'lithos', meaning rock), which, on average, is about 60 km thick. Note that at the lithosphere–asthenosphere boundary there is a change in dynamic properties, and *not* a change in composition.

Remanent magnetism in radiometrically dated old rocks shows that the Earth's magnetic field extends at least 3500 Ma into the past. Indeed, there is plenty of evidence that the Earth has had a hot interior ever since its birth 4600 Ma ago, or not long after. Therefore, if it formed undifferentiated it would soon have become warm enough to differentiate. Thermal models indicate that long-lived radioactive isotopes are the dominant source of energy for the Earth *today*. The compositional structure of the Earth, the chemical affinities of these isotopes and their relative abundances, indicates that they are concentrated into the crust, and to a lesser extent into the mantle. Heat sources long dead, such as the energy from accretion and from short-lived isotopes, still have an effect, because the Earth's interior temperatures remain a good deal higher than they would be if these extinct sources had never been present.

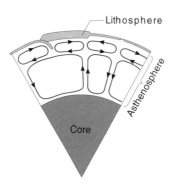

Fig. 5.4 Convection in the Earth's asthenosphere, and the rigid lithosphere

Question 5.1

Suppose that at some time in the Earth's future its outer core solidifies. Describe two observable effects this would have.

5.1.2 Venus

In its size and mean density, Venus is almost the Earth's twin (Figure 4.1). When allowance is made for the lower internal pressures in Venus (because of its slightly smaller mean density and radius) then the uncompressed mean densities are even closer, and these densities are a better basis for comparing compositions. Venus is also our planetary neighbour, and therefore probably had available for its construction much the same sort of materials as the Earth. These few data suggest very strongly that the interior of Venus is not very different from the Earth's interior, and this is reflected in the model in Figure 5.1. Support for this conclusion comes from seven of the Soviet spacecraft that landed on the Venusian surface in the 1970s and 1980s. Six of the craft found the sort of silicates that typify the Earth's ocean basins, and the seventh found silicates of the sort that typify the upper parts of the Earth's continents.

Further support might be expected from seismic, gravitational, and magnetic field data. Unfortunately, we have no seismic data for Venus, and it is so near to having a spherically symmetrical mass distribution that gravitational data provide very weak constraints on the density versus depth (Table 4.2). The spherical symmetry is largely a result of the slow rotation of Venus, the slowest of all the planets, with a sidereal period of 243 days.

❒ Why is this relevant?

There is little rotational flattening—the main cause of departure from spherical symmetry for a planetary body.

The magnetic field of Venus is extremely weak, and only a very small upper limit exists for the magnetic dipole moment (Table 4.2). Does this mean that there is no iron core, or that any iron core is solid? The absence of a liquid iron core would be very surprising in light of Venus's broadly similar size and mass to the Earth.

❒ As well as an electrically conducting liquid, what other ingredients are thought to be required for a magnetic field?

Convection currents, and sufficiently rapid rotation of the body are also thought to be necessary (Section 4.2). The slow rotation of Venus might therefore be the main reason for the absence of a magnetic field. Another reason might be that the cooling of the interior of Venus has been retarded by its hot surface, 735 K versus 288 K for the Earth (because of the massive atmosphere of Venus). If, as a result, no solid inner core has yet formed, then there would be none of the crystallisation of iron that is thought to drive convection in the Earth's outer core. Furthermore, the Earth's solid inner core means that the Earth's conducting fluid is a shell rather than a sphere, and it turns out to be easier to obtain a large magnetic dipole moment from a shell. We can thus retain a liquid iron core in our models of Venus.

Evidence of a hot interior is also provided by the surface of Venus, which displays evidence of recent geological activity and a plastic interior (Section 8.2). Overall, taking everything that we know about the interior of Venus, it seems beyond reasonable doubt that not only does it have a hot interior but also a liquid iron core and a silicate mantle of broadly similar sizes and compositions to those of the Earth. At a more detailed level of modelling there is far more uncertainty.

5.1.3 Mercury

Observational data on Mercury are sparse. We know the mass and the radius and so the mean density can be calculated, and it is between that of Venus and the Earth. However, it is a good deal smaller in radius, and so the internal pressures are considerably less. The uncompressed mean density of Mercury is not much less than the mean planetary value of $5430 \, \text{kg m}^{-3}$, whereas the mean densities of Venus and the Earth are reduced to 4000–$4500 \, \text{kg m}^{-3}$ when uncompressed. Therefore, the uncompressed mean density of Mercury is actually greater than that of the Earth, and is the highest of all the planetary bodies. There is only one abundant, sufficiently dense substance that must predominate—iron. If, like Venus and the Earth, Mercury has an iron core and a silicate mantle, then the iron core must account for a large fraction of its volume, as in Figure 5.1. The surface has not been sampled, but spectroscopic data from Earth-based telescopes indicate a silicate composition.

Gravitational or seismic data could confirm the existence of an iron core. Unfortunately, as for Venus, we have no seismic data, and the gravitational data are comparably unhelpful, and for the same reason—the sidereal rotation period of Mercury is 58.6 days, and such slow rotation has led to inappreciable flattening. The planet is therefore nearly spherically symmetrical, and so we can learn little about the increase of density with depth.

The surface of Mercury indicates no internally driven geological activity since very early in Solar System history—it is covered in impact craters that have accumulated over thousands of millions of years, with little sign of removal. This lack of activity implies that the lithosphere has long been thick, and this is consistent with the high rate of cooling expected from Mercury's small size. Indeed, it could be the case that any iron core is now solid. In this case, Mercury would have a negligible magnetic dipole moment.

In fact, the dipole moment is not negligible (Table 4.2)! Could this mean that at least some of the iron core is liquid? Thermal models indicate that, with a large iron core cooling slowly by conduction through a largely lithospheric mantle, and with a modest abundance of long-lived radioactive isotopes, the inner part of the core might still be molten. But even if this were so, Mercury's slow rotation counts against a strong magnetic field being generated. An important clue is that the magnetic dipole moment is fairly small, and it is plausible that this is consistent with slow rotation. A less likely possibility is that the field is due to iron rich surface rocks magnetised early in Mercury's history, perhaps by a strong field in an iron core that had not then solidified. In that early time Mercury would also have been rotating faster—its rotation has since been slowed by tidal interactions with the Sun. The existence of a liquid core early in Mercury's history is borne out by the surface topography, which indicates contraction early on, such as would result from the solidification of a core.

Question 5.2

State and justify your expectations regarding the existence and extent of an asthenosphere in Venus and in Mercury.

5.1.4 Mars

Mars is somewhat larger than Mercury but has the much smaller uncompressed mean density of 3700–3800 $kg\,m^{-3}$, which is also considerably smaller than that of the Earth. The surface has been sampled, and from this and other data it is clear that iron-rich silicates are common in the crust. Nevertheless, the low uncompressed mean density indicates that, overall, iron is less abundant in Mars than in the Earth. It follows that if Mars has an iron core then it must be small.

Evidence for a small core intrinsically denser than the rest of the planet is provided by the value of C/MR_e^2, given in Table 4.2.

❑ What is the value of C/MR_e^2 for a homogeneous sphere?

For a homogeneous sphere the value is 0.4 (Section 4.1.3). The martian value of 0.365 is sufficiently smaller to be consistent with a small, iron-rich core. The value of C/MR_e^2 has been obtained from the rotation period T, the value of the gravitational coefficient J_2, and the rotation axis precession period (0.1711 Ma), as outlined in Section 4.1.3.

The core is thought to consist of a mixture of iron (with a few per cent nickel) plus less dense materials such as iron sulphide (FeS) and perhaps magnetite (Fe_3O_4). Such mixtures are considered because some thermal models of Mars indicate that differentiation might have been less complete than in the Earth. Incomplete differentiation is also consistent with the iron-rich surface silicates. If FeS is a significant proportion of the core then the mantle would be correspondingly depleted in sulphur. There is some evidence of such depletion from the Martian meteorites. Though they are samples of the crust, they can be used to make inferences about the composition of the mantle, because the crust was derived from the mantle. Figure 5.1 has an Fe–FeS core near the middle of the 1300–1700 km range of sizes for a mantle of composition suggested by the Martian meteorites.

Seismic data for Mars are very limited, confined to measurements attempted in 1976–80 by the Viking 2 Lander. No seismic activity was detected, though the winds were stronger than expected, and would have masked all but the strongest seismic events by terrestrial standards.

Only an upper limit exists for the magnetic dipole moment, and this is less than 0.003% that of the Earth. Given the rapid rotation of Mars, this indicates that the iron-rich core is either solid, or is liquid but not convecting, perhaps because there is no solid inner core forming. Such a completely liquid core is possible if the melting point has been sufficiently lowered by other ingredients. If sulphur accounts for more than about 15% of the core, it would be entirely liquid today. Moreover, it would *always* have been liquid and Mars might never have had a magnetic field. However, weak remanent magnetism has been detected in crustal rocks by Mars Global

Surveyor, notably in the more ancient areas of Mars, indicating that Mars had a magnetic field in the distant past. Therefore, a lower sulphur content and a core that is only now largely or entirely solidified is somewhat preferred on the basis of the magnetic field data.

Thermal models indicate that the interior of Mars started off hot, and that some or all of the core should still be molten. These models also indicate that a thick lithosphere formed early in Martian history, and this is consistent with the gravitational data on a regional scale, and with the surface features (Section 7.3). The polar moment of inertia constrains temperature, because the density of a substance depends not only on pressure but also on temperature, but unless the composition is known accurately the constraint is not strong. Overall, it is concluded that, though cooler than the Earth, the interior of Mars is still fairly warm.

Question 5.3

Explain why we would know much less about the interior of Mars if the planet rotated very slowly.

5.2 Planetary Satellites and Pluto

The large planetary satellites and Pluto fall into two broad groups. The Moon, Io, and Europa are terrestrial bodies in that they are predominantly rocky in composition. Ganymede, Callisto, Titan, Triton, and Pluto differ from the terrestrial bodies in that a substantial proportion of the mass of each of them consists of icy materials, so they are classified as icy–rocky bodies.

5.2.1 The Moon

Next to the Earth, the Moon is the most extensively studied planetary body. This is because it is by far the closest to us. It has been visited by numerous spacecraft, and is the only other body on which humans have walked, during the Apollos 11, 12, 14–17 missions, between 1969 and 1972.

The Moon is sufficiently small that its uncompressed mean density will not be much less than its actual mean density of $3340 \, \text{kg} \, \text{m}^{-3}$. Even so, this is considerably less than the uncompressed mean density of the Earth and even of Mars, strongly indicating that iron is comparatively scarce in the Moon. The value of C/MR_e^2 has been accurately obtained from a combination of J_2 and the Moon's response (as determined by laser ranging) to tidal torques exerted by the Sun and the Earth. The value, 0.394, is slightly but significantly smaller than that of a homogeneous sphere. Because of the slight central compression, this indicates that the Moon is not homogeneous, but has a small dense core. If the core is nearly pure iron then it will be 300–400 km radius, but if it contains a high proportion of iron-rich compounds, such as iron sulphide, then because these are less dense than iron, the core radius will be somewhat larger. Figure 5.5 shows a model of the lunar interior in which the core is assumed to include a fairly low proportion of iron-rich compounds.

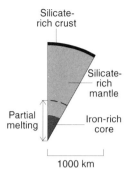

Fig. 5.5 A cross-section through the Moon

A rough indication of present-day internal temperatures has been obtained from measurements of the electric currents induced in the Moon by the magnetic field in the solar wind. These electric currents can be measured through the magnetic fields that they themselves generate. For a given solar wind magnetic field, the higher the internal temperatures, the larger the currents. The temperatures seem to increase with depth, reaching 1000 K or so at 300 km, and 2000 K or so at 1000 km. Further data on internal temperatures were expected from probes that measured the outward heat flow at the Apollo 15 and 17 landing sites. Unfortunately, the data are too sparse and too difficult to interpret to be useful.

We do, however, have seismic data. At the Apollos 11, 12, 14–16 sites, seismometers were set up successfully. The Moon is not as seismically active as the Earth, but 'Moonquakes' (mostly caused by tidal stresses induced by the Earth, or by surface impacts) have yielded the data in Figure 5.6 (seen earlier in Figure 4.9(b)).

S wave speeds are not shown beyond a depth of about 800 km. Such waves are strongly attenuated at greater depths, presumably because the material is warm enough to be plastic—this is consistent with the solar wind temperature data. The rapid increase in P and S wave speeds at shallow depths is consistent with a heavily fractured surface layer that becomes less fractured with depth because of the

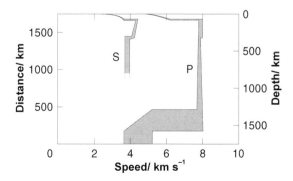

Fig. 5.6 P and S wave speeds versus depth in the Moon

increasing pressure, and is unfractured below a depth of about 20 km. The increase in the speeds at about 60 km is widely believed to be due to a change in composition, from the lower-density silicates found in abundance at the surface, to higher density silicates including the pyroxene and olivine that dominate the Earth's mantle. Samples of the higher-density silicates have also been found on the lunar surface, excavated by huge impacts that created large craters. Thus, in addition to the core, there seems also to be differentiation into a crust and mantle. The core itself is apparent in the seismic data, but its radius is poorly resolved, and its composition poorly constrained by the wide range of possible P wave speeds.

The seismic and thermal data indicate partial melting within a radius of about 700 km of the centre. It is even possible that this region is completely molten. Thus the transition from solid to a partial or total melt could be within the mantle, and not, as in the case of the Earth, at the core–mantle boundary. Thermal models of the Moon, incorporating measured abundances of radioactive isotopes, are consistent with a deep-lying (partially) molten region. These models also indicate that the lithosphere became as much as 60–100 km thick within a few hundred million years of lunar formation. The ancient cratered surface supports this conclusion, because a thin lithosphere would have allowed remoulding of the surface.

The lunar magnetic dipole moment is too small to measure, so we only have an upper limit, about 10^{-6} that of the Earth. Clearly, if any iron-rich core is (partially) liquid, then either it is not convecting today or the rotation of the Moon is too slow for a field to be generated — the sidereal rotation period is 27.3 days. Weak magnetic fields have been detected in the surface rocks. In some cases they seem to have been produced by the effects of impacts, but in other cases exposure to a magnetic field 3500–4000 Ma ago seems to be required. Clearly, one possible source of such a field is the lunar interior, indicating that there might have been a time when the liquid proportion of any iron-rich core was more extensive or more convective. The lunar rotation would also have been more rapid then — it has since been slowed through tidal interactions with the Earth. However, it is very likely that the magnetic dipole moment was always considerably smaller than that of the Earth.

There is a lot of observational evidence that the Moon has a whole-body composition broadly similar to that of the Earth's mantle. This is consistent with the theory of the origin of the Moon outlined in Section 2.2.4, in which an embryo collided with the Earth and ejected the material from which the Moon formed. In this theory, very little of the iron-rich cores of the Earth and the embryo would have been ejected, and so the Moon formed largely from material derived from the mantles of the two bodies. The embryo would have had a similar composition to the Earth, having been formed in the same region of the Solar System, and this is borne out by the oxygen isotope ratios $^{18}O/^{16}O$ and $^{17}O/^{16}O$, which are similar in the Earth and the Moon. Therefore this theory explains the broad similarity between the Moon's overall composition and the composition of the Earth's mantle, and corresponding global depletion of iron. It also explains the Moon's observed depletion in volatile and moderately volatile substances, and the corresponding enrichment in refractory substances — the material sprayed out by the embryo collision would have been hot enough to lose a high proportion of its volatile materials to space.

Question 5.4

Outline the factors that are relevant to explanations of why the interiors of the Moon and Mars are cooler than that of the Earth. You should refer to the general principles in Chapter 4, rather than try to come to any firm conclusions.

5.2.2 Large Icy–Rocky Bodies

The remaining large satellites and Pluto are all in the outer Solar System, and so we have to consider the possibility that icy materials are substantial components. Whether this is the case is indicated by their mean densities, and the values for all but Io and Europa are sufficiently low to indicate that they are icy–rocky bodies, with icy materials accounting for roughly half the mass in the iciest of them. In the solar nebula the icy/rocky mass ratio is thought to have been about three to one, and so it seems that the greater volatility of icy materials has somehow led to their depletion in these bodies, and to an even greater depletion in Io and Europa. Hydrogen and helium are excluded as significant contributors to the mass because only in the giant planets are pressures sufficient to compress these substances to appreciable densities.

Water was undoubtedly the most abundant icy material in the solar nebula, and it condensed in and beyond the Jupiter region. Beyond the Jupiter region the more volatile ices condensed, notably ammonia (NH_3), methane (CH_4), carbon dioxide (CO_2), carbon monoxide (CO), and molecular nitrogen (N_2). If chemical equilibrium was attained, then at the low temperatures of the outer solar nebula, nitrogen would have been present largely as NH_3 and $NH_3.H_2O$, and carbon as CH_4 and $CH_4.7H_2O$. However, the low densities in the outer Solar System disfavoured chemical equilibrium. In the interstellar medium nitrogen is present mainly as N_2 and carbon as CO, and so the outer Solar System could also have had appreciable quantities of CO, N_2, and $N_2.6H_2O$. Near to where a giant planet was forming, the higher densities and raised temperatures might have promoted a shift towards chemical equilibrium. Sufficiently close to a protogiant many icy materials would have been unable to condense.

❐ Why not?

A protogiant was luminous, and so the temperatures in the protosatellite disc (Section 2.3.1) would have been too high.

Recipes for icy–rocky bodies thus have a range of icy ingredients available. The proportions depend on the volatility of the different ices. Water is not only the most abundant icy material, it is also least volatile, and it is stable over a wide range of conditions. Thus, water is expected to be the most common icy material in the interiors of all of the icy–rocky planetary bodies. The more volatile icy materials are expected to be more abundant, the further the body is from the Sun, and they could be concentrated towards the surfaces. This expectation is borne out by what is known of the surface compositions.

The large icy–rocky bodies are Ganymede, Callisto, Titan, Triton, and Pluto. Ganymede and Callisto are two of the four Galilean satellites of Jupiter, and these four will be considered together in the next section. Therefore we start with Titan, the largest satellite of Saturn, and second only to Ganymede for size among the planetary satellites of the Solar System.

Titan has a mean density consistent with 52% of its mass being silicates, and the remainder being predominantly water (as ice). It has a substantial atmosphere, largely of nitrogen but with some methane and other gases. At visible wavelengths atmospheric haze obscures the surface, and the limited data on the surface obtained at other wavelengths have so far failed to provide any useful information about its interior. Differentiation seems likely given the calculated extent of accretional energy and radiogenic heating. If it has a magnetic dipole moment it is extremely small.

Triton is the largest satellite of Neptune. Triton is in a *retrograde* orbit, suggesting that is was therefore captured by Neptune after Triton formed. The tidal energy released upon such capture would probably have melted Triton, in which case it is certainly differentiated, though it is probably solid throughout now, except perhaps for partial melting in the lower part of the icy mantle. Some of the mantle is probably an asthenosphere, perhaps most of it. Ices of N_2, CH_4, CO and CO_2 have been detected on the surface. Observations, particularly from the spacecraft Voyager 2 in 1989, reveal ongoing volcanism involving the very volatile N_2. This is thought to be a lasting consequence of the primordial tidal energy released during capture, and of subsequent radiogenic heating. Volcanism involving icy materials is called **cryovolcanism** (from the Greek 'kruos', meaning icy cold, or frost).

Distant Pluto is small, and has never been visited by a spacecraft. Consequently, we don't know much about its interior. Its mass has been obtained from the orbit of its satellite Charon, which gives the sum of the masses of Pluto and Charon (equation (4.2), Section 4.1.1). Charon is not that much smaller than Pluto so we cannot assume its mass is negligible. The individual masses of Pluto and Charon have therefore been obtained by measuring the position of the centre of mass, and using equation (4.3) to obtain the ratio of the masses. The resulting mean density for Pluto is $2100 \, \text{kg m}^{-3}$ ($\pm 10\%$).

Pluto is thought to be differentiated into a partially hydrated rocky core, an icy mantle mainly of water ice, and an icy crust that spectrometric observations show to be mostly solid nitrogen, with a few per cent methane and some carbon monoxide. If Pluto was not formed ready differentiated, then radiogenic heating plus the tidal heating due to Charon should have been sufficient to cause differentiation. Tidal heating is promoted by the proximity of Charon, only $19\,100 \, \text{km}$ from Pluto, and by the comparatively large mass of Charon—nearly 10% the mass of Pluto. Today, Pluto is expected to be solid throughout, given its small size, though the lower icy mantle might be an asthenosphere.

Figure 5.7 shows a model of Pluto's interior, along with models of Titan, Triton, and the four Galilean satellites. Table 5.3 gives further data from typical models.

Question 5.5

How do solar nebular theories account for Pluto being an icy–rocky body, and being so small?

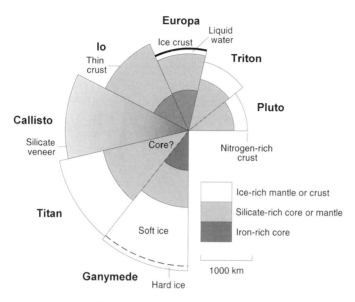

Fig. 5.7 Cross-sections of Pluto and the large satellites in the outer Solar System

Table 5.3 Model pressures at the centres of Pluto and the large satellites of the giant planets, plus some central temperatures and densities[a]

	Io	Europa	Ganymede	Callisto	Titan	Triton	Pluto[b]
ρ^c/kg m^{-3}	5500–8000	5500–8000	5500–8000	?	?	?	2800–3500
T/K	(hot)	(warm)	(warm)	?	?	?	100–600
p/10^9 Pa	6.0	3.2	3.6	2.7	3.3	2	1.1–1.4

[a]A '?' denotes no useful value is available.
[b]If differentiated.
[c]The densities are the estimated values *in situ*, i.e. they are (slightly) compressed densities. The large ranges correspond to various possible compositions for the cores.

5.2.3 The Galilean Satellites of Jupiter

Much has been learned about the Galilean satellites from the flybys of Voyagers 1 and 2 and from the Galileo Orbiter (Table 4.1). Figure 5.7 shows interior models, Table 5.3 gives central temperatures, densities, and pressures, and Figure 1.6 shows their orbits.

Io

Io is the innermost of the Galilean satellites of Jupiter. It is slightly larger than the Moon, and somewhat more dense. The mean density indicates a predominantly silicate plus iron/iron sulphide composition, and the value of C/MR_e^2 of 0.378 is sufficiently less than the uniform-sphere value of 0.4 to require concentration of denser materials into a core. The model in Figure 5.7 has a predominantly iron core extending about 40% of the way to the surface, thus accounting for about 6% of the volume. The Galileo Orbiter obtained inconclusive evidence for a magnetic dipole moment, somewhat larger than that of Mercury, so the core could be (partly) molten.

For the core to be partly molten the interior of Io would have to be hot, and there is dramatic evidence that it is — Io is the most volcanically active of all the planetary bodies in the Solar System! The mantle, throughout most of its volume, is thought to be an asthenosphere in which solid-state convection is occurring, and the mantle is overlain by a thin lithosphere that includes a silicate crust rich in sulphur. Partial melting within the asthenosphere would provide the observed volcanic outflows, which are observed to consist of silicates, sulphur and sulphur dioxide (SO_2).

It came as a surprise to most astronomers that such a small world has a sufficiently hot interior to be so volcanically active.

❐ Why was this surprising?

Equation (4.12) (Section 4.5.4) indicates that small worlds lose energy rapidly, so should now have cool interiors. Indeed, the Moon is comparable in size and mass to Io yet has little or no present-day volcanic activity, and nor does Mercury, which is larger and more massive than Io. We therefore need to find a considerable interior energy source for Io. There is radiogenic heating, but this is far from sufficient. The only plausible additional source is tidal energy.

In Section 4.5.1 you saw that tidal energy will be generated in a body by its rotation, which sweeps the material of the body through the tidal elongation. In the case of Io the rotation period is the same as its orbital period — synchronous rotation — and this is itself a result of tidal effects. Were the orbit circular, Io would keep the same face to Jupiter, as in Figure 5.8(a), and material would not be swept through the tidal elongation — the dot is fixed with respect to the surface. There would then be no tidal heating by this means. Moreover, in a circular orbit Io's

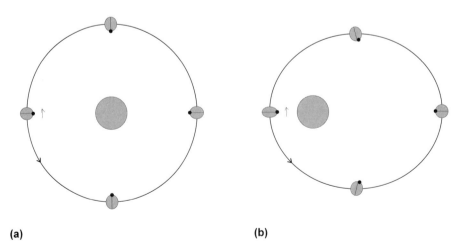

(a)   (b)

Fig. 5.8 Synchronous rotation (a) in a circular orbit and (b) in an elliptical orbit. The dot is fixed with respect to the surface of the small body

distance from Jupiter would be constant, so there would also be no tidal energy generated through variations in the Jupiter–Io distance.

In fact, the orbit of Io is not quite circular. Therefore, there *is* tidal input through its distance from Jupiter varying. Moreover, there is now a contribution from Io's rotation. The rate of rotation is constant, but the rate of motion around the orbit varies, just as the rate of motion of a planet around its elliptical orbit varies in accord with Kepler's second law. Therefore, as seen from Jupiter, Io seems to swing to and fro as it orbits the planet, as shown in Figure 5.8(b), and so material now oscillates through the tidal elongation. Were it not for orbital resonances (Section 3.1.1) with Europa and Ganymede—the orbital periods of Io, Europa, Ganymede are in the ratios 1:2:4—Io's orbit would be much more circular, and tidal heating would be much reduced.

Question 5.6

The Moon is in synchronous rotation in its orbit around the Earth. With the aid of Tables 1.1 and 1.2 and the answer to Question 4.9, show that the tidal field across the Moon is only 0.004 times that across Io. Explain why the greater tidal field across Io must more than offset the greater eccentricity of the lunar orbit.

Europa

Europa, another Moon-sized world, is the next nearest Galilean satellite to Jupiter. Its surface is covered in water ice, though its mean density is only slightly less than that of the Moon, indicating a predominantly silicate composition.

❏ What does the value of C/MR_e^2 in Table 4.2 require?

The value of C/MR_e^2 is less than that of Io, and so requires greater concentration of denser materials into a core, presumably iron-rich, perhaps iron sulphide. The value of C/MR_e^2 also indicates that if the mantle is dry then the shell of water is about 50 km thick, though it is more likely that the boundary between silicates and water is not sharp, and that there is some water in the upper mantle, in which case the shell of water could be as much as 160 km deep. The Galileo Orbiter obtained inconclusive evidence regarding a magnetic dipole moment.

The same tidal input operates as in Io, though the rate at which energy is released in the shell of water and in the interior is roughly twenty times less, because of Europa's greater distance from Jupiter, which more that offsets the larger eccentricity of Europa's orbit. We thus expect Europa to have a cooler interior, though ongoing tidal and radiogenic heating, supplemented by solar radiation and residual primordial heat, should be sufficient to liquefy the lower part of the water shell. There might even be a modest amount of volcanic activity on the ocean floor. The very smooth icy surface indicates an icy crust no more than a few kilometres thick, and though this could be underlain by icy slush rather than by liquid oceans, most astronomers believe that there are widespread oceans at least 5 km deep.

Ganymede and Callisto

The two outer Galilean satellites Ganymede and Callisto have surfaces of water ice, with a thin veneer of silicates covering much of Callisto. Their mean densities require a substantial proportion of ice — these are icy–rocky bodies. The conditions in which Ganymede and Callisto formed, and their observed surface composition, means that the icy component is water ice, with no more than traces of other icy materials.

❐ From Figure 5.7, what is the proportion of the volume of Ganymede that is water ice?

If Ganymede's water is essentially all in its icy mantle, then the proportion that is *not* water ice is the cube of the radius of the silicate-rich mantle divided by the cube of Ganymede's radius. From Figure 5.7 this ratio is 0.16. Therefore, the proportion of the volume that is water ice is 0.84. However, the proportion that is water by mass is not far from 50%, because water is much less dense than silicates and iron-rich substances.

The low value of 0.311 for C/MR_e^2 indicates that Ganymede has the iron-rich core shown in Figure 5.7. It has a magnetic dipole moment about three times that of Mercury, which suggests that the core is at least partly molten and convecting vigorously. Vigorous convection requires greater energy input than at present, and this might have been from greater tides as a result of a more eccentric orbit in the past. It is quite feasible that the orbits of all the Galilean satellites have evolved, and that Ganymede's eccentricity was substantially higher within the last 1000 Ma or so. The surface of Ganymede reveals relatively recent and extensive geological activity, consistent with a warm interior, and there is also evidence for differentiation at some time in the past (Section 7.4.3).

For Callisto, the value of 0.358 for C/MR_e^2 is significantly less than the value of 0.38 that would correspond to self-compression of an undifferentiated ice–rock mixture. However, it is not small enough to indicate full differentiation, and so there is only a modest degree of concentration of rocky materials towards the centre. The surface of Callisto suggests that it has always had a cooler interior than Ganymede, and this is consistent with limited differentiation. A likely reason is negligible tidal heating, a result of the low eccentricity of Callisto's orbit, the lack of orbital resonances, and its greater distance from Jupiter. A further factor is the smaller mass ratio of rocky to icy materials in Callisto, resulting in less radiogenic heating.

❐ Would you expect Callisto to have a magnetic dipole moment?

If Callisto has a cool interior then the magnetic dipole moment should be zero. Callisto is sufficiently far from Jupiter that the Jovian magnetic field is weak, allowing very weak local fields to be detected. This has allowed the Galileo Orbiter to place a very low upper limit on any magnetic dipole moment, entirely consistent with Callisto's other properties.

The Galilean satellites as a group

Among the Galilean satellites, the proportion of icy materials increases as we go out from Jupiter, from zero for Io, to about 55% by mass for Callisto. Unless these satellites formed further out, and migrated inwards, the explanation is that Io formed too close to protoJupiter ever to have had an icy component. Europa acquired a small amount of ice, and Ganymede and Callisto substantial amounts. Only water ice would have been significant — the solar nebula in the Jovian region would have been too warm to allow much of any ices that are more volatile to condense.

Increasing distance from Jupiter also correlates with the inferred cooler thermal histories. This is probably because of the decrease in tidal energy input with increasing distance from Jupiter.

5.2.4 Small Bodies

There remain the host of small satellites, the asteroids, and the comets. The interiors of the asteroids and the comets were discussed briefly in Chapter 3, and I will say no more about them here, but concentrate on the small satellites.

The smallest body we have so far considered is Pluto, which has a radius of 1150 km. Next down in size is Titania, the largest satellite of Uranus. This has a radius of 805 km so this is a real step down, and is a convenient point at which to break off discussion of satellite interiors in any detail. Most of the small satellites that orbit the giants will have a composition roughly the same as the large icy–rocky satellites — by mass about half icy materials and half rocky materials. The densities of some of the innermost satellites of Saturn have been estimated from their effect on the rings and on other satellites, and the values are very low — less than the density of water ice.

❐ What does this indicate?

Such low densities indicate quite a high degree of porosity.

Satellites very close to Jupiter might have been warmed, either tidally or by the luminosity of the young giant, to the extent that they are depleted in icy materials. Just three small satellites do not orbit giants. These are the Martian satellites Phobos and Deimos, rocky bodies that are probably captured asteroids, and Pluto's icy–rocky satellite Charon. Tidal forces are likely to have ensured that Charon is fully differentiated. The smallest satellites of all might not have differentiated even if they have been warmed. This is because in very small bodies gravity is relatively weak, and so the chemical forces between constituents can override the gravitational trend towards differentiation.

The surfaces of some of the small satellites are more remarkable than their interiors, so we shall return to them in Chapter 7.

5.3 The Giant Planets

In arriving at the four giant planets we not only move up an order of magnitude or so in radius but also to worlds that are very different in many other ways. Figure 5.9 shows compositional models of the interiors of Jupiter, Saturn, Uranus, and Neptune that are consistent with the observational data, though

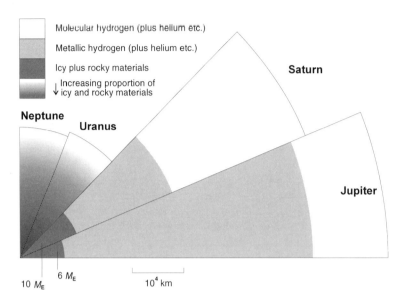

Fig. 5.9 Cross-sections of the giant planets (in their equatorial planes)

some variation is possible and so the models shown are broadly indicative rather than definitive.

The interiors of the giant planets consist largely of hydrogen, helium, and icy materials, with rocky materials making up only a small fraction of the mass. The dominance of hydrogen, helium, and icy materials is required by the low mean densities of the giants (Table 4.2), which range from a 'high' of $1760 \, \text{kg m}^{-3}$ for Neptune to an astonishing low of $700 \, \text{kg m}^{-3}$ for Saturn. These low densities, several times less than typical densities of rocky materials, are even more remarkable given the large size of the giants.

❐ Why is size relevant?

Equation (4.11) (Section 4.4.3) shows that large radii lead to large pressures, and large pressures increase the density of a substance. For Jupiter and Saturn the only models that fit their mean densities consist largely of the intrinsically least dense materials — hydrogen and helium. Uranus and Neptune, being smaller, must have icy materials as the main constituent. On the basis of the relative abundances of the elements in the Solar System, water must be the dominant icy material in the giants.

In the models of all four giants, the outer regions are hydrogen and helium, with small quantities of other substances. The outermost regions are directly observable, and, of course, the models match the observations. The compositional layering in the models is strongly indicated by the gravitational data, though the boundaries between the layers seem to be fuzzy. There is, however, a concentration of icy–rocky materials towards the centre.

❐ Is this in accord with nebular theories of the origin of the Solar System?

Nebular theories predict that the giants started as kernels of icy–rocky materials that grew massive enough to capture hydrogen–helium envelopes and icy–rocky planetesimals (Section 2.2.5). Some central concentration of icy–rocky materials is thus expected.

Table 5.4 lists the temperatures, densities, and pressures at a few depths in the giants — again these are broadly indicative, not definitive. High interior temperatures are indicated by the observed infrared excesses from Jupiter, Saturn, and Neptune (Section 4.5.3). The corresponding outward flows of energy from the interiors are given in Table 4.2. For all the giants, hot interiors today are predicted by thermal models. Furthermore, the large magnetic dipole moment of each giant indicates extensive, convective fluid regions in the interior that are electrically conducting. Convection is predicted to occur throughout much of the interior of each giant. Uranus might be an exception, because of its low infrared excess. This can be explained by a current lack of convection over some range of depths, perhaps due to a composition gradient.

It seems certain that the giants are fluid throughout, in which case hydrostatic behaviour is expected, and the gravitational fields are indeed those expected of rotating hydrostatic bodies.

It is clear that the giants can be divided into two pairs: Jupiter and Saturn; Uranus and Neptune (Question 4.1, Section 4.1.1). Indeed, some people used to restrict the term 'giant' to Jupiter and Saturn, and call Uranus and Neptune 'subgiants'. Let's examine each pair in turn.

5.3.1 Jupiter and Saturn

These two giants are most like the Sun in their composition, though whereas in the Sun the heavy elements (those other than hydrogen and helium) account for only about 2% of the total mass, in most models of Jupiter these elements account for 5–10% of the mass, and for a roughly three times greater proportion in Saturn. Therefore, the actual mass of the heavy elements is about the same in the two planets.

Table 5.4 Model temperatures, densities[a] and pressures in the giant planets

	Jupiter	Saturn	Uranus	Neptune
Envelope[b]				
T (10^5 Pa)/K	165	134	78	71
T (base)/K	~ 6000	~ 4000	—	—
p (base)/Pa	$\sim 2 \times 10^{11}$	$\sim 2 \times 10^{11}$	—	—
ρ (base)/kg m^{-3}	1.1×10^3	1.1×10^3	—	—
Centre				
ρ (centre)/kg m^{-3}	2.3×10^4	1.9×10^4	$\sim 10^4$	$\sim 10^4$
T (centre)/K	2.2×10^4	1.1×10^4	~ 8000	~ 8000
p (centre)/Pa	$\sim 5 \times 10^{12}$	$\sim 2 \times 10^{12}$	$\sim 0.7 \times 10^{12}$	$\sim 1 \times 10^{12}$

[a]The densities are the estimated values *in situ*, i.e. they are compressed densities.
[b]In Uranus and Neptune the envelope has no well-defined base, so values are not given.

Jupiter

To explore the Jovian interior, we will take an imaginary journey to its centre. We start in the atmosphere, which is observed to consist of 75.5% hydrogen by mass, and 23.8% helium. Hydrogen is in the molecular form H_2, and helium, an unreactive inert gas, is in atomic form He. We pass quickly through various cloud layers, the temperature, pressure, and density gradually increasing as we descend, and at a depth of about 10^4 km below the cloud tops the density is $1000 \, kg \, m^{-3}$. This is approximately the density of liquid water at the Earth's surface, and so we are in a hydrogen–helium ocean — and yet we crossed no surface. Figure 5.10 shows the phase diagram of molecular hydrogen (the presence of helium will modify it somewhat, but the story is still the same in all its essentials). Our journey took us along the path shown as a dashed line.

❐ Why was there no sudden transition from the relatively low density gas phase to the much higher density liquid phase?

The path stayed well to the right of the critical point (Cr), and so the density gradually increased.

This absence of a surface makes the Jovian radius a matter of definition. It is defined as the radius at which the atmospheric pressure is 10^5 Pa (standard atmospheric pressure at the Earth's surface is 1.01×10^5 Pa). The same definition is adopted for all the giant planets. In the case of Jupiter, 10^5 Pa is not far below the top of the upper cloud deck.

We do, however, encounter a surface of sorts deeper down, albeit with only about a 10% increase in density across it. This is the transition between liquid molecular hydrogen (plus atomic helium), and liquid *metallic* **hydrogen** (plus atomic helium). At

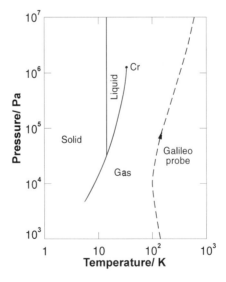

Fig. 5.10 Phase diagram of molecular hydrogen, with Jovian conditions represented by the Galileo probe measurements

pressures around 2×10^{11} Pa, the density of molecular hydrogen is so high that each of the hydrogen atoms in the H_2 molecule is attracted to atoms in neighbouring molecules as much as to its molecular partner. Therefore, the H_2 molecules break up. Moreover, the single electron that orbits the nucleus of each hydrogen atom will also become equally attracted to neighbouring atoms, and so the atoms break up too. The hydrogen will then be a 'gas' of electrons moving in a 'sea' of hydrogen nuclei. Many of the characteristic properties of metals arise from the existence of 'electron gases' within them, so the term 'metallic hydrogen' is appropriate. Metallic hydrogen was once just a theoretical prediction, but then in 1996 sufficiently high pressures were produced in the laboratory for metallic hydrogen to appear as expected. The transition takes place over a range of pressures, and divides the interior into a metallic hydrogen mantle and an overlying molecular hydrogen envelope. The range of depths over which the transition takes place is unknown — it could be very small, as in Figure 5.9.

One of the properties that an electron gas gives to a metal is high electrical conductivity. Thus, electric currents in the metallic hydrogen mantle are an obvious source of the planet's large magnetic dipole moment. The volume of metallic hydrogen, the hot convective interior, and the rapid spin of Jupiter are all consistent with the large dipole moment. A small but significant contribution to the field could arise from liquid iron and other conductors much deeper down. The detailed configuration of the field close to Jupiter is consistent with currents in the metallic hydrogen mantle.

Most models have increasing amounts of rocky and icy materials as depth increases, though they remain minor constituents until we approach the centre. Gravitational data indicate a significant concentration of icy and rocky materials into a dense core, probably with a fuzzier boundary than in Figure 5.9. There might be further differentiation within this core into an inner core dominated by rocky materials, and an outer core dominated by icy materials. Unfortunately, modelling the deep interior of Jupiter is complicated by the significant change of the density of hydrogen over the temperature range thought to exist in the planet — the simplifying approximation cannot be made that density depends only on pressure. On the other hand, the dominance of hydrogen and helium facilitates the development of theoretical equations of state. These are easier to develop for hydrogen and helium than for the iron and the compounds that dominate other bodies. (Moreover, these other bodies consist of uncertain mixtures of a large number of compounds.) It is possible that the heavy elements are several times more abundant in Jupiter and Saturn than they were in the solar nebula. On the basis of relative abundances in the solar nebula the icy/rocky mass ratio should be about three, but there are no observational data that could confirm this is the case in these planets.

Thermal models show that the present-day high temperatures of the Jovian interior can be accounted for by two dominant energy sources — energy from accretion, plus energy from differentiation when the icy–rocky kernel acquired more mass to become the icy–rocky core. An active but small energy source is the settling of helium through the metallic hydrogen mantle. The two major energy sources became inactive about 4500 Ma ago, yet the central temperature of Jupiter is still about 2×10^4 K. The reason is the large size of Jupiter. This has two consequences. First, there would have been a huge amount of accretional energy per unit mass as

Jupiter formed, and so Jupiter would have become extremely hot (Section 4.5.1). Second, Jupiter has a comparatively low ratio of surface area to mass, giving a low rate of internal energy loss (Section 4.5.4) in spite of the efficient outward transfer of energy by convection.

Though convection is expected to occur throughout most of the Jovian interior, calculations also indicate that over the depth range in which the temperatures are 1200–3000 K, energy might be transported outwards by radiation rather than by convection. This would lead to *cooler* interiors than in Table 5.4, and to a deeper transition to the metallic hydrogen phase than in Figure 5.9. This is also the case for Saturn.

Saturn

In many ways, the interior of Saturn is similar to Jupiter. Saturn is a bit smaller, and less dense, leading to lower internal pressures. These lower pressures mean that the transition to metallic hydrogen occurs much nearer the centre, so whereas most of the hydrogen in Jupiter is in metallic form, most of that in Saturn is in molecular form. It is nevertheless the metallic hydrogen in models of Saturn that can account for the planet's magnetic field. With the two planets rotating at about the same rate, and with comparable internal activity, the lower mass of metallic hydrogen in Saturn is predicted to lead to a smaller magnetic dipole moment.

❐ Is this the case?

Table 4.2 shows that the magnetic dipole moment of Saturn is about thirty times less than that of Jupiter.

The smaller size of Saturn also means that its internal temperatures after accretion would have been lower than in Jupiter, and it should have cooled more rapidly. Therefore, the present-day high internal temperatures, indicated by the infrared excess, cannot be accounted for solely by energy of accretion and by past differentiation as the icy–rocky kernel acquired more mass to become the icy–rocky core. An additional source of energy is needed, and this is thought to be the ongoing separation of helium from metallic hydrogen. Initially, the helium in the metallic hydrogen mantle was thoroughly mixed at the atomic level, and random thermal motion prevented any settling of the helium atoms downwards — a tendency arising from the greater mass of the helium atom. As the interior cooled, the miscibility of helium in metallic hydrogen fell, and Saturn is estimated to have reached the point about 2000 Ma ago where helium began to form small liquid droplets. These could not be held by random thermal motions in uniform concentration throughout the metallic hydrogen, so downward separation of the helium droplets began. This is essentially the same process as the separation of vinegar from oil in salad dressing. Convection is believed to have slowed the settling rate, but an outer core of helium is forming, and as it does so energy of differentiation is released. An additional source of energy might be continuing growth of an icy–rocky core.

The removal of helium from the metallic hydrogen would result in helium diffusing down from the molecular hydrogen envelope, so we would expect the observable atmosphere to be depleted in helium compared to Jupiter. And indeed it is! In Jupiter the outer atmosphere is observed to consist of 75.5% hydrogen and 23.8% helium by mass, whereas for Saturn the values are 92.5% hydrogen and about 6% helium, i.e. He/H mass ratios of 0.31 and 0.06 respectively. In Jupiter there is thought to have been much less settling of helium. For both planets viable models have He/H mass ratios for the *whole* planet similar to that estimated for the Sun at its birth — about 0.39.

Question 5.7

As Jupiter's interior cools, a certain energy source will become increasingly powerful. State what this source is, and why it is triggered by cooling. What effect could this source have on Jupiter's subsequent internal temperatures?

5.3.2 Uranus and Neptune

We have noted that the mean densities of Uranus and Neptune show that they are much less dominated by hydrogen and helium than are Jupiter and Saturn. This means that the equations of state for the interiors are less well known, and the consequent range of possible models is larger.

Uranus and Neptune are smaller and less massive than Jupiter and Saturn, and the overall composition of the models can be obtained, very roughly, by stripping away a good deal of the hydrogen and helium from Jupiter or Saturn.

❐ How is this feature explained in nebular theories?

The kernels of Uranus and Neptune formed considerably more slowly than those of Jupiter and Saturn, and so there was less time to capture hydrogen and helium before the protoSun's T Tauri phase drove away the nebular gas. In typical models of Uranus and Neptune, icy plus rocky materials account for about three quarters of the mass of each planet.

It is beyond reasonable doubt that Uranus and Neptune are dominated by icy materials, even though the atmosphere where the composition can be observed consists by mass of 65% hydrogen, and about 23% helium — not very much less than the He/H ratio estimated for the young Sun. As in Jupiter and Saturn, the accessible hydrogen is in molecular form and helium in atomic form. The internal pressures are too low for metallic hydrogen, and so there is no metallic hydrogen mantle. The proportion of icy materials probably increases gradually with depth, and so the transition from the hydrogen–helium rich envelope to an ice-rich core probably occurs over a considerable fraction of the planetary radii, though perhaps more rapidly at about a third of the way to the centre. There is no evidence for any appreciable concentration of rocky materials into rocky inner cores — the gravitational data can probably be explained by wholly or predominantly icy materials becoming increasingly dense as the pressure increases with depth. However, from theories of the origin of the Solar System, rocky materials are certainly expected to be present.

The infrared excess of Neptune indicates a hot, convective interior, though the rate of energy outflow per unit mass is about five times less than for Jupiter and Saturn

(Table 4.2). An interior hot enough to be liquid is also implied by Neptune's large magnetic dipole moment. The detailed configuration of the field indicates that the electric currents are located at no great depth in an ice-rich mantle, rather than in any rocky core. Mixtures of water, methane, and ammonia can become highly electrically conducting at high pressures and high temperatures. The predicted convective interior and the observed rapid rotation of the planet complete the requirements of the dynamo theory. For Neptune to have high internal temperatures today there needs to be an active energy source. This is thought to be differentiation, though uncertainties about the internal distribution of the various icy and rocky materials make the details obscure.

Uranus rotates only slightly more slowly than Neptune and has nearly double Neptune's magnetic dipole moment. The currents are likewise inferred to be located at rather shallow depths in the icy mantle, implying it must be hot. But the infrared excess is barely detectable, corresponding to a power outflow per unit mass at most about half that of Neptune, and probably less. However, the two planets are so similar in so many ways that it is thought that the internal temperatures are roughly the same, and that some process is greatly reducing the rate at which the energy in Uranus is transported to the upper atmosphere, where it would be radiated away to space. It was mentioned earlier that the suppression of convection over some range of depths, due to a composition gradient, might be the reason for the low rate of transfer, though this is far from certain.

Question 5.8

It is believed that radioactive isotopes are only a minor energy source (per unit mass) in the giant planets. Why is this believed to be the case?

5.4 Magnetospheres

This chapter concludes with a short account of a particular consequence of a planetary body having a substantial magnetic field. The magnetic field of the planet then interacts with the solar wind to create what is called a magnetosphere around the planet. Very complex mechanisms are involved in magnetospheres, and so the approach here will be qualitative and brief.

5.4.1 An Idealised Magnetosphere

Figure 5.11 shows a comparatively simple magnetosphere that will introduce the essential features. To the left, the solar wind is flowing in interplanetary space, and it is undisturbed by the planet. The magnetic field in the wind in this particular case is perpendicular to the flow, as also, for simplicity, is the magnetic axis of the planetary field. If the field in the wind were static then the interplanetary field would simply be the sum of the undisturbed wind field and the undisturbed planetary field. But the wind field is entrained in the wind—this is because the wind is a plasma, i.e. it is sufficiently ionised for copious electric currents to flow in it. Entrainment means that the wind carries the magnetic field along with it. As a result, the interaction of the

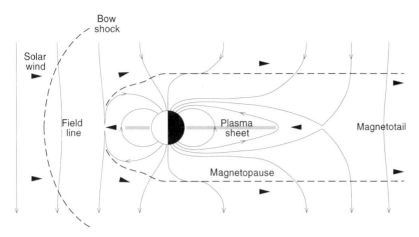

Fig. 5.11 An idealised magnetosphere, with magnetic field lines

wind with the planetary magnetic field gives a different outcome which we will now explore.

The solar wind 'sweeps' interplanetary space clean of the planetary field except in the vicinity of the planet. On the upwind side of the planet (to the left in Figure 5.11) there is a roughly hemispherical boundary outside of which the interplanetary field is dominant. On the downwind side the boundary stretches out into a long magnetotail. Within the boundary the planetary field near the planet is as if there were no wind field, but the planetary field gets more and more distorted as the boundary is approached. The boundary is called the magnetopause, and the volume it encloses is called the **magnetosphere**. The 'sphere' part of the name is to be interpreted as the planet's magnetic sphere of influence, rather than as a description of the shape of the boundary.

There are three sorts of magnetic field lines, and all three are included in Figure 5.11: there are those that start and end on the planetary surface, distorted though they may be; there are those that never encounter the planet; and there are those that start on the planet but, in effect, connect with the wind field and therefore never return to the planet — these are called reconnected lines.

The size of the magnetosphere is characterised as the distance from the centre of the planet to the upwind magnetopause. This distance is proportional to $\mu^{1/3}/n^{1/6}v^{1/3}$, where μ is the magnitude of the magnetic dipole moment of the planet, v is the wind speed, and n is the number density (number per unit volume) of the charged particles in the solar wind (mainly electrons and protons). Though v does not vary much with the heliocentric distance, n diminishes as this distance increases. Also, n varies with solar activity, and so the size of the magnetosphere also varies, being a maximum when n is a minimum.

Because the solar wind approaches the magnetopause at the high speeds of several hundred $\mathrm{km\,s^{-1}}$, it gets a rude shock at a boundary outside the magnetopause, called, appropriately, the bow shock. (The 'bow' is by analogy with a related phenomenon created on the surface of water as the bow of a boat moves through the surface at speeds greater than the speed of the surface ripples.) Between the bow shock and the magnetopause the solar wind is rapidly decelerated, the flow becoming turbulent and the wind plasma strongly heated. The wind flows around the magnetopause, with very little of the plasma entering the magnetosphere. The region between the bow shock and the magnetopause is called the magnetosheath.

The small fraction of the solar wind plasma that enters the magnetosphere is only one source of magnetospheric plasma. Another is cosmic rays (Section 3.3.3). These highly energetic charged atomic particles readily cross the magnetopause, and though most of them pass out again, a small proportion is trapped. Furthermore, a small proportion of the cosmic rays collide with the atmosphere of the planet, or with its surface if it has no atmosphere. This results in the ejection of particles, and these include neutrons that decay into protons and electrons, many of which are then trapped in the magnetosphere. Yet another source of plasma is a slow leak of particles from the planet's upper atmosphere, both in the form of plasma, and in the form of neutral atoms that subsequently become ionised.

Magnetospheric plasma is not uniformly distributed, but becomes concentrated towards the plane of the magnetic equator, where it constitutes the plasma sheet (Figure 5.11). Belts and toruses of plasma surrounding a planet can also occur.

Though the magnetospheric plasma is being added to all the time, there are also losses, outward to interplanetary space and inward to the planet. Among the latter are energetic charged particles that reach the upper atmosphere and excite atoms there. The resulting emission of optical radiation is called an **aurora**. Aurorae are concentrated in a ring around each magnetic pole. Large fluxes of energetic electrons plunging into the upper atmosphere generate radio waves with wavelengths of order 10–100 m. Such decametric radiation emanates from the Earth and from the giant planets, and as early as 1955 indicated that Jupiter has a powerful magnetic field. Aurorae and decametric radiation are intermittent phenomena, depending on the strength of the solar wind. Other radio waves, with wavelengths of the order of 0.1–1 m, are generated by electrons travelling at high speeds in the magnetic fields in the magnetosphere. This is called **synchrotron emission**.

5.4.2 Real Magnetospheres

In the Solar System, the magnetic dipole moments of the Earth and of the giants are far larger than those of any other planetary body (Table 4.2), and correspondingly they have the most extensive magnetospheres. Their magnetic axes are not perpendicular to the solar wind flow, nor on the whole is the magnetic field in the solar wind. Nevertheless, the general form of the magnetosphere in each case is roughly as in Figure 5.11, and there is also a plasma sheet and plasma belts.

Figure 5.12 shows the typical form of the magnetosphere of the Earth. There are two main plasma belts around the Earth — the **Van Allen radiation belts**, named after the American physicist James Alfred Van Allen (1914–) who discovered them in 1958. The inner belt consists largely of protons and electrons. These come from the solar wind, and also from the Earth's upper atmosphere partly through the action of cosmic rays. The outer belt is more tenuous, and the particles are less energetic. It is populated largely by the solar wind. Within the inner belt there is a third belt in which cosmic rays are prevalent.

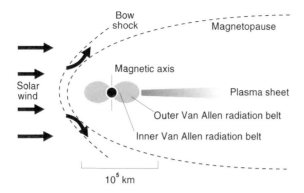

Fig. 5.12 The Earth's magnetosphere

There is also a plasma sheet (Figure 5.12). This has a low density, and is hot, the temperatures being $1–5 \times 10^7$ K. It is fed largely by the solar wind.

❑ What are the main constituents of the plasma sheet?

Being a sample of the solar wind, its main constituents are electrons and protons. Energetic electrons from this sheet find their way by various means into the upper atmosphere, particularly in a ring around the magnetic poles, where reconnected magnetic field lines intersect the ionosphere. There the electrons can give rise to decametric radiation, and also to aurorae — the aurora borealis (Plate 26) in the northern hemisphere, and the aurora australis in the southern hemisphere. When the solar wind is strong, as at times of high solar activity, the ring widens and aurorae are then seen further from the Earth's magnetic poles, down to about 70° or so in magnetic latitude. The magnetic axis is tilted by about 11.5° with respect to the rotation axis (Figure 4.6), and so the corresponding geographical latitude depends on longitude. The auroral displays are at an altitude of only about 100 km, and so the tropics are not the place to go for aurorae!

Figure 5.13 shows Jupiter's magnetosphere.

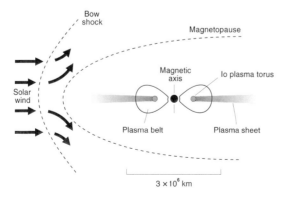

Fig. 5.13 The Jovian magnetosphere

❏ Why is Jupiter's magnetosphere bigger than that of the Earth?

It is bigger because Jupiter is further from the Sun, and so the number density of charged particles in the solar wind is smaller, and because the magnetic dipole moment of Jupiter is 20 000 times larger than that of the Earth. When the solar wind is particularly weak the upwind magnetopause can be about 100 Jovian radii from Jupiter. If we could see such a magnetosphere from Earth with Jupiter at opposition the magnetopause would be appear like a disc with an angular diameter nearly three times that of the full Moon. The magnetotail can extend beyond the orbit of Saturn.

The Jovian magnetosphere is particularly rich in plasma—in the denser regions the human body would quickly receive a lethal dose of ions. This richness is a result of copious internal sources, notably the volcanoes of Io, but also the Jovian upper atmosphere, and the surfaces of Jupiter's satellites and ring particles. Ions and electrons are ejected from the satellites and rings by cosmic rays. As well as being sources of plasma, the satellites and rings also remove plasma particles that collide with them. The solar wind is not an important source of the magnetospheric plasma, except near the magnetopause and far out in the magnetotail. It can be then shown that the plasma sheet must be a result of leakage from the plasma belt. Leakage occurs preferentially at the magnetic equator, where the magnetic containment is weakest. This domination of internal sources of plasma is a result of electric fields in the magnetosphere plus the rapid rotation of Jupiter—the details are beyond our scope.

Beyond Jupiter there are three more planets with extensive magnetospheres—Saturn, Uranus, and Neptune. The magnetosphere of Saturn is intermediate between that of the Earth and Jupiter in extent and plasma content. The considerable magnetospheres of Uranus and Neptune have some peculiarities arising from the large angles between their magnetic and rotation axes (Figure 4.6, Section 4.2), and in the case of Uranus from its large axial inclination, but in terms of the above discussion no new major phenomena are encountered.

Question 5.9

Outline the consequences for the Earth's magnetosphere
(a) If the Earth had no atmosphere
(b) If the speed and number density of charged particles in the solar wind were reduced.

5.5 Summary

The terrestrial bodies are dominated by silicates, and by iron-rich compounds, including iron itself (see Figures 5.1, 5.5, and Io and Europa in Figure 5.7). They are all differentiated, with iron-rich cores and silicate mantles. Europa has a thin icy crust underlain by an ocean of liquid water or slush, whereas the other terrestrial bodies have rocky crusts. The Earth's interior is by far the best known, and it is clear that there is an outer core mainly of iron that is hot enough to be liquid. This outer core is the source of the Earth's magnetic dipole moment. All but the uppermost part of the Earth's mantle is an asthenosphere, and it is probably undergoing solid-state convection. It is thought that the other terrestrial bodies also have warm interiors (Tables 5.2 and 5.3).

Internal temperatures in all of the terrestrial bodies are raised above the values they would have in equilibrium with solar radiation. The temperatures decrease as the size of the body decreases, with the exception of Io, which, in spite of its small size, has an interior hot enough for silicate volcanism. This is because Io has a dominant tidal component in its internal energy sources. By contrast, the other terrestrial bodies have internal temperatures today that are raised mainly through the effects of primordial energy sources and heat from long-lived radioactive isotopes.

Pluto, and the remaining large satellites (Figure 5.7) differ from the terrestrial bodies in having a much larger proportion of icy materials — they are icy–rocky bodies. Callisto is only partially differentiated, but Ganymede is known to be differentiated, and it is presumed that the others are too. Ganymede is also thought to have a partly molten iron-rich core, presumably the result of recent energy input from tides. The interiors of the other icy–rocky bodies are thought to be cool today (Table 5.3).

The remaining satellites have Titania as their largest member. Most of them consist of roughly equal masses of icy and rocky materials. The very smallest satellites are unlikely to be differentiated.

The four giant planets (Figure 5.9) are dominated by hydrogen, helium, and icy materials, the proportion of hydrogen and helium being considerably greater in Jupiter and Saturn than in Uranus and Neptune. For all four bodies the hydrogen–helium ratio for the whole body is thought to be similar to that in the young Sun.

Jupiter and Saturn are each differentiated into an icy–rocky core, a mantle of metallic hydrogen, and an envelope of molecular hydrogen, with helium as the next most abundant component in the mantle and envelope. The boundaries between these regions are fuzzy. Though models of Uranus and Neptune are poorly constrained, it seems that the boundaries are even fuzzier in these subgiants, and because of the lower pressures there is no metallic hydrogen.

The interiors of all four giants are hot (Table 5.4), and they all have very large magnetic dipole moments, located in the metallic hydrogen mantles in Jupiter and Saturn, and in icy mantles in Uranus and Neptune. The infrared excess of Jupiter is largely accounted for by primordial accretion and early differentiation. A substantial supplement of ongoing differentiation is required to account for the infrared excesses of Saturn and Neptune. In Saturn the differentiation is the separation of helium from hydrogen in the metallic phase. This is happening faster in Saturn than in Jupiter because of its lower internal temperatures. In Neptune the nature of the differentiation is unclear. Uranus has a very small infrared excess, and if indeed its interior is hot, then the outward energy transfer rate is somehow being reduced.

A planetary body with a substantial magnetic field will interact with the solar wind to produce a magnetosphere. Well beyond the magnetosphere the solar wind sweeps space clean of the planetary field. Within the magnetosphere the planetary field becomes increasingly dominant as the planet is approached. There will be a variety of plasma in a magnetosphere, much of it concentrated into a plasma sheet and plasma belts or toruses.

6 Surfaces of Planets and Satellites: Methods and Processes

When we wonder about other worlds, it is usually their surfaces to which our thoughts first turn. After all, it is the surface of one particular world upon which we live, and there is a fascination with exploring terrestrial landscapes that differ from those of our own region. How much more so is there an immediate fascination with the landscapes of other worlds. Such individual landscapes will be explored in the two chapters that follow this one. In this chapter I will outline some of the methods of investigating surfaces, and the various processes that have made the surfaces as they are. The giant planets do not have surfaces in the generally accepted sense, so they are excluded from these chapters.

6.1 Some Methods of Investigating Surfaces

The surfaces of Solar System bodies are more accessible than their interiors. The detailed surface form is observable, and the physical and chemical nature of the surface can be determined directly if samples can be obtained, or indirectly from space or from the Earth. We can also observe many surface processes in action, such as volcanism.

But a surface, in the sense of being the outermost layer of a body, distinct from the interior, is not entirely accessible. Therefore, all the methods described in Chapter 4 that are used to investigate the interior are also applied to these shallow depths — seismology, gravitational and magnetic field measurements, and so on. At such depths, as well as a global picture, they also reveal regional and local details.

6.1.1 Surface Mapping in Two and Three Dimensions

Ever since the invention of the telescope around 1600, astronomers have had a growing capability to map the surfaces of planets and satellites. At first this was by looking through the telescope, but later various imaging instruments were added. The best images have been obtained from orbiting or flyby spacecraft (Table 4.1), though ground-based telescopes and telescopes in Earth orbit, such as the Hubble Space Telescope, are yielding images that would astound astronomers of earlier generations. The images are not restricted to visible wavelengths, but are obtained at ultraviolet, infrared and radio wavelengths too. Computer processing can extract every last bit of detail.

A two-dimensional image shows variations *across* a surface, though some altitude information can be obtained, for example from the lengths of shadows cast by a feature, or from evidence that one feature is partially obscuring another. We can, however, obtain a more complete and more accurate picture of the three-dimensional form of a surface — the surface topography — by other means. One way or another we need to measure the altitude at every point on the surface — we need to perform altimetry. For the Earth, altitudes have been measured over the centuries by traditional surveying methods on land and sea, resulting in relief maps of exquisite detail, though the deep oceans were only mapped in the latter half of the twentieth century.

For other planetary bodies the best altimetry data have come from spacecraft in orbit around the body. Altitudes have been obtained by sending a radar or laser pulse from the spacecraft to the surface, and timing the interval for the echo to arrive back at the spacecraft, as illustrated in Figure 6.1. The distance d of the spacecraft from the surface is then $c\Delta t/2$ where c is the speed of light and Δt is the round-trip time of the pulse. One way to obtain the altitude from d is to measure the corresponding distance r of the spacecraft from the centre of mass of the planetary body. This distance can be obtained from the spacecraft's orbit. The altitude of the point on the surface with respect to the centre of mass is then $(r - d)$. From the orbit we also know from where on the surface the pulse was reflected, and so the topographic map is built up.

Another technique using radar is particularly useful in low orbit. This is synthetic aperture radar, in which a series of radar pulses is sent off below or to one side of the spacecraft, as in Figure 6.1. Each pulse illuminates a patch of ground that is much larger than in altimetry. The reflected pulse consists of echoes from every point in the patch, and these points can be distinguished by the different round-trip times and by the changes in wavelength due to the Doppler effect (Section 2.1.2). Points ahead of the spacecraft return shortened wavelengths, and points behind the spacecraft return lengthened wavelengths. Each point in a particular patch is illuminated by several successive pulses as the spacecraft moves in its orbit, and this greatly increases the

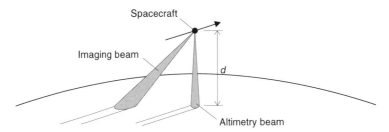

Fig. 6.1 Pulse altimetry, and synthetic aperture radar

spatial resolution. Synthetic aperture radar provides a three-dimensional image, and so the surface topography is obtained.

Though the distance $(r - d)$ reveals surface morphology, it is not the best way of specifying altitude. It is *differences* in altitude that are important, and to emphasise these differences we need to define a zero of altitude that lies within or close to the observed range. In the case of Venus, the zero chosen is the mean equatorial radius of 6051.9 km. The surface of zero altitude on Venus is thus a sphere.

Though a choice of a sphere seems pretty obvious, it's not sensible for the Earth and many other bodies. This is because these bodies are not as spherical as Venus, but are more flattened by their rotations, the faster the rotation, the greater the flattening — Venus rotates slowly, and so is not appreciably flattened. Consider a rotationally flattened planet in the idealised state of hydrostatic equilibrium, as in Figure 6.2. If a sphere centred on the centre of mass were used as the zero of altitude, the equator would be higher than the poles. Yet there is an important sense in which it is *not* downhill from equator to pole, namely, that a plumb line fixed with respect to the rotating surface would hang perpendicular to the surface. Thus, for a real planetary body rotating sufficiently fast to be appreciably flattened, the natural choice of zero altitude is the surface the body would have were it in hydrostatic equilibrium.

We could calculate the exact shape and size of this ideal surface. It is, however, easier to use any fluids that are widespread at the surface of the planetary body. A fluid flows until it is in hydrostatic equilibrium. In the case of the Earth, we have the oceans ('hydro-static' derives from 'stationary water'). For zero altitude we could select any depth of given pressure in the water, but the surface is an obvious choice, so mean sea level is defined as having zero altitude on Earth.

❐ Why the 'mean' in mean sea level?

Sea level changes with tides, winds, atmospheric pressure, currents, and the amount of sea ice, so we have to average out these effects. As far as we know, the Earth is the only planetary body with liquid covering most of its surface, but several rotationally flattened bodies are totally covered by a different sort of fluid — an atmosphere. The zero of altitude can then be defined by some value of atmospheric pressure. In the

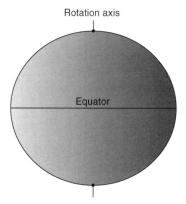

Fig. 6.2 A rotationally flattened planetary body in hydrostatic equilibrium

case of Mars zero altitude is where the mean atmospheric pressure is 610 Pa — the triple point pressure of water (Section 4.4.3). In the case of the giant planets, the choice is 10^5 Pa — close to mean atmospheric pressure on the Earth's surface.

Some bodies, such as the Moon, have surfaces that are not closely approximated by a sphere nor by the effects of rotational flattening, nor do they have an atmosphere. In this case zero altitude is defined by what is called a reference ellipsoid. This is a 'distorted sphere' characterised by three different radii at right angles.

Question 6.1

Saturn's large satellite Titan rotates slowly and has a massive atmosphere. How could you define its zero of altitude?

6.1.2 Analysis of Electromagnetic Radiation Reflected or Emitted by a Surface

The radiation from the surface of a body includes thermal emission and reflected solar radiation. The radiation can by analysed by the techniques of photometry and reflectance spectrometry outlined in Section 3.1.5 in relation to asteroids. Just as for asteroids, each technique yields information about the surface composition and surface roughness. For example, Figure 6.3 shows the reflectance spectra of Europa and of coarse-grained water ice. The correspondence is close enough to conclude that water ice is the dominant constituent of Europa's surface.

With radar we provide our own illumination, at wavelengths in the approximate range 8–700 mm (microwaves). However, radar data are not easy to interpret. The strength of the reflection depends on the angle at which the incident pulse strikes the surface, on the roughness of the surface at the scale of the wavelength used, and on the composition of the surface down to a depth of order 10 wavelengths. On the other hand, further information can be obtained if the radar pulse is circularly

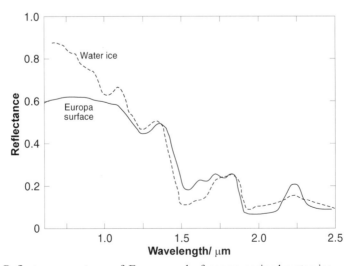

Fig. 6.3 Reflectance spectrum of Europa and of coarse-grained water ice

polarised. Circular polarisation is possible in any wave in which the oscillatory motion is perpendicular to the direction of travel, i.e. if it is a transverse wave. The S wave in Figure 4.7 (Section 4.3.1) is transverse, and so too is electromagnetic radiation. A transverse wave is circularly polarised if the oscillations rotate around the direction of travel of the wave to create something like a corkscrew. For radar, if the pulse sent is circularly polarised, then the echo will consist of a component rotating in the same direction as that sent, and a component rotating in the opposite direction. The ratio of the strength of these two reflected components provides further information on the surface.

6.1.3 Sample Analysis

The most direct way of establishing surface composition is to examine samples from known locations. Such samples can also be radiometrically dated. The Earth has, of course, been extensively sampled, and from volcanic activity we even have samples of the upper mantle. For the Moon, we have some meteorites, plus samples brought back to Earth by various missions or analysed *in situ* by landers. In the case of Mars we have *in situ* analyses plus a few meteorites of probable Martian origin, and for Venus we have *in situ* analysis. Meteorites and micrometeorites also provide samples of asteroids and comets. For all the other bodies in the Solar System we have no samples at all. Samples are analysed by a great battery of chemical and physical techniques, but the details are beyond our scope.

Question 6.2

Venus is covered in clouds that are opaque to visible and infrared radiation. Describe how a spacecraft above the clouds could determine the surface morphology and composition.

6.2 Processes that Mould the Surfaces of Planetary Bodies

There are many processes that mould the surface of a planetary body. Here we shall concentrate on those that are widespread and of particular importance. Most of the examples will be for rocky materials, though most of the processes act on icy materials too.

6.2.1 Impact Cratering

You will recall from Chapter 3 that interplanetary space is populated with large numbers of small bodies in orbits that can intersect those of planetary bodies. It is therefore to be expected that the surfaces of planetary bodies will bear the scars of individual impacts. Indeed they do, in the form of impact craters. The lunar craters are a familiar consequence of impact (Figure 6.4(a)), and about 120 impact craters have been identified on Earth, the best known being Barringer Crater in Arizona (Figure 6.4(b)).

(a)

(b)

Fig. 6.4   (a) The lunar impact crater Copernicus, about 90 km in diameter. (NASA/NSSDC AS17-151-23260) (b) Barringer Crater in Arizona, about 1.2 km diameter. (USGS, D. R. Roddy, 1cT)

An impact crater is produced when a projectile strikes the surface of a planetary body with sufficient kinetic energy to excavate a hole. The kinetic energy of the projectile is given by

$$E_k = \tfrac{1}{2}mv^2 \tag{6.1}$$

where m is the mass of the projectile and v is its speed with respect to the surface. The value of v depends on the relative motions of the projectile and the planetary body, and on any changes in projectile speed during the encounter, notably through its acceleration by the planetary body's gravity. Orbital speeds and the acquired speeds are each about $10\,\mathrm{km\,s^{-1}}$, to an order of magnitude. Therefore, to an order of magnitude v is $10\,\mathrm{km\,s^{-1}}$. In the case of the Earth, the impact speeds are in the range 5–$70\,\mathrm{km\,s^{-1}}$. Such high impact speeds mean that even a small mass can excavate a considerable crater.

❐ What is the kinetic energy of a small body with a mass of $10^8\,\mathrm{kg}$ and with a speed of $15\,\mathrm{km\,s^{-1}}$?

From equation (6.1) the kinetic energy is

$$E_\mathrm{k} = \tfrac{1}{2}\,10^8\,\mathrm{kg}\,(15\,000\,\mathrm{m\,s^{-1}})^2 \approx 10^{16}\,\text{joules}$$

This is the energy that would be liberated by the explosion of about 3 million tons of TNT, and it is also about the energy that was required to excavate the $1.2\,\mathrm{km}$ diameter Barringer Crater. This crater was excavated by an iron meteorite, and so the radius of the meteor at the Earth's surface would have only been about $15\,\mathrm{m}$ if it had been travelling at $15\,\mathrm{km\,s^{-1}}$.

Figure 6.5 illustrates the stages in the formation of an impact crater. In stage 1 (Figure 6.5(a)) the projectile has struck the surface and has penetrated perhaps only two or three times its diameter before being brought to rest. It does not get far because the speed of the projectile at impact exceeds the speed of seismic waves in the surface material, which at most will be about $4\,\mathrm{km\,s^{-1}}$. Therefore, the material ahead of the projectile gets no 'advanced warning' of the impact, and so cannot move away. This leads to the piling up of a sudden and enormous compression — a **shock wave**. The pressures generated exceed the strength of the materials by factors of 10^3–10^4, and so the surface materials and the projectile are highly fractured. At the same time nearly all the kinetic energy of the projectile goes into heating the projectile and its immediate surroundings. Most of this material is vaporised and this produces a violent explosion. During stage 1 surface materials are ejected at high speed.

In stage 2 (Figure 6.5(b)) shock waves spread out from the site of impact, fracturing and melting subsurface layers and throwing huge quantities of material outwards. The immediate result (stage 3, Figure 6.5(c)) is a hole with a volume greatly exceeding the volume of the projectile itself. The hole is rimmed with the distinguishing characteristic of overturned rock strata. Subsequently, modifications can occur by a variety of processes, as you will see.

The direction of the projectile has little effect on the shape of the crater unless it impacts at grazing incidence — the explosion and the shock waves spread out uniformly from the point of impact. Craters are thus roughly circular, unless produced by a near grazing impact, when the crater will be elongated in the direction in which the projectile was travelling.

The volume of a crater is roughly proportional to the kinetic energy of the projectile. It also depends on the gravitational field of the planetary body, and on the strength and density of its surface layers. For example, for a surface of given strength and density, the crater size diminishes as the gravitational field increases. This is because the surface materials are held down more strongly. Also, there is a big difference between rocky surfaces and icy surfaces, partly because of the lower densities of icy materials, and partly because of the lower strength of icy materials,

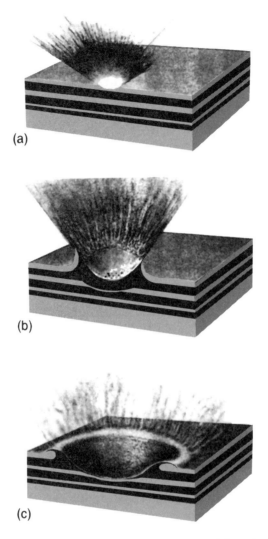

(a)

(b)

(c)

Fig. 6.5   Three stages in the formation of an impact crater. (Adapted with permission from Kluwer, from Figure 3.1 of *Planetary Landscapes*, R. Greeley, Chapman and Hall, 1994)

particularly if they are not far below their melting points. For a specific projectile kinetic energy and surface gravitational field, a crater produced in an icy surface will be roughly double the diameter of one produced in a rocky surface.

The morphology of a crater depends on its size. The reasons are complicated, but the outcome is illustrated in Figure 6.6, where the diameter ranges are for the Moon. Small craters are simple, bowl-shaped depressions (Figure 6.6(a)). The bowl is flatter at larger diameters, and at yet larger sizes a central peak will be present (Figure 6.6(b)), probably the combined result of floor rebound immediately after the excavation is complete and the reflection of shock waves from any deep interfaces. At about this size there might also be slumping of the crater walls soon after the impact, forming terraces. In yet larger craters the central peak is replaced by a cluster of peaks, or by a ring of peaks, called a

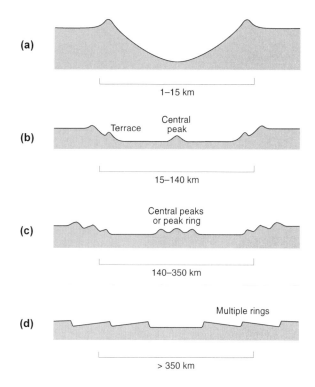

Fig. 6.6   Crater morphologies. The diameter ranges are for the Moon, and are approximate

peak ring (Figure 6.6(c)). The largest craters are complex multi-ringed structures, called multiring basins (Figure 6.6(d)), that are probably the result of a combination of slumping and waves of surface motion. The larger craters can become flooded by lavas, though this might not happen for millions of years. The larger craters are also modified by isostatic adjustment (Section 4.1.4).

❐ What effect do you think this will have?

A crater is a deficit of material, so isostatic adjustment will cause uplift within the crater at the surface (and horizontal motion deeper down). This reduces the depth-to-diameter ratio to 1/20 or less. Craters of all sizes are continuously subject to erosion and partial infill by some material or other.

The actual size ranges of the different types of crater depend on the gravitational field and on the strength and density of the surface layers, and so vary from body to body. For example, on the Earth, which has a rocky surface like the Moon but a larger gravitational field, central peaks occur in craters as small as 6 km diameter. On the icy surface of Ganymede, which has a comparable gravitational field to the Moon, they occur in craters down to about 10 km diameter.

As well as the crater itself, there will also be evidence of the ejected surface material, notably rays, ejecta blankets, and secondary craters, i.e. craters produced by the impact of ejecta thrown out by the primary impact (Figure 6.7). These

Fig. 6.7 Rays, secondary craters (very small), and an ejecta blanket from the lunar crater Euler (27 km in diameter). (NASA/NSSDC AS17-2923)

features, like the crater itself, are subject to modification through erosion and through deposition of material.

6.2.2 Craters as Chronometers

Impact craters can be used to determine the age of a surface. The older a surface, the greater the number of impacts per unit area it will have accumulated. Therefore, a heavily cratered surface on a planetary body must be older than a lightly cratered surface. Figure 6.8 shows a heavily cratered region on the Moon adjacent to a much more lightly cratered region. Clearly the latter is the younger surface.

For the Moon as a whole, Figure 6.9 shows the number densities of craters of different sizes, i.e. the number of craters per unit area in defined ranges of diameters. Two graphs are shown, the one averaged over the lunar highlands, the other averaged over the lunar maria, which are the smooth, dark areas in Plate 7. ('Maria' is the plural of 'mare' (ma-ray), Latin for 'sea', though we now know that there are no seas on the Moon.) The graphs clearly show that for the Moon as a whole the maria are younger than the highlands. Regional studies show that even the oldest mare is younger than the youngest highland area. Both graphs show that the smaller the crater, the greater their number density. Therefore, the smaller the kinetic energy of a projectile, the more numerous they are. Among the great range of projectile masses, impact speed is uncorrelated with mass and so, broadly speaking, Figure 6.9 shows that massive projectiles in the Solar System have been fewer in number than

Fig. 6.8 Lunar craters in the highlands and the adjacent Oceanus Procellarum. The frame is about 160 km wide. (NASA/NSSDC AS15-2483)

less massive ones. This is in accord with our understanding of the size distribution of planetesimals (Section 2.2.3) and with the subsequent evolution of the population of left-overs, as asteroids and comets.

In what sense is the age of a surface indicated by craters? On a given planetary body, one surface will have a lower crater density than another because of some resurfacing event that obliterated some or all of the existing craters. An obvious example is a flood of lava. Another is melting of the surface. In such cases, the age of the surface indicated by its crater density is the time in the past when such resurfacing occurred. Craters can also be obliterated by various erosional or depositional processes. These will obliterate small craters much more rapidly than large ones. This is a different sort of resurfacing, continuous in time rather than concentrated near to a particular time, and it is important to avoid confusing its effects with those that result from lava floods and the like. For example, suppose that erosion is more powerful near the poles of a planet than near its equator. The crater density at the poles, particularly for small craters, will consequently be lower than at the equator for surfaces of equal age.

❏ How could this be misinterpreted?

This could be misinterpreted to mean that the polar regions had been resurfaced by lava or by melting more recently than surfaces elsewhere. It is therefore common to exclude from counts those craters with diameters less than a few kilometres.

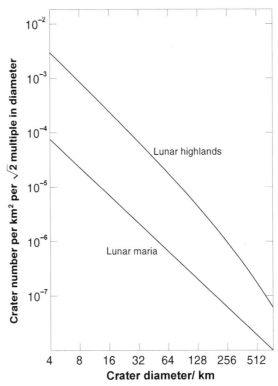

Fig. 6.9 Numbers densities of lunar craters of different sizes, averaged over the lunar highlands and over the lunar maria

Further complications arise from the need to avoid counting craters of volcanic origin and secondary craters. Fortunately, most volcanic and secondary craters have morphologies and spatial distributions that betray their origin. Again it is wise to exclude small craters, many of which will be secondaries.

A final complication is the phenomenon of saturation. A surface becomes saturated with impact craters when further impacts, on average, obliterate as many craters as they create. It is then not possible to distinguish between older and younger saturated surfaces on the basis of impact crater densities alone. The most heavily cratered parts of the lunar highlands might be close to saturation, as might be regions on some of the icy–rocky satellites, though definitively saturated surfaces in the Solar System are rare. One measure of saturation is the degree of randomness of the spatial distribution of craters. Simulations show that as an area approaches saturation, the random distribution that characterises subsaturated areas becomes more uniform as sparsely cratered subareas acquire more, and subareas near to saturation are little changed.

With care, crater densities can be used to place the various surfaces on a single planetary body in the order of their age since some widespread resurfacing. This can also be done for the surfaces on *separate* planetary bodies, provided that we know of any differences in

- The bombardment history of the two bodies
- Their surface properties
- Rates of erosion and deposition.

Unfortunately, such differences are rarely well known, particularly the first. The bombardment history depends (among other things) on the local population of potential projectiles. Unfortunately, differences between local populations are poorly known. For example, Mars is close to the asteroid belt, and so will surely have been more heavily bombarded than the Moon. But estimates of exactly how much more are highly uncertain. One estimate is that a (unsaturated) surface of given age on Mars has received twice the number of impacts as a surface of similar age on the Moon. If this is the case then we can estimate whether the surface on Mars is older or younger than a particular surface on the Moon. But there is much uncertainty.

To obtain *absolute* ages we need to know in absolute terms the cratering history of at least one body, and how to apply the data to other bodies. Only for the Moon do we have the absolute ages for surfaces with *widely* different crater densities. These ages have been obtained by radiometric dating, and they have enabled us to deduce the rate of impact cratering on the Moon almost as far back as the birth of the Solar System. The crater densities and corresponding absolute surface ages are shown in Figure 6.10 for the total number of craters larger than 4 km in diameter. (Smaller craters are excluded to avoid the problems noted earlier.) The limited data for the Earth have been included (corrected for Earth–Moon differences). Terrestrial craters are few because much of the surface is young, and because erosion and deposition are severe.

Figure 6.10 also shows an impact rate inferred from the crater densities. These densities are not as well known as the crisp line in Figure 6.10 indicates, and so the

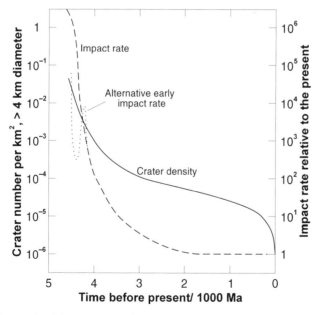

Fig. 6.10 Crater densities versus surface age in the Earth–Moon system, and an inferred impact rate

impact rate is correspondingly uncertain. Nevertheless, it is certain that the rate was initially very high and then declined rapidly. Before about 4200 Ma ago two possibilities are shown. In the one there is a monotonic decrease in bombardment, but in the other there is a peak. In both cases the overall rapid decline presumably reflects the final stages of mopping up of interplanetary debris left over from the formation of the Solar System. The period from the end of accretion about 4500 Ma ago to about 3900 Ma ago is called the heavy bombardment. Any peak during this time is the *late* heavy bombardment.

In applying the data in Figure 6.10 to other bodies we have the same uncertainties that we encounter in placing surfaces in age order, particularly the differences in local populations of projectiles through the ages. The only feature that presumably has roughly the same age everywhere is the decline of the heavy bombardment. Therefore, any near-saturated surface is likely to be older than about 3900 Ma. This reveals an interesting difference between the inner Solar System and the outer Solar System. On the most densely cratered surfaces, the number densities of craters of different sizes are rather different in these two regions, which indicates that the populations of heavy bombardment projectiles were also different. This also seems to have been the case subsequently, and this makes it impossible to place absolute ages on outer Solar System surfaces that post-date the heavy bombardment.

Question 6.3

A plain called Chryse Planitia on Mars has a crater density of 2.2×10^{-4} per km^2 for craters greater than 4 km diameter.
(a) How many such craters are there on a typical area of 10^5 km^2 in Chryse Planitia?
(b) Assuming that data for the Moon can be applied to Mars, estimate how long ago this area was resurfaced. Why is your value an upper limit to the age?

6.2.3 Melting, Fractional Crystallisation, and Partial Melting

The composition of the surface of a planetary body depends on the composition of the materials from which the planetary body formed, but it also depends on the subsequent processing of that material. An important process is melting.

Consider first those planetary bodies dominated by silicates plus iron or iron-rich compounds. The body will either have an iron-rich core from the start, or one will start to grow by differentiation as the interior temperatures rise. In either case there will be a silicate mantle more or less rich in iron. In Section 5.1.1 you saw that in the case of the Earth the mantle is predominantly peridotite, which consists of the minerals olivine $((Mg, Fe)_2SiO_4)$ and pyroxene $((Ca, Fe, Mg)_2Si_2O_6$, where the metals can less commonly be Na, Al, or Ti). Table 6.1 lists these minerals, along with some others we will shortly encounter.

Consider a peridotite mantle in the late stage of accretion of a planetary body. It is possible that during this stage the upper mantle becomes completely molten. A magma ocean is then said to have formed, **magma** being partially or wholly molten material (rocky here, but icy too). The ocean will consist of a mixture of minerals and, as long as no crystals form, the random thermal motions keep the magma well mixed. The ocean gradually cools and the first minerals begin to crystallise. Among these will

Table 6.1 Important igneous rocks and minerals in the crusts and mantles of terrestrial bodies

Rock	Mineral content (major components)	Where found on the Earth and the Moon
Peridotite	Pyroxene + olivine	Earth: mantle
Basaltic–gabbroic rocks		
● Basalt (extrusive)	Feldspar + pyroxene	Earth: oceanic crust
		Moon: maria
● Gabbro (intrusive)	Feldspar + (Fe, Mg)-rich silicates such as pyroxene	Earth: oceanic crust
		Moon: mantle (with olivine)
● Anorthositic gabbro/ anorthosite (intrusive)	Calcium-rich feldspar	Moon: highlands
Granitic–rhyolitic rocks		
● Granite (intrusive)	Feldspar + quartz	Earth: upper continental crust
● Rhyolite (extrusive)	Feldspar + quartz	Earth: upper continental crust
Andesite	Intermediate between basaltic–gabbroic and granitic–rhyolitic	Earth: continental crust

Pyroxene	$(Ca, Fe, Mg)_2Si_2O_6$	(rarely Na, Al or Ti in the brackets)
Olivine	$(Mg, Fe)_2SiO_4$	
Feldspar	$(K, Na, Ca)AlSi_3O_8$	
Quartz	SiO_2 (a particular crystalline form of silica)	

be silicates rich in magnesium and iron, such as olivine. These are dense silicates and so tend to sink downwards. Other minerals that crystallise at the comparatively high temperatures at this stage are calcium–aluminium rich silicates that are rich also in silicon and oxygen. Feldspars are an important example — $(K, Na, Ca)AlSi_3O_8$. Such silicates have comparatively low densities and so tend to float upwards. The separate crystallisation of different minerals is called **fractional crystallisation**.

The formation of feldspars, particularly those rich in calcium, is moderated by the effect of pressure. At pressures in excess of about 1.2×10^9 Pa, calcium and aluminium crystallise within denser minerals that do not float upwards. This means that feldspar would not emerge from depths greater than those at which the pressure reaches 1.2×10^9 Pa. Such pressures are reached at depths ranging from 40 km in the Earth to 250 km in the Moon, so the larger terrestrial bodies would have been more vulnerable than the smaller bodies to pressure moderation of feldspar formation.

Complete differentiation might well not occur. Partial differentiation would yield a surface layer rich in feldspar, plus other silicates rich in iron and magnesium, notably pyroxene. Nevertheless, in all cases the formation and subsequent cooling of a magma ocean would result in the surface layer being depleted in iron and magnesium, and enriched in silicon, oxygen and in metals such as calcium and aluminium. The upper mantle would then consist of peridotite correspondingly somewhat depleted in elements such as calcium and aluminium. A crust–mantle distinction would thus be established, as in Figure 6.11.

Fractional crystallisation is not the only way of creating a crust. This can also occur through the partial melting of a solid outer mantle consequent upon a rise in internal temperatures, or a reduction in pressure. Partial melting can occur whenever there is a *mixture* of minerals. As the temperature rises, or pressure falls, there will come a point where a proportion of the mixture will melt.

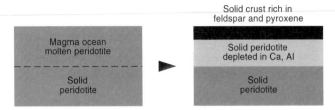

Fig. 6.11   Formation of a crust by differentiation in a peridotite mantle

❏ Why does pressure reduction promote melting (see Figure 4.11, Section 4.4.3)?

This is because when the pressure is reduced the melting point of a substance is also reduced. If the liquid from partial melting is less dense than its surroundings then it will be buoyant and will tend to rise and form a crust with a composition distinct from that of the solid from which it was derived. Once a crust is established, further partial melting can occur within the crust to produce further differentiation. The outcome is more than one crustal type, and a rich variety of surface rocks.

The simple mixture of water (H_2O) and ammonia (NH_3) can serve to illustrate an important feature of partial melting. A simplified phase diagram of various mixtures with up to 50% ammonia is shown in Figure 6.12, which corresponds to 10^5 Pa. (At the higher pressures below a planetary surface, but not far below, the diagram will not be very different.) At temperatures below 176 K there is a solid mixture of water (H_2O) and various ammonia hydrates (e.g. $NH_3.H_2O$). At temperatures above 176 K liquids of various sorts appear, depending on the proportion of ammonia. The important point is that 176 K is lower than the melting temperature of ammonia (195 K) and of water (273 K). In general, mixtures of any sort, icy or rocky, partially melt at lower temperatures than their pure constituents.

The discovery at the surface of a planetary body of materials that could have resulted from partial or total melting is evidence that such melting has occurred in the past. This indicates that interior temperatures were sufficiently high to cause the melting.

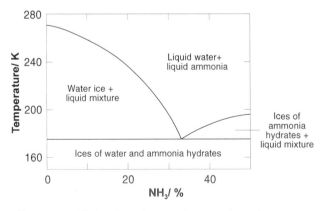

Fig. 6.12   Phase diagram at 10^5 Pa of a mixture of ammonia and water

Question 6.4

In one sentence, state why silica (SiO_2) is not subject to partial melting or partial crystallization.

6.2.4 Volcanism and Magmatic Processes

Volcanism covers all processes by which gases, liquids, or solids are expelled from the interior of a body into the atmosphere or onto the surface. If icy materals are involved the processes are called cryovolcanism.

Volcanism and cryovolcanism start with partial melting in the interior, of rocky materials and icy materials respectively. The magma, being less dense on average than the surrounding materials, finds its way to the surface through fissures. *En route*, rocky magma will dissolve volatile substances such as water, and the magma might partially solidify as it approaches the surface. Volcanism and cryovolcanism lead to further chemical differentiation.

The exact composition of the magma, its temperature, and volatile content, determine what is called the *style* of volcanism. There are two extreme styles. In explosive volcanism the eruption is very violent, because the magma is very viscous and rich in dissolved volatiles. With rocky materials, rock fragments and ash are erupted at high speeds. By contrast, in effusive volcanism the volatile content is low, or has become low by the time the magma reaches the surface, and the magma also has a low viscosity. Consequently there is a surface flow of molten rock (or icy materials), called **lava**. Viscosity (which depends on composition and temperature) and volatile content are continuous and somewhat independent variables, so there are mixed styles too, including volatile-rich low viscosity magma that erupts explosively yet leads to lava flows. The details of each style are influenced by the rate of eruption and by the gravitational field. In explosive volcanism the density of any atmosphere and the winds help determine the spatial distribution of the products.

Given so many different factors, it is not surprising that a wide range of volcanic features is found. The 'classic' volcanic cone (Figure 6.13(a)) is the result of explosive volcanism where the rate of eruption is modest. At modest eruption rates and low viscosities, a sequence of separate lava flows can create **shield volcanoes** (Figure 6.13(b)), which derive their name from their resemblance to a warrior's shield. Less familiar are vast volcanic plains created by low-viscosity lavas flowing at high rates from channels and tubes that radiate out from a vent or fissure.

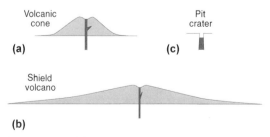

Fig. 6.13 (a) Volcanic cone. (b) Shield volcano with summit caldera. (c) Volcanic pit

There are many other volcanic landforms, but we need only consider volcanic craters, because of the need to distinguish them from impact craters. Many volcanoes have summit craters (Figure 6.13(a), (b)) called summit **calderas**. There are also volcanic pits and depressions that do not sit on mountains (Figure 6.13(c)). Some of these forms bear a superficial resemblance to impact craters, but in most cases a more careful morphological examination will reveal their volcanic origin. You can get some sense of this if you compare Figure 6.13 with Figure 6.6.

Sometimes, magma moving up through a fissure does not reach the surface, but spreads out sideways into other fissures and solidifies. Erosion can subsequently reveal these **intrusive rocks** as they are called. Lavas and products of explosive volcanism give rise to **extrusive rocks**. Widespread examples of extrusive–intrusive pairs are **basaltic–gabbroic rocks**, basalt being the extrusive form and gabbro the intrusive form, and **granitic–rhyolitic rocks**, where the extrusive and intrusive forms are, respectively, rhyolite and granite. The extrusive and intrusive forms have much the same mineral content, determined by partial melting. Thus, basalt and gabbro are each mixtures of the minerals feldspar and pyroxene, and rhyolite and granite are each mixtures of feldspar and quartz. The intrusive forms are coarser grained, i.e. have larger crystals. This is because they solidified more slowly in their underground environments than did the rocks extruded onto the surface.

Rocks produced from magma, whether intrusive or extrusive, constitute what are called **igneous rocks**. Table 6.1 lists the important igneous rocks and their corresponding mineral content, and where they are commonly found in the Earth and the Moon.

Question 6.5

Why could you conclude from its location that the summit crater in Figure 6.13(b) is unlikely to be of impact origin? How could you confirm your conclusion by surveying the composition of the surface (assume a rocky world), and any layering around the crater perimeter?

6.2.5 Tectonic Processes

Tectonic processes are those that cause relative motion or distortion of the lithosphere, and they derive their name from the Greek word for carpenter — 'tekton'. Volcanism can be associated with tectonic processes, but this need not be the case. Tectonic features can vary in size from a few kilometres to planetary scale.

Faults are a common tectonic feature. Figure 6.14 illustrates three of the many kinds of fault. The normal fault in Figure 6.14(a) usually arises from stretching of the lithosphere — the lithosphere is in tension and therefore cracks down to some depth, followed by relative vertical motion. Some rift valleys, or **grabens**, are the result of slumping between two parallel normal faults (Figure 6.14(b)). The Great Rift Valley that extends for about 6000 km from Syria to southern Africa, is a huge example. It is the longest rift valley on Earth, and has a typical width of about 50 km. Part of one of its walls is shown in Figure 6.15. A graben can also result from the collapse above a linear disturbance below the surface, such as a vertical sheet of intrusive rock (a dyke). A thrust fault, or reverse fault (Figure 6.14(c)) usually results from

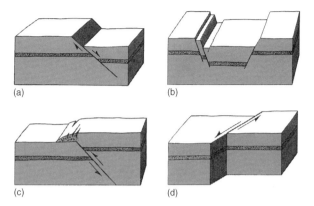

Fig. 6.14 Three kinds of fault. (a) A normal fault. (b) A graben (rift valley). (c) A thrust fault. (d) A strike-slip fault. (Adapted with permission from Figure 11–16 of *Moons and Planets*, W. K. Hartmann, Wadsworth, 1993, © Wadsworth Inc.)

lithospheric compression. A strike-slip fault (Figure 6.14(d)) can arise from compression or tension, but is characterised by motion that is predominantly horizontal and not vertical as in the other examples.

Compression can also cause bending and folding of the lithosphere, as in Figure 6.16(a). On the largest scale the result is the fold mountains that snake across the Earth, such as the Alpine chain in Europe, and the Himalayas in Asia. You might think that the highest bits of folded lithosphere always correspond to the anticlines in

Fig. 6.15 Part of the eastern wall of the Great Rift Valley. (The author)

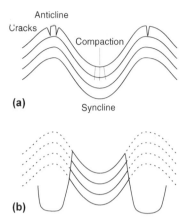

Fig. 6.16 (a) Folding of the lithosphere. (b) The effect of subsequent erosion of the anticlines

Figure 6.16(a), and indeed this is the case in young folds. But notice the cracks that have opened up at the tops of the anticlines. These make the anticlines more prone to erosion. Conversely, the compaction that has occurred in the synclines makes them less prone to erosion. As a result, the anticlines can be eroded to below the level of the synclines, as in Figure 6.16(b).

Planetary bodies differ greatly in the extent to which the lithosphere has experienced tectonic processes, and whether the processes have been local, or form a coherent planet-wide system. For example, the Moon shows few tectonic features, whereas the Earth shows many, largely the result of a planet-wide system called plate tectonics (Section 8.1.2). The extent to which a lithosphere has experienced tectonic processes depends on its composition and its thermal history. Conversely, the extent and nature of tectonism provides important information on the thermal evolution of the body. Further discussion of individual planetary bodies is in Chapters 7 and 8.

❐ Would you expect icy lithospheres to be subject to tectonic processes?

Yes indeed, though their expression will reflect differences between the behaviour of ices and rocks.

6.2.6 Gradation

Gradation covers all processes by which material is eroded from a surface, and then transported and deposited elsewhere, perhaps in a different mineralogical form. During deposition, the gravitational field of the planetary body tends to cause settling to the lowest available altitude, and therefore gradation tends to level off a landscape. There are several distinct gradational processes.

Mass wasting

Mass wasting is the downslope movement of materials under the influence of gravity. Material can be loosened physically in many ways—for example, by seismic waves, the day–night thermal cycle, the impacts of micrometeorites, and the action of liquid water or ice. It can be loosened by chemical reactions at the surface, particularly if there is an atmosphere. Material can also be made subject to mass wasting by the removal of adjacent material by other gradational processes. Once sufficiently loosened, gravity does the rest. The distinctive feature of mass wasting is that transportation is downhill directly from where the material lay to its new location — there is no long-range transportation. One of many possible outcomes is shown in Figure 6.17, where mass wasting has produced the fan-shaped deposits.

Aeolian processes

Whereas mass wasting can occur on the surface of *any* planetary body, aeolian processes require an atmosphere in which winds blow. Wind moves solid particles such as dust and sand, and it can move them over large distances before they are deposited. Small particles tend to be carried in the wind, and larger particles tend to bounce or creep along the surface. Wind-borne particles cause further erosion (Figure 6.18). Wind can also sculpt a sandy surface to create sand dunes, and these can creep over the landscape like slow waves. Accumulations of small particles are called sediments, and deposition from winds is one way of producing them.

Hydrological processes

One of the more familiar results of hydrological gradation is the terrestrial system of river valleys resulting from surface drainage. The very existence of such systems is

Fig. 6.17 An example of mass wasting, in Wyoming. (USGS, T. A. Jaggar 120)

Fig. 6.18   An effect of wind erosion in Chile. (USGS, K. Segerstrom 554)

testimony to the large quantities of material moved downstream by the surface run-off of liquid water. Motion of ground water near the surface and of water deeper underground also results in erosion and transportation. Valley glaciers and ice caps remove huge quantities of material, but rather more sedately. When glaciers and ice caps recede, characteristic landforms are left behind, such as valleys with U-shaped cross-sections. As well as such large-scale erosional features, ice can also cause widespread erosion on a small scale. For example, if the day–night temperature cycle is such that melting and freezing of water repeatedly occurs, then the break-up of water-bearing rocks or wet rocks is accelerated by the expansion of water on freezing.

Lakes and oceans also erode materials, transport and deposit them, as can be seen in landforms like cliffs and beaches, and in extensive deposits of sediments, particularly on ocean floors. Sediments deposited in water can subsequently be exposed if the oceans recede, or if a shallow lake dries up.

Water moves material not only mechanically but also chemically by dissolving some or all of the minerals in rocks. Limestone, which is mainly calcite ($CaCO_3$) and dolomite ($CaMg(CO_3)_2$), is particularly susceptible, especially if the water contains dissolved carbon dioxide. The removal of limestone from under the Earth's surface has produced many magnificent underground caverns. Another chemical effect of water is to *modify* the minerals present. **Clay minerals** are a common outcome. For example, the clay mineral montmorillonite has a large molecule with the formula (Al, Mg)$_8$(Si$_4$O$_{10}$)$_3$(OH)$_{10}$.12H$_2$O. This formula shows the effect of water on silicates not only through the attached water molecules H_2O but also through the presence of the water molecule fragment OH (hydroxyl).

Hydrological and aeolian processes are important aspects of *weathering*.

Question 6.6

List the gradational processes to which a planetary surface is subject if the planet has an atmosphere but no water.

6.2.7 Formation of Sedimentary and Metamorphic Rocks

A sediment, however formed, and wherever it lies, can form a **sedimentary rock**. Chemical cementation of the solid particles is an important part of the process, though modest pressure helps through the consolidation it produces. On Earth, shale is a particularly abundant sedimentary rock, making up an estimated 4% of the upper 6 km of the Earth's crust. It consists of very fine grains derived from consolidated sediments. These sediments consisted of clays and silts, dominated by clay minerals in both cases but distinguished on the basis of size — clay particles are defined as those smaller than 1/256 mm across, and silt particles as those in the range 1/256–1/16 mm. Limestone is another type of sedimentary rock, consisting mainly of cemented carbonate particles. A **carbonate** is a compound that contains the chemical unit CO_3, a common terrestrial example being calcium carbonate, $CaCO_3$.

In addition to igneous and sedimentary rocks, there is just one other major rock type — metamorphic rock. A **metamorphic rock** is an igneous or sedimentary rock that has been modified but not completely remelted. Metamorphosis can result from any combination of raised pressure, raised temperature, or a change in the chemical environment. For example, if shales are subject to a combination of raised pressure and temperature, then slate is a possible outcome. If granite is subjected to high pressure and temperature then granite–gneiss ('nice') is formed, with a prominent banded structure.

6.3 Summary

Surfaces are investigated through mapping that establishes the topography. Mapping can be using images, altimetry, and synthetic aperture radar. For bodies that are rotationally flattened, the zero of altitude is defined by a surface of hydrostatic equilibrium, such as mean sea level on Earth.

The composition and other characteristics of a surface are investigated by photometry and reflectance spectrometry, and through the use of radar. Direct analysis of surface samples has only been achieved for the Earth, the Moon, Venus, and Mars. Meteorites and micrometeorites provide samples of asteroids and comets, and also of the Moon and Mars.

There are several different types of process that create and modify the surface of a planetary body:

- Melting, fractional crystallisation, and partial melting
- Impact cratering — in which bodies from interplanetary space impact the surface
- (Cryo)volcanism — all processes by which gases, liquids, or solids are expelled from the interior
- Tectonic processes — all processes that cause relative motion or distortion of the lithosphere

- Gradation — all processes by which material is eroded from a surface, and transported and deposited elsewhere
- The formation of sedimentary and metamorphic rocks.

Impact craters can be used to place surfaces in a sequence of ages, the greater the crater density, the older the surface, i.e. the longer ago it was last resurfaced by lava or by melting. Care has to be taken to allow for gradational effects, and to exclude volcanic craters and secondary impact craters. Absolute ages of surfaces with widely different crater densities have been obtained for the Moon. In comparing one body with another we have to allow for any differences in the bombardment history of the two bodies, their surface properties, and in the rates of gradation.

7 Surfaces of Planets and Satellites: Inactive Surfaces

Except for the Sun and the giant planets, which have no surfaces in the generally accepted sense, the surfaces of all bodies in the Solar System are still subject to impact cratering and gradation, but only some of them now experience (cryo)volcanism or tectonic processes to any significant extent, and in this sense are active. This suggests an obvious grouping of the surfaces into two sorts, and this is the basis on which this and the following chapter are organised. Thus this chapter describes surfaces no longer active, and Chapter 8 describes those that are. The bodies with active surfaces are the Earth and Venus—the two largest terrestrial planets—and some of the larger satellites. The bodies with inactive surfaces range in size from Mars down to the smallest bodies in the Solar System. I shall concentrate on the larger ones. Broadly speaking, a body has an inactive surface because its interior has cooled to the point where its lithosphere is now too thick to allow (cryo)volcanism or tectonic processes to occur.

7.1 The Moon

The Moon is the only body on which we can see surface features with the unaided eye (Plate 7). The dark areas are the maria (singular, mare). These are relatively smooth, and they lie amid more rugged highland terrain that constitutes most of the lunar surface. The highlands reach up to 16 km above the lowest-lying regions, and are dominated by impact craters.

The Moon is in synchronous rotation around the Earth. Therefore one side—the near side—always faces towards the Earth, and the other side—the far side—always faces away from it. However, this does not mean that one face of the Moon never sees the Sun. When we see the Moon as less than full then part of the near side is in darkness, and consequently part of the far side must be in sunlight. At any point on the lunar surface the time between successive noons—the lunar 'day'—is 29.53 days. With no atmosphere to moderate them, the temperatures at the equator reach about 400 K at noon only to plunge to about 100 K at midnight.

With the Moon in synchronous rotation we would see just 50% of the lunar surface if every one of the following conditions was met (you do not need to dwell on these):

- Its orbit around the Earth were circular
- Its rotation axis were perpendicular to the plane of its orbit around the Earth
- The lunar orbit did not precess about an axis perpendicular to the ecliptic plane
- The radius of the Earth were negligible (so we would always observe the Moon from the same vantage point as the Earth rotates).

None of these conditions is met, and as a result, from any point on the Earth's surface the Moon appears to rock slowly to and fro (east–west) and nod up and down (north–south). These are called geometric librations, and they allow us to see 59% of the lunar surface over a period of about 30 years. In addition, because of tidal forces, the lunar rotation rate oscillates, and consequently there is a physical libration that allows us to see a tiny bit more. Nevertheless, about 40% of the lunar surface was hidden from us until 1959, when the spacecraft Lunik III provided us with our first images of the far side. A more recent image of part of the far side is shown in Figure 7.1.

7.1.1 Impact Basins, and Maria

Impact craters are the dominant lunar landform. Much of the highlands is near to saturation, indicating great age. The largest impact craters are the impact basins, many of which have subsequently been partially filled to create the maria, which constitute about 17% of the lunar surface. The maria are concentrated on the near side (Figure 7.1). The largest impact basin of all, the South Pole-Aitken basin, is on the far side near the south pole. It has a maximum depth of 8.2 km below the reference ellipsoid that defines zero altitude, and a diameter of 2250 km, making it the largest known impact basin in the Solar System. There is very little infill. The impact might have excavated the upper mantle, and image analysis is consistent with the presence of silicates richer in iron and magnesium than the crust.

There is plenty of evidence that the maria are partially filled impact basins. The circular nature of the mare boundaries (Plate 7) and arc-shaped mountain ranges within the maria surely indicate a multiring impact basin beneath. Further evidence is gravitational field measurements which show that coinciding with many maria there is excess mass — a mass concentration called a **mascon**. One possible cause is the upward bowing of relatively dense mantle immediately after the impact, which would have greatly reduced the mass deficit after excavation, but still left a depression. Any subsequent infill of the basin would create the mascon, provided that the lithosphere had by then become too rigid to achieve isostatic adjustment (Section 4.1.4). In some cases, the upward bowing itself could have been excessive, due to rebound held in place by a rigid lithosphere, and so created a mascon before any infill.

Studies of impact basins, filled and unfilled, indicate that the Moon has long had a thick lithosphere that has prevented isostatic equilibrium from being achieved.

Fig. 7.1 Lunar far side (left half) and near side (right half), imaged by the Galileo spacecraft in 1990 en route to Jupiter. (NASA/JPL P37327)

7.1.2 The Nature of the Mare Infill

There is a good deal of evidence that the mare infill is mainly lava, and not debris from later impacts, nor migrating dust. For example, mare samples have a basaltic composition, in sharp contrast to the pyroxene-poor composition of the surrounding highlands (Table 6.1). Also, shallow channels, called sinuous rilles, snake across the maria — these could be the remnants of lava supply channels or collapsed lava tubes. Some linear rilles might be grabens created by the extraction of subsurface magma. More obvious signs, such as volcanoes, are represented by only a handful of small features, and so it is presumed that the upwelling of the lava was mainly through fissures that the lava itself buried — there are terrestrial examples. Had the lava erupted all in one go, the resulting fluid would have filled the basin to the level required by isostatic adjustment and there would be no mascon. It is therefore necessary to suppose that the infill was in a series of sheets, each a few tens of metres thick. There is evidence for such sheets on Mare Imbrium, in the form of scarps on the surface (Figure 7.2).

Fig. 7.2 Part of the lunar Mare Imbrium, showing what are probably thin sheets of lava. The frame is about 60 km across the base of this oblique view (NASA/NSSDC AS17-155-23714)

It is widely believed that the mare lavas are derived from the partial melting of the mantle a few hundred kilometres beneath the lunar surface, the melting being aided by the release of pressure as the mantle material ascended. Detailed studies of large basins of different ages suggest that this infill was available more readily early in lunar history than later, and this indicates that the lithosphere gradually thickened as the interior cooled, making it more difficult for lava to be released after later large impacts. Even so, the magma always took some time to reach the surface. Radiometric dating shows that several hundred Ma separated maria basin formation from maria infill.

❐ What other evidence is there that the maria surfaces are younger than the rest of the lunar surface?

The low density of impact craters on the maria surfaces also indicates relative youth. There is thus plenty of evidence for the delay in mare infill necessary to explain the maria mascons.

7.1.3 Two Contrasting Hemispheres

On the far side there is less infill of impact basins. The second largest of the far-side basins, Mare Orientale, about 900 km across, is in the centre of Figure 7.1. It is only slightly filled, and the gravitational field shows a deficit of mass — a 'negative mascon'. The lack of infill on the far side is a possible outcome of the observed higher altitudes of the basin floors there than on the near side, which could have placed the far-side basins beyond the reach of magma. The exception is the South Pole-Aitken basin, which is so deep that the lack of infill indicates significant regional differences in the properties of the crust and upper mantle.

Another striking difference between the near side and the far side is the crustal thickness. This has a global mean value of 61 km, but gravitational and topographic measurements, particularly by the Clementine orbiter (Table 4.1), have shown that the mean thickness on the far side is about 12 km greater than on the near side. On a regional rather than hemispherical scale, the thickness of the crust varies over the approximate range 20–120 km, though it is much thinner than 20 km in the South Pole-Aitken basin. These regional variations must be due in part to the transport of crustal materials across the surface of the Moon by giant impacts, and this could explain the difference on the hemispherical scale. Another regional factor with hemispherical consequences might be major variations in the extent of melting of the lunar exterior. Additional possibilities are that erosion by mantle convection early in lunar history moved crustal material from the near side to the far side, or that mantle convection has re-incorporated into the mantle more of the near side crust than the far side crust.

7.1.4 Other Lunar Surface Features

Tectonic features include faults and ridges, largely confined to the maria, and some of the linear rilles — these occur in the highlands and on the maria. These features can be explained by a combination of crustal tension early in lunar history, crustal deformation around impact basins, and cooling of lava.

Gradation is confined to mass wasting and impact-related events. There is no evidence at all that the Moon ever had oceans, lakes, rivers, or an atmosphere of any significance, though water ice seems to be present in areas near the poles that have been shaded from sunlight for the last 1000 Ma or so. The Clementine orbiter found indirect evidence of ice at the south pole, and subsequently the orbiter Lunar Prospector, using a different technique, found indirect evidence of ice at both poles. The order of 10^{13} kg in total lie near the surface in various craters. The water must have migrated there from elsewhere on the Moon, and has survived because the sublimation rate from permanently shadowed areas is very low, though dust protection is necessary to shield any water from photodissociation by UV radiation from starlight. The Moon must have been so dry at birth that the water is likely to have been delivered subsequently, by volatile-rich impactors such as comets.

7.1.5 Crustal and Mantle Materials

Lunar samples have been returned to Earth by the six Apollo manned landings and by three Soviet Luna robotic missions, all on the near side. Samples have also been analysed at the surface by several robotic landers, again on the near side. In all cases the samples comprise small rocks found lying on the surface, pieces chipped off larger rocks, and samples from cylindrical tubes that have penetrated down to 2.4 m below the surface.

The surface of the Moon is covered in fine dust, called lunar fines or 'soil'. Figure 7.3 is a typical view of the dusty surface. The fines is a complicated mixture of silicate rock fragments and glassy particles with much the same composition at the nine sites at which it has been sampled. On the maria the fines have albedos of only 5–8%, whereas in the highlands the values are in the range 9–12%, which is why the highlands look brighter. There are four sources of the fines: recondensed minerals that were melted or vaporised by impacts, fine ejecta, surface rocks fractured by micrometeorites, and dust infall from space. As well as the fines there are lots of small rocks, but few large ones. These small rocks are mainly breccias ('bretchy-ars')—the result of the pressure and temperature welding of rock fragments by impacts. Fines plus breccias dominate throughout the length of the core samples. The whole assemblage of fines and pieces of rock is called the **regolith**.

What lies beneath the regolith? Seismic data from landers, all on the near side, are shown in Figure 7.4, and indicate that beneath the regolith there is broken rock that becomes fully compacted at a depth of about 20 km.

❐ What does the sharp increase in speed at the greater depth of 60 km indicate?

The sharp increase in speed at about 60 km (at least on the near side) is widely interpreted as a change in composition at a crust–mantle interface.

We turn now to the mineralogy of the lunar rocks. The highland rocks are dominated by anorthosite, an intrusive igneous rock consisting largely of the mineral feldspar (Table 6.1), and in particular the calcium-rich variety plagioclase feldspar. Anorthosite in this quantity could be produced by strong differentiation in a

Fig. 7.3 A typical view of the lunar surface, dominated by dust with few large rocks. This is at the Apollo 17 landing site, on the edge of Mare Serenitatis. (NASA/NSSDC AS17-145-22165)

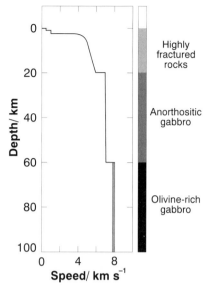

Fig. 7.4   Seismic P wave speeds versus depth for the outer part of the Moon

widespread magma ocean created by impact melting in the later stages of lunar accretion. To account for the anorthosite it would have needed to be a few hundred kilometres deep, with a peridotite composition similar to that of the Earth's mantle, though depleted in the more volatile elements and modified by the formation of a small iron-rich lunar core. Fractional crystallisaton would have established a lunar crust rich in anorthosite, underlain by a mantle rich in olivine and gabbro. It is this change in composition that is held to be responsible for the increase in seismic speed at about 60 km in Figure 7.4. The seismic speeds are consistent with anorthosite above this level and olivine plus gabbro below it. It is possible that the heat-producing radioactive isotopes of uranium, thorium, and potassium are concentrated into the crust, and this would help explain why the Moon's interior seems to have cooled early in its history (Section 4.5.1).

The mare rocks differ from the highland rocks through being dominated by the extrusive igneous rock basalt — just the sort of rock that would result from partial differentiation of the mantle after crust formation. Though the maria cover about 17% of the lunar surface, the mare infill is at most only a kilometre or so deep, and so the basalts comprise less than 1% of a lunar crust of mean thickness 61 km.

Taking the crust-plus-mantle as a whole, analyses show that, compared to the Earth, the lunar crust-plus-mantle is enriched in refractory compounds but heavily depleted in iron and iron-rich compounds and in siderophile elements such as magnesium. The lunar crust-plus-mantle is also heavily depleted in volatiles such as water, carbon dioxide, and hydrocarbons (compounds of carbon and hydrogen). It is also depleted in the more volatile silicates, such as those rich in potassium. Metallic iron particles in the lunar rocks indicate that the lunar surface has *never* been exposed to oxygen-rich volatiles to much extent — otherwise they would have been dissociated and oxidised all the metallic iron.

❐ Does any theory of the origin of the Moon account for these differences in crust-plus-mantle compositions between the Earth and the Moon?

As explained in Section 5.2.1, these differences can be explained by the collision theory of the origin of the Moon that was outlined in Section 2.2.4.

7.1.6 Radiometric Dating of Lunar Events

Two types of event have been radiometrically dated for the lunar rock samples. For the mare basalts and many highland rocks we have the age of solidification; for the breccias we have the time at which the rock fragments were impact-welded.

The oldest date is a solidification age of 4600 Ma for a highland sample. This is about the time the Moon formed. Its composition suggests that it might have been excavated by a giant impact from the solid mantle *below* the magma ocean. Most other highland rocks solidified in the range of ages 3800–4300 Ma ago, which is in accord with the great antiquity of the highlands inferred from the high-impact crater densities there. Mare basalt solidification ages are significantly younger. Breccia ages range down to 3100 Ma ago, except for an age of only about 900 Ma for breccia from the impact that formed the crater Copernicus (Figure 6.4(a), Section 6.2.1). The fresh-looking crater Tycho is even younger, perhaps as little as 100 Ma.

The older breccias help us to date the formation of the mare basin impacts, and the mare basalt solidification ages help us to date the mare infills. The brecchia ages range from 4000 Ma for Maria Serenitatis, Nectaris, and Humorum, to 3900 Ma for Mare Imbrium. Other basins, such as Mare Tranquillitatis, are inferred to be older, because that have been modified by the earliest dated basins. Other ancient basins have presumably been entirely obliterated. Table 7.1 gives the ages of some lunar basins, along with the mare infill basalt solidification ages. As noted earlier, you can see that infill was delayed for several hundred Ma, the most recent, that of Mare Procellarum, being completed about 3200 Ma ago. It is not known what caused such delays. There are no radiometric dates for the lunar far side, but Mare Orientale, from ejecta relationships, seems to be a bit younger than Mare Imbrium, and so has an estimated age of 3800 Ma.

A small fraction of the lunar highlands is somewhat depleted in craters. This is largely due to blanketing by ejecta from huge impacts, though in some regions there might be ancient lava flows, over 4200 Ma old, since broken up by impacts and

Table 7.1   Ages of some lunar basins and mare infill

Basin	Basin age/Ma	Infill age/Ma
Tranquillitatis	Before 4000	3600
Fecunditatis	Before 4000	3400
Serenitatis	4000	3800
Crisium	3900	3400
Imbrium	3900	3300
Procellarum	?	3200
Orientale	About 3800	?

A '?' denotes an uncertain value.

mixed with ejecta. Some features here and there on the Moon are consistent with more recent volcanism. For example, crater densities on parts of Mare Imbrium are so low that volcanism might have persisted here and there until about 2500 Ma ago. Even today there is the rare observation of possible gas or ash emission in tiny quantities. However, these transient lunar phenomena have many possible causes, for example meteor strikes and moonquakes—volcanism is not required. Overall, from about 4200 Ma ago lunar volcanism was dominated by lava infill of mare basins, and this was almost entirely over by 3200 Ma.

The lunar cratering rate through lunar history

It was explained in Section 6.2.2 how the radiometric ages of lunar surfaces with different crater densities have been used to deduce the cratering rate throughout its history, with the outcome shown in Figure 6.10. The cratering rate was very high early on, and as a result most of the lunar surface became nearly saturated. There has since been little resurfacing in what we see now as the highland regions. The cratering rate declined steeply as the supply of debris left over from planetary formation diminished, with 3900 Ma ago marking the end of the heavy bombardment. The rate continued to decline until about 2000 Ma ago, since when it has not varied much.

7.1.7 Lunar Evolution

We can summarise a plausible lunar evolution in broad terms as follows, on the assumption that just after its formation the Moon had a uniform composition broadly similar to that of the Earth's mantle today, though depleted in volatiles and enriched in refractories.

(1) Early in lunar history, a small core formed, depleting the mantle in siderophile elements.
(2) Impact melting helped create a magma ocean several hundred kilometres deep. A thin skin solidified and gradually thickened, but not before a crust rich in anorthosite formed, overlying a mantle rich in olivine and gabbro.
(3) The magma ocean solidified by about 4000 Ma ago, with the whole surface nearly saturated with impact craters, including some large basins. Further impact basins formed, but with a rapidly declining impact rate even the youngest (Imbrium and Orientale) are about 3800 Ma old.
(4) By about 4000 Ma ago the heat from radioactive decay had raised temperatures to the point (about 2000 K) where there was an extensive asthenosphere. Isolated pockets of partial melt formed in the rising legs of convection cells, and the asthenosphere migrated deeper as radiogenic heating declined.
(5) The partial melt supplied basalt lava to infill some of the large impact basins, notably on the near side where the basins are at lower altitude. Because of slow, deep convection the magma did not reach the basins until hundreds of Ma after they formed.

(6) As the rate of radioactive heat generation subsided, the interior cooled and the lithosphere thickened, to about 300 km by perhaps 3600 Ma ago, and it continued thickening until lava became only very rarely available from about 3200 Ma ago.

Note that the magma ocean might not have been planet-wide. The degree of isostatic compensation varies considerably across the Moon in a manner indicating that the crust and upper mantle might have become rigid at different times in different regions.

Question 7.1

If the lunar highlands had a peridotite composition, how would this modify our view of lunar evolution?

Question 7.2

(a) If the mare infill had been derived from the highlands through gradation, how would the composition of the infill differ from that observed?
(b) Were a large impact basin to be excavated in the Moon today, why is it likely that any subsequent infill would only be through gradation?

7.2 Mercury

Mercury, the planet closest to the Sun, is a small world almost entirely devoid of atmosphere. Its proximity to the Sun means that at perihelion, when it is only 0.31 AU from the Sun, the equatorial temperature at noon is about 825 K, though just before dawn the absence of an atmosphere leads to a frigid 90 K. Mercury rotates three times during two orbits of the Sun, giving a mean solar rotation period (Mercury's solar 'day') of 176 days. It is the tidal force of the Sun that has slowed Mercury's rotation to the point where it is now in this 3:2 resonance.

Mercury is much less well explored than the Moon. The only spacecraft to have visited the planet was the flyby mission of Mariner 10, twice in 1974 and again in 1975. About half of the surface was imaged, with a resolution of about 100 m — about the same as that of the Moon in the larger Earth-based telescopes. Overall, the altitude range is a few kilometres.

We have no samples of the Mercurian surface, though the remotely sensed properties of the surface from Mariner 10 and from the Earth are consistent with dusty basaltic silicates everywhere, plus iron sulphide. This makes it rather like the lunar fines, except that Mercury is more uniform in its albedo, not displaying the maria–highlands contrast of the Moon.

7.2.1 Mercurian Craters and Associated Terrain

The surface of Mercury is dominated by impact craters (Plate 4), ranging from small bowls at the limit of resolution up to huge impact basins. Mercurian impact craters are broadly similar to those on the Moon, the main differences being attributable to the higher gravitational field at the surface of Mercury, $3.7\,\mathrm{m\,s^{-2}}$ compared to $1.6\,\mathrm{m\,s^{-2}}$ on the Moon.

❏ What effects on the craters should the higher field have?

Among the expected effects of higher gravity are that, for surfaces with similar compositions and compaction, the craters will be smaller for given projectile kinetic energy, and ejecta will be flung less far. Also, the diameter ranges of the various crater morphologies in Figure 6.6 (Section 6.2.1) will be different. Though these expectations are borne out, closer inspection reveals further differences between Mercurian and lunar craters, presumably because the two surfaces have somewhat different properties.

Gradation of Mercury's craters has occurred in various ways. Ejecta from more recent craters partially obscures neighbouring older craters, and many craters show evidence of infill that has created plains—in many cases this infill might be lava. Larger craters show evidence of isostatic adjustment. There has also been mass wasting, presumably a result in part of seismic activity. The paucity of craters greater than about 50 km in diameter suggests that the crust might have been rather soft early in Mercury's lifetime, leading to viscous relaxation. This occurs in any solid substance warm enough to flow slowly under its own weight, and the effect is exaggerated if the lithosphere is thin. The number of large basins (300–1000 km diameter) is much the same as on the Moon, but they are more subdued because of burial and infill.

The largest basin, indeed the largest structure on the surface, is Caloris, a multiring basin 1300 km across, with ejecta reaching as far again beyond the outer rim (Figure 7.5(a)). The infill is a mixture of ejecta and impact melt. Caloris also has smooth plains thought to be lava flows, an interpretation supported by the possible presence of lava channels. The smooth plains seem to have been modified by isostatic adjustment, by magma withdrawal, and by tectonic processes. Diametrically opposite the Caloris basin is a unique region of hilly terrain criss-crossed by linear features (Figure 7.5(b)). On a smaller scale, radar studies from Earth show this region to be very blocky, or fractured. It is thought to be the result of the focusing by a large and dense planetary core of seismic waves from the Caloris impact. This is further evidence for the large iron core postulated in Section 5.1.3.

The type of smooth plain found in the Caloris basin is also found elsewhere on Mercury, notably in other basins and also in large craters (like Petrarch in Figure 7.5(b)). The less cratered half of Figure 7.6 is another example. As the name suggests, smooth plains are flat and sparsely cratered. The cratering density is similar on all such plains, suggesting that all the smooth plains are of much the same age. The sparseness of the cratering indicates comparative youth, borne out by the relationship of smooth plains to adjacent material, which shows that the smooth plains came later. Indeed, the smooth plains are the youngest surfaces on Mercury, and all of them could be lava flows, a suggestion supported by sinuous lobed ridges that could be the edges of lava sheets. The smooth plains seem to be the Mercurian equivalent of the lunar maria.

Mercury also has an equivalent to the lunar highlands, heavily cratered terrain, the more heavily cratered areas in Figure 7.6 and Plate 4 being examples.

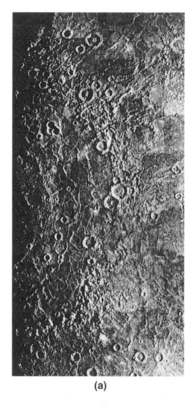

(a)

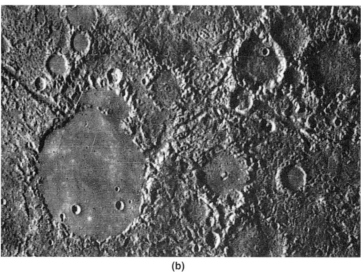

(b)

Fig. 7.5 (a) The Caloris basin on Mercury, displaying ejecta, multiple rings, infill, smooth plains. The frame height is about 1800 km. (NASA/NSSDC AoM F21) (b) Hilly terrain diametrically opposite the Caloris basin. The largest crater, Petrarch, is about 150 km across. (NASA/NSSDC FDS27370)

Fig. 7.6 The surface of Mercury, showing smooth plains (upper half) and heavily cratered terrain. The frame is about 490 km wide. (NASA/NSSDC P15427)

⊓ What could have caused this type of terrain?

This terrain is considered to be left over from the terminal phase of the heavy bombardment, as in the case of the lunar highlands.

The most common type of surface on Mercury has no lunar equivalent. These are the rolling intercrater plains, so called because they lie between the larger craters (greater than 20 km diameter) and in low-resolution images seemed to have few craters. The less cratered areas of Plate 4 are intercrater plains. In fact, the crater density on the intercrater plains is much higher than on the smooth plains, the population being dominated by craters smaller than 10 km diameter. Many of the craters are secondaries, as indicated by their shallow, elongated forms and tendency to form clusters and chains. Spatial relationships and morphological features suggest that the intercrater plains formed at various times throughout a substantial fraction of Mercury's early history. Volcanic resurfacing is probably responsible for them.

7.2.2 Other Surface Features on Mercury

A few craters have dark halos, as have a few craters on the Moon. Such craters might be volcanic, but in the case of Mercury better images are required to investigate this possibility.

Scarps and ridges are very common, and cross all types of terrain (Figure 7.7). Taken together, they suggest planet-wide crustal compression. A 1–2 km decrease in

Fig. 7.7 The scarp Discovery Rupes on Mercury. The largest crater is Rameau, 55 km diameter. (NASA FDS528881-4, 27398-9, 27386, 27393, R. G. Strom)

the radius of Mercury could account for them, though some might be the result of more local tectonism and others could be the result of lava flows. Local crustal *tension* is indicated by a few graben and a few strike-slip faults.

Radar studies from Earth have indicated that there might be water ice in permanently shadowed areas near the poles, presumably protected by dust, though perhaps a more likely interpretation of the radar data is sulphur. In either case, the infall of volatile-rich bodies and the migration of volatiles to the polar cold traps is a likely explanation of the deposits, as it is for the polar ice on the Moon.

There are no absolute ages for the surface, though if lunar cratering rates apply, then there has been little volcanic or tectonic activity on Mercury in the past 3000 Ma.

7.2.3 The Evolution of Mercury

On the basis of the limited observations, including those that relate to the interior, a plausible (but not unique) picture of the evolution of Mercury is as follows.

(1) Any early magma ocean would have solidified by 4000 Ma ago.
(2) Though we have no radiometric dating for Mercurian surfaces, models indicate that the bombardment history of Mercury has been broadly similar to that of the Moon, and so the heavy bombardment of Mercury would have ended about 3900 Ma ago.
(3) The existence of heavily crater terrain indicates that a lithosphere formed before the end of the heavy bombardment. It was not very rigid, and therefore allowed some of the craters, particularly those around 50 km diameter, to relax into oblivion, thus creating some of the intercrater plains.
(4) The formation of the massive iron core and radiogenic heating warmed the interior sufficiently for an asthenosphere to extend up to the comparatively shallow depth of about 50 km, thus making lava available to the surface. Any early expansion of the crust during this heating phase is not seen in surface cracks, so it would have ended before the end of the heavy bombardment.
(5) Crustal contraction created fractures, scarps and ridges. Lava flows created more intercrater plains. The contraction could have resulted from cooling and shrinkage of an iron core, aided by lithospheric contraction. Fracturing could also have resulted from the stresses produced by the tidal slowing of Mercury's rotation by the Sun. If at this early time there was a molten iron core, then it must still be largely molten today, because the large crustal shrinkage on solidification would be preserved but is not seen.
(6) The final stage of heavy impact cratering created Caloris and its associated features, rather as it created Imbrium and Orientale on the Moon.
(7) Subsequently, when the impact rate had declined to a low level, smooth plains formed through lava flows, and the impact rate has continued at a low rate up to the present.
(8) The lithosphere gradually thickened, to about 200 km by 2000 Ma ago, and it continued to thicken thereafter. There might have been little volcanic or tectonic activity on Mercury in the past 3000 Ma.

Question 7.3

Discuss the factors relating to energy gains and losses that have led scientists to conclude that the lithosphere of Mercury is thicker today than those of the Earth and Venus.

7.3 Mars

Mars is somewhat larger than Mercury. On average it is 1.5 times as far from the Sun as we are, and therefore under its thin atmosphere its surface is cold, with temperatures rarely above 273 K, and at night plunging to as low as 150 K in the coldest regions. The day on Mars is nearly the same length as on Earth, the mean solar rotation period being 24 hours 39 minutes 34.7 seconds, compared with 24 hours for the Earth. Mars takes longer than the Earth to orbit the Sun, so there are nearly 669 Martian solar days in a Martian year.

Mars has for a long time been a favourite target of investigation, with its surface markings that exhibit seasonal changes. For more than 300 years it has been scrutinised from Earth, and in the Space Age there have been many successful spacecraft missions, some of which are listed in Table 4.1.

7.3.1 Albedo Features

From Earth, the most obvious features on Mars are the white polar caps, the dark markings, and the widespread light red background (Plate 8). The polar caps consist of ices of carbon dioxide and water. The light red regions are fairly uniformly covered in dust, whereas the dark regions are more streaky, and are darker because of a higher proportion of dark dust, and because of the exposure of dark underlying terrain.

These albedo features exhibit seasonal changes — a result of Mars's axial inclination of 25.2°. The orbit of Mars is fairly eccentric, with a perihelion distance of 1.38 AU and an aphelion distance of 1.67 AU. Midsummer in the southern hemisphere occurs near perihelion, and midwinter near aphelion (the same as for the Earth). Therefore the seasonal changes in the southern hemisphere are more extreme than in the north. The most obvious seasonal change is the cyclical growth and retreat of the polar caps, visible from Earth even in a modest telescope. We now know that this cycle is the result of carbon dioxide condensing at the poles in the autumn and subliming back into the atmosphere in the spring.

There are also seasonal changes in the shape of the dark areas, and in their contrast against the light areas. The agent is atmospheric winds. These mobilise dust, and as a consequence streaks are created downwind of obstacles such as craters — light streaks where light dust has been deposited on dark terrain, and dark streaks where dark dust has been deposited on light terrain. Winds also create dark streaks by scouring away overlying light dust to reveal dark terrain beneath. The dark areas are only weakly correlated with large-scale topographic features, but are strongly correlated with small-scale topography (such as small craters) and with wind strength and direction. It is seasonal changes in the winds that cause the seasonal changes in the dark areas, for example by scouring light dust from some areas and depositing it in others. Furthermore, changes in the winds on time scales of decades cause changes in the dark areas on a similar time scale.

The most dramatic manifestation of wind-raised dust is huge dust storms. These are most frequent near to perihelion, and sometimes cover almost the whole planet in yellow-tinted clouds that consist predominantly of the light dust. Spectrometric studies of these clouds and of the light areas of the surface itself indicate basaltic minerals mixed with various clay minerals, notably montmorillonite (Section 6.2.6). The dark material seems to be dominated by basaltic silicates rich in iron and magnesium. The red tint of both the light and the dark areas is the result of iron-rich minerals. It is thought that the light dust is derived from the dark material by various physical and chemical processes.

❒ How are clay minerals produced?

Clay minerals are the result of the aqueous alteration of silicates. Direct evidence for the action of water is a 500 km diameter area near the equator which is rich in an iron oxide called hematite. Such a concentrated deposit of hematite suggests that it was formed in a body of liquid water, now long gone.

7.3.2 Two Different Hemispheres

Spacecraft images show that the surface of Mars consists of two contrasting hemispheres, divided approximately by a great circle inclined at 30° to the equator (Figure 7.8). The mean altitude of the hemisphere north of this line, the northerly hemisphere, is 3 km lower than that of the southerly hemisphere, and it is generally much flatter. It is not known why it is so flat, but its lower altitude is usually attributed to a thinner crust. The possible reasons are the same as those given in Section 7.1.3 for the differences in crustal thickness of the two hemispheres of the Moon. In spite of the north–south differences, there is no tendency for altitudes on Mars to cluster around two separate mean altitudes. This is shown by the frequency of occurrence of altitudes across the whole surface. This **hypsometric distribution** is included in Figure 7.8, and you can see that there is only one maximum i.e. it is unimodal.

The hemispheres also differ in other ways. The southerly hemisphere is dominated by impact craters and impact basins, whereas the northerly hemisphere is dominated by plains, domes, volcanoes, and huge grabens and pits. The scarcity of impact craters in the northerly hemisphere shows that it is the younger hemisphere, and this is also indicated by abundant evidence that it has been resurfaced by lava flows. Its relative youth is consistent with a thinner crust, because magma would then have had easier access to the surface.

Radiometric dating has not been carried out at the Martian surface. We therefore have to obtain absolute ages by adapting to Mars the lunar cratering rate in Figure 6.10 (Section 6.2.2), making due allowance for Mars's different gravity, its proximity to the asteroid belt, the different nature of the Martian surface, and the greater degradation on Mars. Unfortunately, this cannot be done with much precision, and consequently the absolute ages are poorly known. Relative ages from crater densities are used to define three epochs: the Noachian, with near-saturation crater densities,

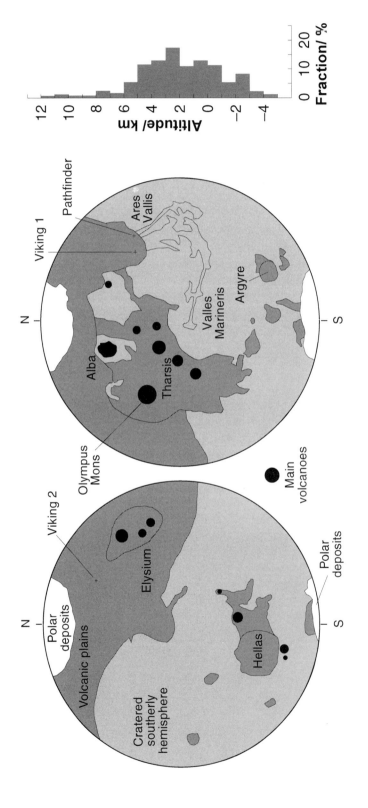

Fig. 7.8 A simplified topographic map of Mars, and the Martian hypsometric distribution. (Adapted with permission from Figure 22–13 of *Foundations of Astronomy*, M. A. Seeds, Wadsworth, 1994, ©Wadsworth Inc.)

is the time of the heavy bombardment; the Hesperian, with moderate crater densities, is the time immediately following the Noachian; the Amazonian, with light cratering, immediately follows the Hesperian, and extends right down to the present. Different cratering models lead to very different absolute ages for the boundaries between these epochs, 3500–3800 Ma for the Noachian–Hesperian boundary and 1800–3550 Ma for the Hesperian–Amazonian boundary.

Most of the southerly hemisphere has ages greater than the mid-Hesperian, and most of the northerly hemisphere has ages less than this. In absolute terms the age that divides the two hemispheres is probably somewhere in the range 3100–3800 Ma. The ages of the very youngest areas on Mars might only be the order of 1 Ma, though it is possible that about 85% of the whole Martian surface is more than 2500 Ma old. Let's look more closely at the two hemispheres.

7.3.3 The Northerly Hemisphere

Figure 7.8 shows the main topographic features of the northerly hemisphere. The plains that constitute a large fraction of this hemisphere are young and thought to be largely volcanic. The numerous volcanic features support this conclusion. Spectrometric evidence from orbiting spacecraft and evidence from surface analyses by the two Viking Landers is consistent with the presence of iron-rich basaltic materials such as would comprise silicate lava flows. Radar reflectivities and the rapid response of the temperatures of some areas to changes in insolation are consistent with a covering of volcanic ash. This ash could have been carried by the wind from the sites of explosive volcanic eruptions.

Though the northerly plains are rather flat, superimposed on them are two large raised domes, the Elysium and Tharsis regions. The Tharsis region is the more dramatic, a broad dome about 5000 km across, rising to 10 km above the surrounding plains, and to about 6 km above the zero of altitude. Various processes could have contributed to the formation of these domes — convective plumes in the mantle, isostatic adjustment, intrusive and extrusive igneous activity. Different regions of each dome have different ages, indicating protracted formation. Whatever single or combined process created them, the gravitational fields in their vicinity show that there is excess mass present near the surface, perhaps because the underlying lithosphere is so thick that it takes hundreds or thousands of Ma for isostatic equilibrium to be achieved, or perhaps because the domes are still being borne aloft by convective plumes in the mantle.

Each of these domes is associated with features arising from crustal tension that would result from uplift, or perhaps in some cases from the effect of the intrusion of dykes of magma. Most spectacular is Valles Marineris, a system of cracks and grabens enlarged and modified by slumping, land slides, wind erosion and magma withdrawal, and perhaps also modified by water flow at and below the surface. It is 4000 km long (Figure 7.8), up to several hundred kilometres across, and up to 8 km deep. A short section of it is shown in Figure 7.9. Mars Global Surveyor images have revealed layering in the walls right down to the floor. The nature of the layers is unknown. At the west end of Valles Marineris is Noctis Labyrinthus, consisting of short grabens running in all directions. At its east end is an example of Martian chaotic terrain — jumbled, heavily mass wasted, with wide channels leading downhill.

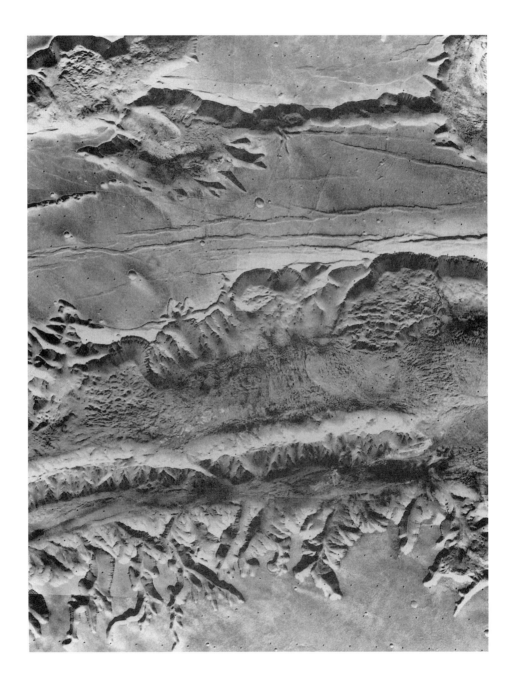

Fig. 7.9 A short section of Valles Marineris on Mars. The frame width is about 110 km. (NASA/NSSDC P17872)

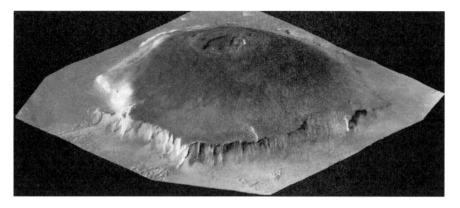

Fig. 7.10 Olympus Mons on Mars. The steep scarp is about 550 km across, and the summit is 25 km above the adjacent plains. The vertical scale is exaggerated. (NASA/JPL)

The domes also bear huge shield volcanoes. One old volcano, Alba, on the Tharsis dome, has a lopsided topography that suggests it predates the Tharsis dome, and was then tilted as the dome grew. At the other extreme, some volcanoes are so unweathered that they could be dormant or recently extinct, and could have contributed lava flows to dome building. The least eroded and youngest volcanoes are four on the eastern half of the Tharsis dome, on what is called the Tharsis ridge. These resemble terrestrial shield volcanoes such as Mauna Loa in Hawaii, though the Martian examples are much larger than the terrestrial ones. Olympus Mons is the largest of all (Figure 7.10), rising to an altitude of 27 km (25 km above the adjacent plains), and 550 km across. Lava has flowed from a huge complex of summit calderas and from numerous vents on its flanks. The expected basaltic composition of the lava is borne out by orbital spectrometry. The great size of each Martian shield volcano could result from the combined effects of a magma source in a fixed position with respect to the crust, and a thick lithosphere retarding isostatic adjustment.

Some volcanoes seem to have erupted ash rather than lava.

❒ What does the magma need to contain to form ash?

Ash indicates the presence of dissolved volatiles in the magma. Other volcanic features include fissures from which lava emerged, and channels along which lava has flowed. Low-rise lobes are common, consistent with lava flows, though the source vents in many cases are now hidden, as in the case of the lunar maria. Some of the layering in the walls of Valles Marineris might be lava flows that helped build up the northerly plains.

In spite of these extensive indications of volcanism, detailed studies show that the accumulated volcanic activity on Mars is considerably less than on the Earth. This might be because crustal formation from a magma ocean concentrated the heat-producing radioactive elements in the crust, as might be the case for the Moon too. Mars, however, is bigger than the Moon, so volcanism and tectonic activity has certainly been more widespread and persisted longer.

7.3.4 The Southerly Hemisphere

The southerly hemisphere is dominated by impact craters, much of it near to saturation. The largest crater is the impact basin Hellas, within which is found the lowest elevation on Mars, 4 km below zero altitude. The interior of Hellas is about 1800 km across, making it second only to the Moon's South Pole-Aitken basin among impact basins in the Solar System. Beyond the main ring of uplifted mountains are fractures, with associated volcanism. Within the Hellas basin is the plain Hellas Planitia, built of lava and wind-blown dust. The next largest impact basin is Argyre (Figure 7.11), 900 km in diameter, also filled in the manner of Hellas, as indeed are many smaller craters.

On the most heavily cratered parts of the southerly hemisphere, many of the craters are heavily weathered, and smaller craters have been totally eradicated. On slightly less cratered and therefore slightly younger terrain the smaller craters still survive, implying a decrease in weathering rate at some time in the distant past, before 3000 Ma ago. A decline in atmospheric mass would account for such a decrease—more on this in Chapter 10.

Apart from impact craters, the southerly hemisphere displays a few heavily weathered, almost obliterated shield volcanoes, and lava channels that indicate long-

Fig. 7.11 Part of the Argyre basin on Mars, visible in the lower left quarter of this image. The basin is about 900 km across. (C. J. Hamilton and NASA (P17022 is similar))

extinct volcanic activity. There are not many such features, and so the activity was not as widespread as in the northerly hemisphere.

The boundary between the hemispheres

The boundary between the two hemispheres is ragged, with hummocky outliers of high ground in the northerly hemisphere, some of them surrounded by landslides. This is called fretted terrain, and it could either be the result of retreating scarps or the weathering of a once-sharp boundary, or the removal of underground magma, or some combination of all three.

7.3.5 The Polar Regions

The seasonal polar caps in the northern and southern hemispheres extend in midwinter to latitudes of about 65° and 50–60° respectively. In both hemispheres the caps undoubtedly consist largely of carbon dioxide ice. This is the dominant constituent of the Martian atmosphere, and the temperature of the seasonal caps, about 150 K, is that at which carbon dioxide would condense (gas to ice) at Martian surface pressures. The temperature remains near to 150 K throughout much of the long winter, sustained by the latent heat released on carbon dioxide condensation. Some of this condensation occurs at ground level, but some occurs in the polar regions of the atmosphere, where it forms carbon dioxide clouds from which powdery snow falls. The depth of the seasonal caps has been estimated from their rates of advance and retreat, a metre or so topping the range of values.

By high summer the seasonal cap has retreated, leaving behind the residual cap (Figure 7.12). The north pole residual cap reaches temperatures of about 240 K, far too high for it to consist of carbon dioxide. These temperatures, at Martian surface pressures of water vapour, correspond to the 'Solid + gas' line on the phase diagram of water, and so the residual northern cap is presumed to be water ice. Reflectivity

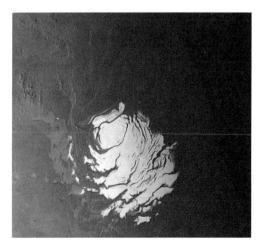

Fig. 7.12   Residual Martian cap at the south pole. The cap is about 400 km across. (NASA/ NSSDC and USGS MG90-0s000-407B)

measurements by the Galileo Orbiter indicate that dust is also present, in accord with earlier findings. At the south pole the residual cap seems be carbon dioxide (plus dust), though presumably underlain by dusty water ice.

❐ If seasonal changes are more extreme in the southern hemisphere, how could the residual cap be carbon dioxide, when in the northern hemisphere it is water ice?

The south pole is at the higher altitudes characteristic of the southerly hemisphere, and so is colder than the lower-lying north pole.

The north cap is generally smooth, but both caps are scarred by wind-scoured pits and valleys. These show that the caps overlie extensive, nearly horizontal layered sediments a few km deep. Each layer is 10–50 m thick and is richer in dust than the residual caps. The layering is thought to arise from a variation in the dust/water ice ratio from layer to layer, and it is estimated that each layer represents many years of deposition. Much thinner annual layers could exist within those resolved. These layered sediments overlie unlayered and largely ice-free bright dust that is several hundred metres thick near the poles. It extends further towards the equator than the layers, thinning as latitude decreases, and petering out at mid-latitudes. Nearer to the poles, and encircling the residual caps, are dark polar collars that consist of dunes of dark grains. The polar deposits are all very lightly cratered, but though Amazonian they could still have ages in excess of a few hundred Ma. The pits and valleys extend down to the underlying terrain, which in the north is seen to be the lightly cratered plains typical of the hemisphere. The north cap reaches an altitude of at least 2500 m above its surroundings. Beneath the polar deposits at the south pole lies the heavily cratered terrain that is typical of that hemisphere.

The dust has presumably been transported to the polar regions by winds, each dust particle having acted as a nucleus on which carbon dioxide and water condensed from the atmosphere, the water adding to any water content the grain initially had. The particle with its icy mantle was then precipitated at the pole. Subsequently, at least some of the carbon dioxide content was lost through sublimation. For so much dust to be present at the poles, evidence of denudation elsewhere is a reasonable expectation, and such evidence is found. For example, Valles Marineris seems to have suffered wind erosion, as have various hummocky terrains.

A scenario that can account for many of the polar features has as its central feature switches from net deposition to net erosion. First, the thick unlayered deposit is laid down over many years, perhaps as a set of layers no longer preserved. Net deposition then switches to net erosion, cutting into the deposit. Then there is a switch back to net deposition, resulting in the oldest layers that *are* preserved today. The layered deposits are then partially eroded and this is followed by another deposition episode. Evidence for several layering episodes is provided by what is called an unconformity, a mismatch in orientation between a set of layers and another set above it. Net deposition then switches to net erosion again, to give the pits and valleys that we see today. It is not known whether there has recently been a switch back to net deposition. A major factor in these switches is probably quasi-periodic changes in the axial inclination of Mars — more on this in Chapter 10.

Fig. 7.13 The Martian crater Arandas, with an ejecta blanket suggesting a surface flow of water plus entrained rocky materials. The crater itself is about 28 km across. (NASA/ NSSDC P17138)

7.3.6 Water-related Features

One of the most intriguing types of feature on Mars are those that seem to require the involvement of liquid water. The intrigue is that liquid water is not evident anywhere on the Martian surface today and that the conditions everywhere on the surface would result in rapid freezing. What are these intriguing features?

A small proportion of Martian impact craters have unusual ejecta blankets— Figure 7.13 shows one example. Such blankets are readily explained by the impact-generated melting and surface flow of liquid water with entrained rocky materials, and so they are evidence that the Martian surface contains water in some form. (A less favoured explanation is that such blankets were produced by dry ejecta interacting with the thin atmosphere.) The lack of impact craters on the blankets indicates their youth, and so the water is presumably still there. Older blankets of this sort that would now bear craters are not seen, presumably because such low relief features are readily eroded away by the atmosphere.

Even more intriguing are Martian channels. Figure 7.14 shows three sorts of channels, all of which suggest the flow of liquid water. Figure 7.14(a) shows an example of an outflow channel. These are of order 10^3 km long, 10^2 km wide and a few kilometres deep, and their morphology strongly suggests the sudden outflow of liquid water, in many cases more than once. Downstream there are features that indicate ice flow as well as liquid flow, presumably because some of the water froze. Lava flows would have generated quite different morphologies.

At the head of the channel in Figure 7.14(a) is an example of Martian chaotic terrain. Such terrain is found at the head of most outflow channels.

❐ What mechanism does this suggest for the formation of chaotic terrain?

A ready explanation is that chaotic terrain is the result of collapse following the removal of the water that created the channel. The outflow channels indicate that the removal was rapid, and this is thought to be due to the release of groundwater from

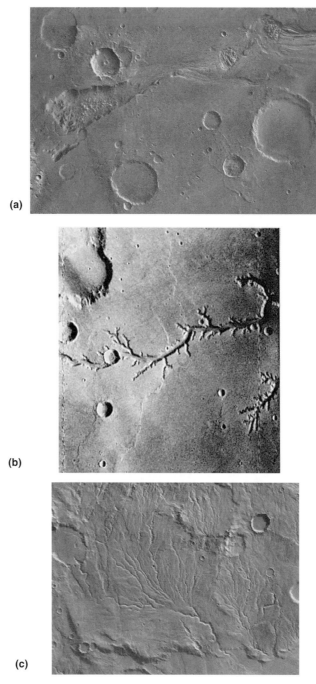

Fig. 7.14   (a) The Martian outflow channel at the head of Simud Vallis. The width of this frame is about 300 km. (C. J. Hamilton and NASA (P16983 is similar).) (b) The fretted channel Nirgal Vallis on Mars. The frame is about 80 km across. (NASA/NSSDC P466A50.) (c) A valley network on Mars. The frame is about 130 km across. (C. J. Hamilton and NASA (from 63A09))

an aquifer that was sealed beneath a permafrost layer until the layer was suddenly disrupted. This disruption could have been caused by an impact, direct or nearby, by fault movement, or by volcanic activity. Another possibility is the bursting forth of water from the aquifer as the pressure of water in it exceeded a critical value. The terracing of channel walls is one indication that there have been repeated floods as the aquifer refilled and emptied. In this case there might be further outflows in the future, though most of the channels are on Hesperian terrain and so were formed in that early epoch.

The remaining outflow channels originate in canyons. Groundwater seepage into these canyons could have created lakes with ice-covered surfaces. Subsurface seepage of water could have eroded and weakened the canyon walls leading to catastrophic break-out of the water beneath the ice, and the formation of the outflow channel.

Most of the outflow channels flow into low-lying plains in the northerly hemisphere and into the Hellas basin in the southerly hemisphere. Lakes must have formed on these plains, and there are features in the northerly plains that are consistent with this possibility, for example possible shorelines and layered sediments. It is even conceivable that much of the northerly hemisphere was once covered by an ocean. It is not clear where the lake or ocean water went. It might still be present as ice deposits hidden by dust, or it might have seeped into the groundwater system. At equatorial latitudes under present conditions, surface ice can quickly be lost by sublimation.

Fretted channels (Figure 7.14(b)) are narrower than outflow channels and are more sinuous. Another distinguishing feature is short, stubby tributaries. They occur along parts of the boundary between the northerly and southerly hemispheres, and stretch hundreds of kilometres into the uplands. Valley networks (Figure 7.14(c)) consist of channels that are narrower still, with typical widths 1–10 km, and typical depths of 100–200 m, and they have better-developed tributary systems. Most of them occur in the southerly hemisphere in areas of Noachian age, particularly at high altitudes and low latitudes. It is widely accepted that the fretted channels and the valley networks were carved by the flow of liquid water. But this leaves open the question of whether the water was supplied by precipitation or by groundwater sapping. The detailed form of the fretted channels and of most of the networks is similar to that of terrestrial systems that were created by sapping, though some Martian networks might have been carved by precipitation. In either case, in order for liquid water to survive for hundreds of kilometres along narrow channels it is necessary for atmospheric temperatures and pressures to have been higher in the Noachian than they are today.

A minority view is that the fretted channels and networks were carved by mass wasting aided by groundwater at the base of the debris, in which case most of the material that flowed was rocky material from the channel itself. This explanation avoids the need for substantially different conditions in the distant past, and it is supported by evidence of mass wasting in the fretted channels, for example debris aprons at their mouths. But it has difficulties explaining the networks, not the least of which is how debris could be transported over hundreds of kilometres in channels only a few kilometres wide. We shall return to Martian atmospheric conditions in Chapter 10.

7.3.7 Observations at the Martian Surface

In 1976 the Viking Landers became the first two spacecraft to land on the Martian surface and carry out a long and successful series of investigations. They landed in well-separated areas in lowland plains in the northerly hemisphere (Figure 7.8). The view from both sites was broadly similar, a gently undulating, dusty landscape strewn with boulders in the 0.01–1 m size range, most of which were probably ejected from nearby impact craters. The limited photometry that the Lander cameras could perform on the boulders is consistent with basalt. Bedrock might be visible here and there. At the Viking Lander 2 site some of the surface has the same crusty appearance as so-called duricrust on the Earth, where it is the result of dust grains becoming bound together by materials precipitated in pools of water as they evaporate. In the case of Mars, the water could have come from upward percolation, or from the melting of permafrost.

At both sites the dust was sampled and analysed. The relative abundances of chemical elements and isotopes were measured, rather than chemical compounds being detected, and so the mineralogy has to be inferred from the elemental analysis and from sparse data from other observations. Broadly, the data can be matched by dust consisting largely of clays rich in iron and magnesium. On Earth such clays result from the action of water on iron- and magnesium-rich basalts, and on both planets such basalts are to be expected in volcanic lavas. Orbital spectrometry provides evidence that iron- and magnesium-rich clay minerals are common in the bright regions. The dust in general has a high iron content. Oxidation, among other things, produces hematite, and this gives Mars its red appearance, though the reason for the oxidation is unknown. The dust is rich in sulphur, a possible result of sulphates left behind by water evaporation. The action of water is also expected to produce carbonates. The Landers had no means of detecting carbonates, but they might have been detected in Earth-based studies.

Evidence for organic compounds was not found. This is consistent with the negative results of the Lander experiments to detect life. But even without the activities of living organisms, organic compounds must continuously be delivered to the surface of Mars.

❐ What sorts of bodies deliver organic compounds?

The bodies delivering these compounds are comets and carbonaceous meteorites. The absence of any traces of organic compounds is thought to be the result of their destruction by peroxides produced by solar UV radiation. Such radiation reaches the Martian surface almost unattenuated, because the atmosphere is largely devoid of the ozone that protects the Earth's surface.

In July 1997 Mars Pathfinder landed on Mars near the mouth of the outflow channel Ares Vallis, just inside the northerly hemisphere (Figure 7.8). The view to the west is shown in Plate 9. Particularly noticeable are the two peaks on the horizon, about 1 km away. These are 30–35 m tall, and, unsurprisingly, have been dubbed Twin Peaks. Though the most recent outflow from Ares Vallis might have been as long as 2000 Ma ago, the outflow has been copious, and the expected evidence of water flow at the Pathfinder site is found in topographical forms such as ridges, troughs, and a distant streamlined island, and also in the boulders — the assortment of types, their size distribution, the roundness of some of them, and the way that some are stacked. One or two boulders, and also the more southerly of Twin Peaks might be layered, suggesting that they are sedimentary.

As well as evidence of water flow, there is abundant evidence at the Pathfinder site, as elsewhere on Mars, of aeolian processes — dunes, ripples, moats, wind tails, and centimetre-sized scours on rocks. In the thin atmosphere of Mars it takes a very long time to produce such features, and their prominence is further evidence of the low level of the volcanic, tectonic, and fluvial processes that would erase the aeolian features. Further evidence of wind action is air-borne dust. At the Pathfinder site samples were collected and analysed, and a high magnetic content was found, at least partly due to maghemite, an iron oxide that on Earth is formed by precipitation from water that is rich in iron compounds. The surface dust was also analysed and found to be much the same as at the two Viking Lander sites. Bright dust was more common than dark dust, the latter predominantly being found where winds would have swept the bright, finer dust away.

A crucial aspect of the Pathfinder mission was a briefcase-sized rover, called Sojourner, that made several trips up to 12 m from Pathfinder to measure the relative abundances of the elements in six boulders — the Viking Landers could only perform such analyses on the dust. These analyses were supplemented by photometry in more wavelength bands than the Viking Landers could perform, covering the range 0.44– 1.0 μm. The somewhat surprising result is that some of the boulders are rather andesitic in composition, indicating chemical differentiation in basaltic–gabbroic crustal rock (Table 6.1). If so, then rather more extensive differentiation has occurred in the Martian crust than previously thought. However, it is possible that the andesitic material is only a veneer produced by weathering.

In broad terms, the surface composition of Martian boulders and dust is consistent with volcanically produced basalts, perhaps with some subsequent modification by partial melting and by the action of water. Accumulation of organic compounds has been prevented by solar UV radiation.

7.3.8 Martian Meteorites

The Martian samples examined by the various landers are probably not the only samples of the Martian surface that we have scrutinised at close range.

❐ In Chapter 3 you met others. What are they?

There are (so far) thirteen meteorites that are widely regarded as having come from Mars. Twelve have been thoroughly analysed. They are broadly basaltic–gabbroic in composition, and so have presumably crystallised from magma at and below the Martian surface. Basalt predominates in several of them, thus providing further evidence that basalts are common on the Martian surface. The radiometric solidification ages vary from one to another. The oldest (ALH84001) has a solidification age of 4500 Ma; among the other eleven it ranges from 1300 Ma to as recent as 180 Ma. Since solidification, the Martian meteorites have been modified by liquid water, and weathering products in them adds to the evidence for clays on Mars.

These meteorites also indicate that the mantle is depleted in siderophile elements and in elements that preferentially enter sulphide melts. This indicates an iron-rich core in which FeS is prominent. By piecing together the history of the meteorites it is possible to infer that Martian core formation occurred 4400–4600 Ma ago.

7.3.9 The Evolution of Mars

A plausible picture of the history of Mars resembles lunar history, with differences because of the lower rate of Martian heat loss per unit mass, consequent upon its greater size, and because of differences in composition.

(1) Towards the end of accretion there was a magma ocean created by impact melting (perhaps aided by a thick blanketing atmosphere). This led to a crust with a more basaltic/gabbroic composition than the mantle, perhaps with a concentration of the heat-producing radioactive elements in the crust. The magma ocean had solidified by the end of the heavy bombardment.

(2) Ancient impact basins became places of crustal weakness that were sites of subsequent volcanic activity that flooded these basins with lava, and in some cases led to later volcanic activity.

(3) Probably before a well-developed lithosphere was established, a core formed consisting largely of iron and iron sulphide.

(4) As the heavy bombardment was coming to an end, the northerly/southerly divide was somehow created, and this made the northerly hemisphere more prone to volcanic resurfacing.

(5) The uplift of the Tharsis bulge started, producing extensive fracturing. This uplift might have been the result of a convective plume in the mantle. Radioactive heating increased the temperature of the interior, and the subsequent crustal expansion resulted in further fractures and in volcanic activity, particularly in the northerly hemisphere.

(6) An early, warm, dense atmosphere lasted long enough to allow the valley networks to form.

(7) Volcanism persisted throughout much of Martian history, focused in later times on the Tharsis and Elysium regions. Volcanism declined as the interior cooled and the lithosphere thickened. The lithosphere is at least 100 km thick today, and the asthenosphere might now have vanished completely.

(8) Volcanic and tectonic activity might rarely still occur. Impact cratering and degradation are certainly continuing, with the thin Martian atmosphere playing a prominent role.

Question 7.4

Make a brief case for transferring this section on Mars to the next chapter — on active surfaces.

Question 7.5

Outline how the surface of Mars might have been different if its crust had everywhere been as thin as it is thought to be in the northerly hemisphere.

7.4 Icy Surfaces

Pluto, its satellite Charon, and most of the satellites of the giant planets have icy surfaces. This is revealed by spectroscopy, and by the low mean densities of those bodies for which data are available. Icy surfaces are expected in the outer regions of the Solar System because of the low temperatures, and you have seen that, because of its high cosmic abundance and low volatility among icy materials, water is expected to be dominant. Beyond Jupiter other ices can be significant components of surfaces, such as ammonia, methane, and nitrogen. The near absence of these icy materials from the satellites of Jupiter is due to the higher temperatures closer to the Sun, and to infrared radiation from the formation of this massive planet.

Among the icy surfaces, only those of Europa, Enceladus, and Triton are active. The rest are inactive, and we shall concentrate on the five largest of these. Table 7.2

Table 7.2 Distinguishing surface features of the inactive intermediate size icy satellites

Object	Orbit semimajor axis/10^3 km	Radius/km	Surface features
Saturn			
Mimas	187	195	Heavily cratered, though craters > 30 km are rare, perhaps due to partial resurfacing before the end of the heavy bombardment
Tethys	295	530	Heavily cratered, except for a region of early cryovolcanic resurfacing
Dione	378	560	Cratered, though with considerable cryovolcanic resurfacing near the end of the heavy bombardment
Rhea	526	765	Imaged hemisphere dominated by the heavy bombardment. A later, lighter bombardment confined near to the north pole. Little resurfacing
Iapetus	3561	730	Heavily cratered. One hemisphere is much darker than the other, perhaps due to 'soot' impact eroded from the outer satellite Phoebe
Uranus[a]			
Miranda	130	243	Heavily cratered, plus three regions resurfaced after the heavy bombardment — due to collisional disruption followed by tectonics and cryovolcanism?
Ariel	191	580	Surface cratered after the heavy bombardment with global tectonic features and associated volcanism lasting to about 2000 Ma ago
Umbriel	266	595	Heavily cratered. A particularly low albedo (0.19), perhaps from a surface veneer additional to effects of methane decomposition
Titania	436	805	Similar to Ariel, and perhaps with a similar surface history
Oberon	583	775	Heavily cratered, rather like Umbriel
Neptune[a]			
Proteus	118	210	Poorly imaged. Craters and tectonic features present, plus a large depression
Nereid	5510	170	No good images or other data, but presumably icy. Some evidence of surface markings

[a]Many of the satellites of Uranus and Neptune are darkened by the cosmic ray decomposition of methane in the surface ices.

highlights the distinguishing features of the largest of the remainder — the icy satellites of intermediate size. Not included are the very smallest satellites, icy or rocky. Very few of these have in any case had their surfaces explored to any significant extent.

7.4.1 Pluto and Charon

With a radius of 1150 km Pluto is by far the smallest planet, and because of its great distance from the Sun its surface temperatures never exceed about 60 K. Its high albedo is consistent with fairly pure icy materials, and Earth-based infrared spectroscopy has identified nitrogen (N_2) ice on the surface, plus smaller quantities of methane (CH_4) ice and a trace of carbon monoxide (CO) ice. Hubble Space Telescope images have revealed a bright polar cap and bright patches elsewhere, presumably consisting of clean ice. The rest of the surface could be ice mixed with rocky materials, or with organic compounds generated by the action of UV radiation on ices. The distribution of albedos over the surface exhibits changes, perhaps due to atmospheric transport of frosts — there is a very thin, slightly hazy atmosphere with a surface density of 2×10^{-5} kg m^{-3}. CH_4 is certainly present in the atmosphere, and N_2 and CO are presumed to be components also. It is thought that a layer of methane about 10 km thick could have been lost to space over Pluto's lifetime, in which case the surface will be depleted in impact craters up to about 10 km in diameter. No spacecraft has yet visited Pluto so we cannot confirm this prediction.

An icy surface is in accord with the model of the interior in Figure 5.7 (Section 5.2.2). Water ice is the main constituent of the icy mantle, but this is expected to be topped by the more volatile ices that have been detected at the surface.

Pluto's satellite, Charon, has a radius of 600 km, about half that of Pluto. Spectroscopy indicates that methane might be absent from its surface, though water ice seems to be present. Any methane crust might therefore be lost, and if this happened very early in Charon's history then it might bear more impact craters than Pluto.

It is expected that Pluto and Charon are in mutual synchronous rotation, and there is some observational evidence for this from periodic brightness variations thought to arise from a combination of rotation and non-uniform surface albedo. Radiogenic and tidal heating might have been appreciable, but it is very unlikely that volcanic or tectonic processes still operate.

❐ Why is this?

They are small bodies, so cool rapidly.

7.4.2 Titan

The surface of the huge world Titan, larger than Mercury and second only to Ganymede among the planetary satellites, is so obscured by the cloud and haze in its massive atmosphere that the images returned by Voyagers 1 and 2 in 1980 and 1981 respectively, revealed nothing of the surface (Plate 17). The atmosphere is, however, less obscuring at certain near-infrared wavelengths, and observations at these wavelengths have now been made from the Earth's surface and from the Hubble Space Telescope. Water ice is dominant, but the surface is not uniform in brightness, and so variable proportions of other substances are presumed to be there, notably hydrocarbons.

The low density of Titan, and the likelihood of at least some degree of differentiation, had earlier led astronomers to expect a surface rich in water ice. The atmospheric composition and surface temperature had also indicated that there might be a surface veneer of methane and ethane (C_2H_6), plus smaller quantities of larger hydrocarbon molecules. The possibility of extensive oceans of hydrocarbons had been ruled out by Voyager 2, and also by Earth-based radar investigations, the atmosphere being transparent at wavelengths greater than about 1 cm. The radar reflectivities are consistent with regions of bright ice, probably a mixture of water and ammonia, and darker regions of solid and liquid hydrocarbons.

We will know a lot more about the surface of Titan in 2004, if the Huyghens probe, part of the Cassini mission, lands successfully on its surface.

7.4.3 Ganymede and Callisto

Of the four Galilean satellites of Jupiter, the outer two, Ganymede and Callisto, are no longer tectonically or volcanically active. Ganymede and Callisto were investigated by Voyager 1 and Voyager 2 in 1979 and by the Galileo Orbiter from 1996 onwards. The outer regions are dominated by water ice, and hydrated minerals are also present, along with small quantities of so far unidentified substances. These satellites are too close to the Sun for the more volatile ices to be present as more than traces.

Callisto is the third largest satellite in the Solar System, slightly smaller than Titan. It is very heavily cratered (Plate 15), indicating little by way of resurfacing since the heavy bombardment, except for the partial burial of small craters by dark, smooth material. Indeed, among the icy satellites, Callisto is one of the darkest, with an albedo of only about 20%. This might be a consequence of the great age of the surface — loss of ice through sublimation or through bombardment by charged particles in the magnetosphere could have created a surface enriched in the rocky materials that were originally a minor component. A significant proportion of rocky dust in the ice is also indicated by the manner of gradation of some crater walls and by the low radar transparency of the ice.

❐ Where would you expect to see high albedos on Callisto?

If there is purer ice beneath the surface, then young craters and the ejecta from them should have high albedos. This is observed to be the case.

Large craters, greater than about 60 km diameter, are scarce, and many craters show evidence of viscous relaxation. The effect on large craters is greater than on smaller ones, and as a result many of the oldest larger craters must have vanished. Under present conditions on Callisto this might not happen even in 4600 Ma. Therefore, a thinner lithosphere and warmer subsurface conditions are indicated for the past, and this is consistent with thermal models of Callisto. The higher interior temperatures in the distant past were a result of residual heat from accretion and differentiation, tidal slow-down of the rotation (to yield the present synchronous rotation), and greater radiogenic heat generation from the radioactive isotopes in the rocky materials. A thin lithosphere and viscous relaxation in the past are also indicated by several multiring basins with flat areas in the centre. Today, a comparable impact would not generate rings around the impact basin, and the central region would be bowl-shaped. However, Callisto might not be entirely solid today. The Galileo Orbiter has detected evidence of electric currents near the surface, such as could be carried by liquid water.

The other inactive Galilean satellite, Ganymede, is the largest satellite in the Solar System. It has a radius 1.08 times that of Mercury, though it has only 45% of Mercury's mass.

❒ What does this indicate about the composition of Ganymede?

This indicates a large proportion of icy materials (water), as in Figure 5.7 (Section 5.2.2). Its surface displays two distinct terrains, each accounting for about half the surface, though the two are intermingled (Plate 14). There are regions of heavily cratered darker terrain, and bands of less heavily cratered brighter terrain that here and there makes wedge-shaped incursions into the darker terrain.

On the darker terrain the crater density is about a third of that on Callisto, with a notable scarcity of craters larger than about 100 km diameter. The craters show evidence of even greater viscous relaxation than on Callisto and this could account for their smaller number, though cryovolcanic resurfacing before the end of the heavy bombardment is another possibility. Though large craters are scarce on Ganymede there are flat circular features called palimpsests that could be almost fully relaxed impact basins, or impacts that penetrated a thin lithosphere and released slushy ice.

The brighter terrain is criss-crossed by belts of ridges and grooves that suggest several episodes of formation and a consequent range of ages for this type of terrain (Figure 7.15). This conclusion is borne out by the variation in crater densities from place to place on it. Overall, the crater density on the brighter terrain is less than on the darker terrain, indicating that the brighter terrain is younger, though the

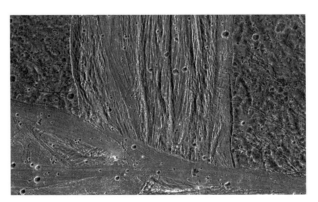

Fig. 7.15 Two swathes of brighter terrain on Ganymede, down the centre, and across the bottom of this 950 km wide frame. (NASA/JPL P50037)

high-resolution Galileo Orbiter images have confused this simple picture by revealing that one dark area has been reworked relatively recently, and that one bright area has fine grooves plus numerous small impact craters that can be explained by crustal expansion long ago.

The most likely origin of the bright terrain is from the thermal expansion of Ganymede consequent upon differentiation, or expansion upon freezing of an icy mantle. This cracked the lithosphere to form grabens that filled with slushy ice. This ice froze, and the ridges and grooves could be the result of cracks and subsidence, or of cryovolcanism along cracks. In this model the infill is brighter because as a partial melt it is relatively free of dust and rock fragments. This also makes it less dense than the overlying material, which helps it to rise to the surface. Episodes of tidal heating, with the most recent perhaps about 1000 Ma ago, can explain the range of ages in the bright terrain. Such episodes could arise from orbital evolution (Section 5.2.3).

Thus, compared to Callisto, Ganymede has been more extensively resurfaced. This is consistent with Ganymede's greater size (slower cooling rate), its slightly greater ratio of rocky to icy materials (greater radiogenic heating), and a possible history of significant tidal heating. This in turn is consistent with the clear evidence that Ganymede is differentiated more completely than Callisto (Figure 5.7).

Question 7.6

State, with justifications, how you would modify each of the orbits of Pluto and Ganymede to cause cryovolcanism on these bodies.

7.5 Summary

The solid surfaces in the Solar System can be divided into those that today are subject only to impact cratering and degradation, and those that are additionally subject to a significant level of volcanic or tectonic activity. Broadly speaking, the larger the body, the more likely it is to have a surfaces that falls into the latter category, though tidal heating blurs this size-based distinction.

The largest bodies that are no longer subject to volcanic or tectonic activity comprise the Moon, Mercury, and Mars, which have rocky surfaces, and Pluto, Charon, Titan, Ganymede, and Callisto, which have icy surfaces. In general, these surfaces are now dominated by impact craters, though they all bear evidence of volcanic and tectonic activity early in their history, the details varying from one body to another.

The surface of Mercury has been modified by crustal shrinkage and by viscous relaxation early in its history, and subsequently by lava flows. This has produced a variety of terrains that differ in their crater densities, from heavily cratered terrain, through intercrater plains, to smooth plains.

The Moon has a predominantly anorthositic crust, thought to be derived from a peridotite mantle. The impact basins on the near side have been filled by basalt lavas to form the maria—areas with much lower crater densities than the nearly saturated highlands that dominate much of the Moon. The basins on the far side, because they are at higher altitudes, are only partially filled. The crust on the far side is generally thicker than that on the near side.

The surface of Mars, like that of the Moon, also has two distinct hemispheres. The southerly hemisphere is dominated by impact craters, and has an abundance of channels that have been created by flowing water. The northerly hemisphere has a lower altitude, perhaps because the crust is thinner there. There are few impact craters, but extensive evidence of volcanic and tectonic activity, with some as recently as 180 Ma ago, though mostly much older. Unlike Mercury and the Moon, there are extensive deposits of solid water and carbon dioxide, particularly at the poles. The surface is basaltic, presumably derived from a magma ocean and volcanism. Subsequent alteration by various processes has occurred, including those involving water, and solar ultraviolet radiation. Further crustal differentiation might (somehow) have taken place, to produce rocks more andesitic in composition.

Pluto, and many of the larger icy–rocky satellites, have a long history of surface inactivity, and are known or presumed to be heavily impact cratered. Their surfaces are not pure water ice, but contain rocky dust and, beyond Jupiter, other ices, in some cases as a thin veneer. They all show some variation in composition across the surface. Callisto and Ganymede shows evidence of viscous relaxation in the distant past. Ganymede also seems to have experienced some resurfacing due to enhanced tidal heating, most recently about 1000 Ma ago.

Except for Enceladus, the smaller bodies all lack volcanic and tectonic activity today. The surface characteristics of some of them are summarised in Table 7.2.

8 Surfaces of Planets and Satellites: Active Surfaces

In this chapter we discuss the surfaces that are still volcanically and tectonically active—the surfaces of the Earth, Venus, Io, Europa, Enceladus, and Triton.

8.1 The Earth

The Earth is the largest of the terrestrial bodies, slightly larger than Venus, and for obvious reasons its surface is the best known in the Solar System. About 70% of the surface is covered in oceans, but here, as well as the continents, we are concerned with the solid surface under the oceans, postponing a discussion of the oceans to Chapter 10.

The Earth is both tectonically and volcanically very active, largely because of its high internal temperatures and thin lithosphere. The high temperatures are a consequence of primordial heat plus the heat from long-lived radioactive isotopes, and a comparatively low rate of heat loss. The Earth is unique in that it has a *global* tectonic system called plate tectonics. This has sculpted the large-scale features, and has given the Earth a predominantly youthful surface almost devoid of impact craters. To understand plate tectonics we first need to look at the Earth's lithosphere in more detail.

8.1.1 The Earth's Lithosphere

On the basis of rock samples, seismic, gravitational, and other evidence, a typical section through the Earth's lithosphere is known to be as in Figure 8.1. The concentric layering that characterises the structure of the deep interior is absent. Instead, the structure of the lithosphere varies from one region to another. It is mostly 50–70 km thick, and consists of the Earth's crust plus the uppermost part of the mantle. The mantle consists of peridotite, which you will recall is a mixture of iron- and magnesium-rich silicates, notably pyroxene and olivine (Table 6.1). The crust consists of less dense silicates that are not so rich in iron and magnesium. The

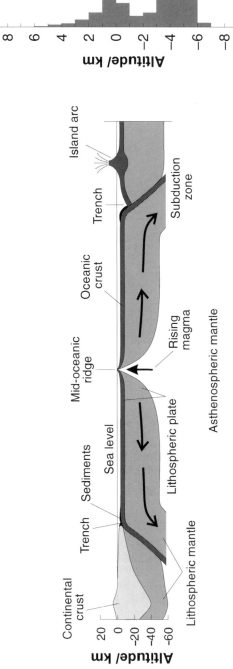

Fig. 8.1 The Earth's lithosphere, and the Earth's hypsometric distribution

crust is subdivided into oceanic crust and continental crust. As their names imply, these are found under the oceans and on the continents respectively. However, though continental crust reaches higher altitudes than oceanic crust and thus accounts for most of the dry land, and though oceanic crust with rare exceptions lies beneath the sea, the seashore is rarely the boundary between the two crustal types. In many regions continental crust only gives way to oceanic crust some distance offshore. Only if the volume of water in the oceans were somewhat reduced would the seashore predominantly coincide with the boundary.

Continental crust not only reaches higher altitudes than oceanic crust, it also occupies a largely separate range of altitudes. This is shown by the Earth's hypsometric distribution in Figure 8.1. You can see that there are two peaks — it is bimodal. The lower peak corresponds mainly to oceanic crust, and the higher peak mainly to continental crust. Oceanic crust is 5–10 km thick, has an average density of about $2900 \, kg \, m^{-3}$, and is dominated by basaltic–gabbroic rocks. Continental crust is 20–90 km thick, has an average density of about $2600 \, kg \, m^{-3}$, and has an andesitic composition, which is intermediate between basaltic–gabbroic rocks and granitic–rhyolitic rocks (Table 6.1). In its upper reaches it is more granitic–rhyolitic. Much of the Earth's continental crust is covered by a thin veneer of soil. This is derived from rocks through gradation plus extensive modification by the Earth's biosphere. Much of the oceanic crust is covered in sediments.

❐ What general process could give rise to the formation of a crust on the mantle, and the separation of the crust into two types?

Partial melting created the crust and then separated it into two types. This process continues today.

In Figure 8.1 you can see that crustal elevations are mirrored by deep 'roots'. This is a consequence of isostatic equilibrium, evident in gravitational data, and common over the Earth's surface. Such extensive isostasy implies the existence of a plastic region — an asthenosphere — for which there is much evidence (Section 5.1.1). Were oceanic crust as thick as continental crust then, because of the greater density of oceanic crust, isostasy would still lead to it 'floating' lower on the asthenosphere than does the continental crust. The smaller thickness of oceanic crust increases the altitude difference, as does the weight of water lying on top of it.

8.1.2 Plate Tectonics

Most of the large-scale features of the lithosphere, and the volcanic and tectonic processes that mould it can be explained by the theory of **plate tectonics**. This theory developed from seeds sown in the early years of the twentieth century, and it had its full flowering and wide acceptance in the 1960s.

According to the theory, the lithosphere of the Earth is divided into fairly rigid plates in motion relative to each other. The continents are carried on these plates, and therefore the modern map of the world is transitory. There are seven large plates, as shown in Figure 8.2, and a greater number of smaller ones, only a few of which are shown. Their relative motions define three kinds of boundaries between the plates. The first kind is where two plates are sliding past each other, side by side.

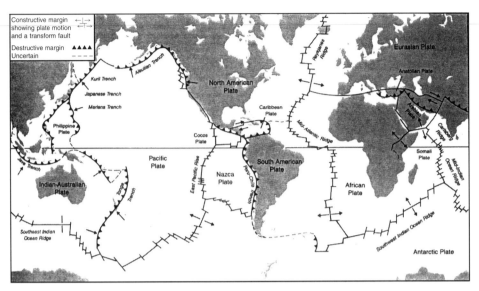

Fig. 8.2   The Earth's major lithospheric plates

This kind is called a conservative margin, and the boundary is a transform fault where the predominant displacement is horizontal. Figure 8.2 shows many examples.

The second kind of plate boundary is the constructive margin. The plates on each side of a constructive margin are continuously created from the upper mantle by partial melting of the mantle beneath mid-oceanic ridges (Figures 8.1 and 8.2). This partial melting is the result of pressure release as material ascends, and it creates a magma of basalt–gabbro which has a lower density than mantle peridotite and therefore rises to form basaltic–gabbroic crust as the upper layer of the plate. This is how oceanic crust is created. The rest of the oceanic plate consists of mantle peridotite somewhat depleted in the elements such as aluminium that are enriched in basalt–gabbro. The elevation of the ridges is largely due to isostasy associated with their higher temperatures and consequent lower densities. The plates slowly spread away from the ridges and, because the Earth is not expanding, this must lead to plate collisions elsewhere.

To consider collisions it is necessary to distinguish between oceanic plate, where the crust is oceanic, and continental plate where it is continental. Consider first the collision between two *oceanic* plates, as shown to the right in Figure 8.1. This is the first kind of destructive margin. You can see that one plate dives beneath the other forming an oceanic trench, the diving plate being reincorporated into the mantle in what is called a subduction zone. The heat from the friction between the two plates as the one moves over the other, plus heat conducted into the diving plate from the mantle, gives rise to partial melting and volcanism that builds a chains of islands called an island arc. The melting is, however, fairly complete and so the island arcs consist of material not very different from oceanic crust.

In Figure 8.1 you can also see the collision between an oceanic part of one plate and a continental part of another plate — this is the other kind of destructive margin. The oceanic plate is shown diving below the continental plate, to give another subduction zone, again producing volcanism for the same reasons as before. The magma consists of partial melts of both plates. The net effect is the creation of intrusive and extrusive material of andesitic composition, i.e. a growth in the volume of continental crust. The associated volcanic activity builds volcanoes. This, plus any uplift of the continental plate as it over-rides the oceanic plate, results in a range of mountains bordering the ocean. An example is the Andes. The partial melting of the continental crust at destructive margins leads to some differentiation within the crust, with granite intrusions near the surface, and rhyolite extrusions on the surface. Note that in this kind of collision it is almost always the oceanic plate that does the diving.

❏ Why is this?

Continental crust is less dense than oceanic crust, and consequently a continental plate is less dense than an oceanic plate. Therefore, the continental plate is more buoyant. Continental sediments on the oceanic crust resist subduction, and so tend to collect in the subduction region.

  If the diving plate also carries continental crust, then, as the subduction continues, this crust can ultimately meet the continental crust of the other plate. This will uplift any sedimentary deposits and bits of oceanic crust trapped between the two continents. This uplift, plus any crumpling of the continental crusts, raises mountain ranges that do not border oceans, such as the Himalayas. There is also a thickening of the continental crusts from the compression of the two plates. It has been estimated that gradation at the rates that we know have been operating over the past few hundred Ma can convert all the continental crust above sea level into ocean sediments in only a few Ma! Therefore, some opposing mechanism must have been operating, and this thickening of continental crust is a good one.

  The collision of continental plates will halt the relative motion of the plates towards each other, and this can trigger the creation of a new constructive margin elsewhere. This is a major means by which plate motion is changed. A new constructive margin can form within a continent where in its initial stages it will create a rift valley, such as parts of the Great Rift Valley in Africa (Figure 6.15, Section 6.2.5). Therefore, plate motion not only moves continents and joins continents, it can also disrupt them.

  According to plate tectonics, oceanic crust is created at constructive margins and destroyed at destructive margins — hence their names. Continental crust is created at destructive margins between oceanic and continental crust, and is destroyed in small quantities through the subduction of a small fraction of the ocean floor sediments from the continents. But is it created and destroyed elsewhere? The answer to this important question is uncertain, but a widely accepted hypothesis involves processes that are presumed to act on the underside of continental crust. It is thought that magmas from the mantle rise and spread along the underside of the continental crust and melt its base. This results in differentiation within the crust to produce a somewhat andesitic upper crust and a more basaltic lower crust. Water seems to be essential for this differentiation. By this process the crust thickens. It also thickens in collisions between continental plates. In each case the lower lithosphere reaches depths where the temperatures are high enough for melting to occur, and subduction then carries the material deep into the mantle. This process is called **delamination**. It is believed to be the main way in which continental crust is destroyed.

The initial source of continental crust was presumably island arc material that gradually became more andesitic through further partial melting. This might not, however, have been the only initial source of continental crust. During the later stages of the formation of the Earth there was probably a magma ocean from which fractional crystallisation generated a planet-wide basaltic–gabbroic crust, rather like the present oceanic crust. Subsequent impacts in this primeval crust might have led to local partial melting that resulted in regional differences in crustal composition. Later, plate tectonics could have worked on these differences to produce the first volumes of the two types of crust that we have today.

From sea level records it is possible to conclude that the volume of continental crust has changed little during the past 3000 Ma, and so on average the rate of creation of continental crust must have roughly equalled its rate of destruction. From various measurements it is estimated that over the past 3000 Ma the average rate of cycling of continental crust is about 2.5 km^3 per year, though variations in the rate have occurred and the rate of creation has not always exactly equalled the rate of destruction.

8.1.3 The Success of Plate Tectonics

You have probably already noted many features of the Earth's surface that can be accounted for by plate tectonics. Thus, it accounts for the two types of crust — these arise from the partial melting at constructive and destructive margins and through delamination. It also accounts for the hypsometric distribution — the bimodal form of this distribution arises from a combination of the greater density of the oceanic crust and the greater thickness of the continental crust, with both crusts in isostatic equilibrium over most of their areas. Plate tectonics also accounts for the greater thickness of the continental crust.

One of the earliest indications of the existence of mobile plates was the fit of the continents, notably on the opposite sides of the Atlantic Ocean. Down the centre of the Atlantic there is a mid-oceanic ridge, roughly equidistant from Europe and Africa on one side, and the Americas on the other side. This was explained by the creation of new oceanic plate material and the consequent spreading of oceanic plates away from the ridge, carrying the pre-existing continents further apart. Today there are measurements of the rate at which oceanic plates spread from mid-oceanic ridges — typically a few centimetres per year. For example, the South Atlantic Ocean is presently widening at about 3 cm per year.

❐ The South Atlantic Ocean is about 5000 km wide in the direction of widening. If it has always been widening at about 3 cm per year, how long ago were South America and Africa in contact?

At this rate of widening, the Americas were in contact with Europe and Africa about 200 Ma ago. Figure 8.3 shows an estimate of where all of the continents were at that time. Further back they were somewhat more scattered again.

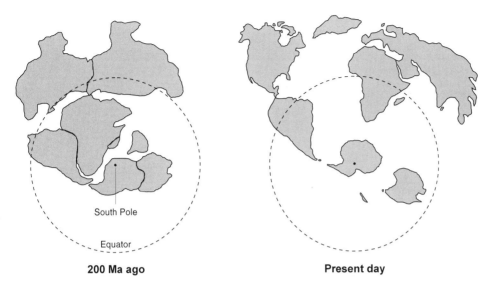

Fig. 8.3   The Earth's continents today and 200 Ma ago

It is clear that, according to the theory, oceanic crust near a constructive margin is young, and that it is older the further it is from the margin. This is just what is found by radiometric dating. Oceanic crust near mid-oceanic ridges is less than 1 Ma old, increasing up to about 200 Ma near subduction zones. As expected, most continental crust is older than this, some of it more than 2600 Ma old.

As well as mid-oceanic ridges, many other types of expected landform exist, and in the right places. For example, many mountain ranges border a coastline with a deep ocean trench that has every appearance of a subduction zone. Moreover, volcanic activity in such ranges generates just the sort of andesitic materials expected from the partial melting of basaltic and upper mantle materials (indeed, the name 'andesite' is derived from one such range — the Andes). Transform faults also exist, and some of these are in motion, the most notorious being the San Andreas Fault in California.

Earth's volcanic and tectonic activity is concentrated at transform faults, mid-oceanic ridges, the mountain ranges that border ocean trenches, and at island arcs. In plate tectonics these are the plate boundaries, and the theory predicts that volcanic and tectonic activity will indeed be concentrated at such boundaries. Also, in the region of mountain ranges and mid-oceanic ridges, there are departures from isostasy of the sort expected from continuing vertical lithospheric motions at plate boundaries.

The explanatory power of plate tectonics is vast, and you can explore it further in Question 8.1.

Question 8.1

In terms of plate tectonics, account for each of the following features of the Earth's surface: (a) the scarcity of impact craters, (b) a rift valley that crosses Iceland, (c) the

Urals (the mountain range east of Russia), (d) seismic activity at shallow depths near mid-oceanic ridges and ocean trenches, (e) the Aleutian Islands.

8.1.4 The Causes of Plate Motion

The immediate cause of plate motion is convection in the mantle. Most of the mantle is solid, so it is solid-state convection that is predominant, though the partially molten zone just below much of the lithosphere assists the motion of the plates. A possible pattern of mantle convection was shown in Figure 5.4 (Section 5.1.1).

A constructive margin is where motion in the convection cells is upwards, and a destructive margin is where it is downwards. The sideways motion that carries a plate away from a constructive margin is assisted by the slight downhill gradient — rather like snow sliding off a roof, though the plate gradient is much shallower and the process very much slower. At destructive margins the plates are cool and dense, and though heated as they descend they also become compressed, the net outcome being greater densities. They therefore tend to sink and this also aids plate motion by pulling on the horizontal part of the plate. At some unknown depth of at least several hundred kilometres there is horizontal motion back towards the constructive margin to complete the convective cycle. A crucial ingredient is the comparative lack of rigidity of the continental plates. Without this flexibility the global system of plates would be too rigid to permit motion — it would lock up.

As well as the large-scale convection that drives the plates it is known that there are rising columns of hot mantle material underneath plates, well away from any margins, called plumes. Therefore, though convection offers an explanation of plate motion, there is no simple relationship between convection patterns and plate boundaries.

Plumes account for strings of volcanic islands well away from plate boundaries (unlike island arcs). The Hawaiian Islands are one such chain. The sequence of ages when the different volcanoes became extinct indicate that plate motion over the plume generates the chain. Plumes are also a possible source of the huge quantities of basalt the have occasionally flooded parts of the Earth. As the plume head rises the pressure decreases, and in the asthenosphere the pressure can decline to the point where extensive partial melting occurs, creating large volumes of basalt that spill on to the Earth's surface to create volcanic plains.

Of course, the deeper question is — what causes the convection?

❐ Recall from Section 5.1.1 what the cause might be.

Models of the Earth's interior indicate that heat from long-lived radioactive isotopes, plus the residual effect of primordial heat, can sustain an adiabatic temperature gradient in the mantle, and that the pressures and temperatures are sufficiently high for solid-state convection to occur from the base of the lithosphere to great depth, perhaps as far as the core boundary.

Plate cycling accounts for about 70% of the energy that reaches the crust from the Earth's interior, as hot plates are created at conservative margins and cold plates descend at subduction zones. Plumes and delamination probably account for a large proportion of the remaining 30%.

8.1.5 The Evolution of the Earth

A plausible history is as follows, starting with a roughly uniform composition.

(1) Early on, radioactive heating, perhaps assisted by energy from accretion and other heat sources, raised temperatures to the melting point of iron, and so partial melting occurred with the appearance of liquid iron (plus some minor constituents such as nickel). This was denser than the remaining material and so the iron drained downwards, a process aided by the plasticity of the surrounding material. The iron core was thus formed, and heat of differentiation was released. The overlying mantle then had the composition of peridotite.

(2) The final giant impact by a body of broadly terrestrial composition created the Moon.

(3) Impact melting (perhaps assisted by a dense atmosphere) created a magma ocean several hundred kilometres deep. Fractional crystallisation led to the formation of a chemically distinct oceanic-type crust of basaltic–gabbroic composition, and a corresponding depletion in the upper mantle of elements (like aluminium) that are enriched in the crust. Subsequent impacts in this primeval crust might have led to regional differences that aided the subsequent formation of continental crust. The magma ocean froze throughout its depth before 4000 Ma ago.

(4) Meanwhile at the surface, plate tectonics was already established, creating and destroying basaltic–gabbroic crust, but increasing the volume of continental crust up to about 3000 Ma ago, since when the volume of continental crust has been roughly constant. Even without a magma ocean, plate tectonics would ultimately have created both types of crust.

(5) The asthenosphere initially extended to the molten iron core, and might well do so today, though partial melting is confined to its upper reaches.

(6) Plate tectonics continues, driven by mantle convection, and throughout Earth history has been responsible for the continuing high level of tectonic and volcanic activity, and for the sculpting of the Earth's surface.

8.2 Venus

Venus has a volume only 15% smaller than that of the Earth, and its global mean density is 5% less than the terrestrial value. As noted in Section 5.1.2, there is little doubt that its interior broadly resembles that of the Earth, though Venus is not the Earth's twin, and this is particularly true when it comes to its surface.

Venus has been extensively explored by orbiters and by landers. The landers were from the USSR, and they reached Venus in the 1970s and 1980s. Seven of them analysed the surface, and four of the seven returned images—a considerable achievement given the mean surface temperature of 735 K. None carried out seismic measurements. The many orbital missions have so far culminated in NASA's Magellan Orbiter, that between September 1990 and September 1992 mapped 99% of the surface using radar altimetry and synthetic aperture radar to 'see' through the 100% cloud cover. The images obtained with synthetic aperture radar have spatial resolutions down to a few tens of metres. The Magellan Orbiter subsequently mapped the gravitational field, in an extension of the mission with the spacecraft in a lower orbit that provided gravitational detail.

8.2.1 Major Topological Divisions

The surface of Venus can be divided broadly into three topographic domains. About 10% of the surface is highlands, with altitudes greater than 2 km (the zero of altitude is the mean equatorial radius, 6051.9 km). The mesolands have altitudes in the approximate range 0–2 km, and constitute somewhat more than half the surface — they roughly correspond to the 'rolling plains' of pre-Magellan times. The rest of the surface is called lowlands or plains, and generally lies below zero altitude. The highest altitude is 10.7 km in the mountainous area called Maxwell Montes, and the lowest is about 4 km below zero in the trough Diana Chasma. The overall altitude range is thus about 15 km, rather less than the Earth's 20 km range. Figure 8.4 is a simplified topographic map of Venus.

Figure 8.4 also shows the hypsometric distribution. It is unimodal, whereas the Earth's (Figure 8.1) is bimodal.

❑ What does this indicate?

The bimodal distribution on Earth arises from the existence of two types of crust, continental and oceanic. The unimodal distribution on Venus thus indicates that there is one crustal type, or that any other type is scarce. Comparison of the two distributions also shows that the altitudes cluster around the one maximum on Venus more that they do around either maximum on Earth, and therefore on a large scale Venus is a good deal smoother, probably the result of its higher surface temperature — this reduces rock strength.

8.2.2 Radar Reflectivity

Almost all the imaging of the Venusian surface has been done by orbiting radar, at wavelengths of the order of 10 cm. At such wavelengths the radar reflectivity of the surface correlates with altitude — broadly, the greater the altitude, the higher the reflectivity. This suggests that on a scale of a centimetre or so the lowlands are smooth whereas the highlands are rough. This is thought to result from gradation, in particular, the erosion of highland material to form fine dust that collects on the lowlands. The generation of dust presumably owes much to the high atmospheric pressures and temperatures, and to traces of corrosive gases. Further evidence for gradation is wind streaks behind obstacles, and the existence of some dune fields and land slides.

At altitudes above 3.5 km the reflectivity abruptly becomes considerably higher, and this might be due to a thin veneer of various metal chlorides, fluorides, and sulphides. These are somewhat volatile, and therefore condense at the lower surface temperatures in the highlands — the temperature decreases by about 8.5 °C for every kilometre increase in altitude. These substances would be emitted in small but sufficient quantities by the volcanoes of Venus. The very highest altitude summits

257

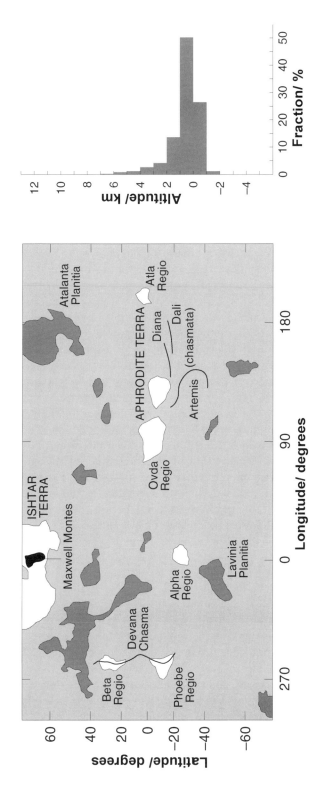

Fig. 8.4 A simplified topographic map of Venus, showing only the highest and lowest altitudes, the lighter the shade, the higher the altitude, except for Maxwell Montes. The hypsometric distribution is also shown

have low reflectivities, perhaps because of a veneer of materials that are even more volatile.

Radar shows that about 75% of the surface is bare rock, and the four landers that returned images each showed much bare rock, with only patches of gravel and finer material. This is one indication that gradation is not as powerful on Venus as on the Earth. This is not surprising given the lack of any precipitation and the low surface wind speeds — less than about $2\,m\,s^{-1}$.

8.2.3 Impact Craters, and Possible Global Resurfacing

The surface of Venus is lightly covered in impact craters, ranging from simple bowls a few kilometres across to multiringed basins with diameters of more than 100 km (Figure 8.5). Really small craters are absent because Venus has long had a dense atmosphere that has disrupted small projectiles before they reached the ground; any surviving fragments would produce small pits well beyond Magellan image resolution. The fragments of larger projectiles are presumably responsible for the observed crater clusters, and for shallow surface scars.

Many craters have radar-dark halos. These are presumed to be smooth, perhaps because the ejecta blanket contains much fine dust, or perhaps as a result of surface pulverisation by an atmospheric shock wave caused by the projectile — such a shock wave would be very powerful in the dense atmosphere. Flows extend from some craters as far as 150 km. These could be ejecta that was confined near to the surface by the dense atmosphere. Alternatively, or additionally, the flows could be magma released by the impact.

The spatial distribution of impact craters is not very different from random, with a global surface density indicating a surface age of about 300 Ma. Moreover, many of the craters have suffered little degradation. One interpretation is that most of the Venusian surface was wiped clean of craters about 300 Ma ago and that there has been little resurfacing since. Another view is that the resurfacing has been more extended in time, and that 300 Ma is an average retention age for craters. There has, however, been some more recent resurfacing, because areas of particularly low crater density correlate with volcanic and tectonic features. Overall, it is likely that there has been a decline in the resurfacing over the past few hundred Ma, but the details are unclear.

8.2.4 Volcanic Features

Volcanic features are abundant on Venus. Particularly widespread are volcanic domes from a few kilometres to several tens of kilometres across. The largest are flat-topped and only about 1 km high, and are consequently called pancake domes (Figure 8.6). Domes also occur on Earth, where they are the result of viscous lava oozing from a central vent, and though on Earth they are only up to a few kilometres in diameter, they are thought to have the same origin on both planets. There are also Venusian shield volcanoes (Plate 5), built from low-viscosity lavas. They are up to hundreds of kilometres in diameter, several kilometres high, and tend to be on broad regional rises. The larger shields are the same sort of size as the largest shields on Earth.

(a)

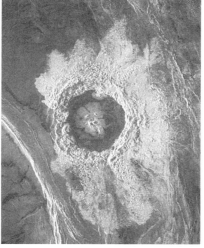

(b)

Fig. 8.5 Radar images of two Venusian impact craters. (a) Meitner, about 150 km across the outer rim. (NASA, part of F-MIDR55S319, R. Greeley.) (b) Dickinson, 69 km in diameter. (NASA/NSSDC P37916)

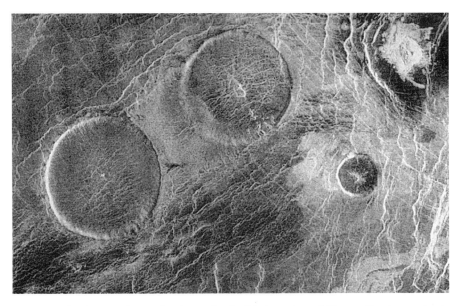

Fig. 8.6 Radar image of pancake domes in the Eistla region of Venus. The largest is about 65 km across. (NASA/NSSDC P38388)

Lava floodplains cover about 75% of the surface. Some of these have very few craters and so must be recent. Presumably the lava erupted from fissures, and ran along the channels that are seen to meander for up to thousands of kilometres. Some of these channels now run uphill, presumably as a result of uplift since the lava flowed. The vast floodplains and long channels indicate lava with low viscosity.

❐ What else do these features indicate?

The lava must also have remained liquid for a long time. The high surface temperature of Venus would facilitate longevity and low viscosity, but a distinct composition is also required. One possibility is lava with a high sulphur content. Another is carbonatite lavas. These are rich in igneous (not sedimentary) carbonates and in simple metal compounds such as sodium chloride, and have melting points not much above the Venusian surface temperature. Their sources would be near the surface. Carbonatite lavas are rare on the Earth because the generally lower surface temperatures prevent the extensive melting of such substances.

Low lava viscosity also helps to account for the rarity of explosive volcanism. A further factor would be any dryness of the Venusian crust and mantle, and the consequent low volatile content of the magma (Section 10.4). For any explosive volcanism that did occur, the high surface atmospheric density would have confined the explosively erupted material close to its source.

Volcanism might continue today. There is indirect evidence in the existence of fresh-looking lava flows, and also in fluctuations in atmospheric sulphur dioxide (SO_2) that could indicate variations in volcanic activity. However, as yet there have been no direct observations of active volcanism on Venus.

8.2.5 Surface Analyses and Surface Images

Direct analyses of surface materials have been carried out by seven of the USSR landers — Veneras 8, 9, 10, 13, 14, and VEGAs 1 and 2. These all landed in regions of volcanic flows. Venera 8 landed 5000 km east of Phoebe Regio (Figure 8.4), and the rest of the Veneras landed on the flanks of Beta Regio. The VEGAs landed on Rusalka Planitia, which is on the northern flanks of Aphrodite Terra, a predominantly highland region with several large volcanoes. Except for Venera 8, the analyses suggest basaltic rocks. The rocks analysed by Venera 8 were more granitic. This suggests that some division into two crustal types has taken place, though the hypsometric distribution indicates that this has not gone very far.

Veneras 9, 10, 13, and 14 returned images from the Venusian surface, and those from Veneras 9 and 14 are shown in Figure 8.7. Venera 9 seems to have landed on a hillside of boulders and gravel, whereas Venera 14 landed among protruding slabs of flat rock plus some gravel. Veneras 10 and 13 landed on terrain not very different from that around Venera 14, with smooth sheets of rock flush with the ground, plus some patches of dust. The sheets and slabs of rock could have been formed from flows of fluid lava.

8.2.6 Tectonic Features

Tectonic features are abundant on Venus. Some mountain ranges have complex folds and faults resulting from crustal compression and tension. Among these is Ishtar Terra (Figure 8.4), which includes the mountain range Maxwell Montes plus three other ranges, and a high plateau. As well as mountain ranges, the highlands also bear rugged tracts of faulted and folded terrain called tesserae. Alpha Regio, a rugged plain with a mean altitude of about 2 km and about 1000 km across, is a major tessera. Some of the fracturing in the tesserae might be due to a process that degrades all elevated features on

(a)

ВЕНЕРА-9 22.10.1975   ОБРАБОТКА ИППИ АН СССР 28.2.1976

ВЕНЕРА-14 ОБРАБОТКА ИППИ АН СССР И ЦДКС
(b)

Fig. 8.7 Images from two of the four Venus landers that returned images. (a) Venera 9. (b) Venera 14. (Via NASA/NSSDC)

Venus This is the plastic flow of rocks at the surface, a result of the high surface temperatures. The process is much more rapid on Venus than on the Earth, but it still takes tens of millions of years to have a significant effect.

Other tectonic features are belts of ridges and grooves hundreds of kilometres long. There are also chasms — chasmata. Some of these seem to have formed through crustal tension. One such example is Devana Chasma which, at 3000 km long, is of Africa's Great Rift Valley dimensions, and divides the young volcanic regional rise Beta Regio and also the neighbouring rise Phoebe Regio (Figure 8.4). Other chasmata, such as those in Figure 8.8(a), have profiles that suggest they might be places where one slab of crust is being forced under an adjacent slab, somewhat like subduction on Earth. Another type of tectonic feature is the coronae. These are

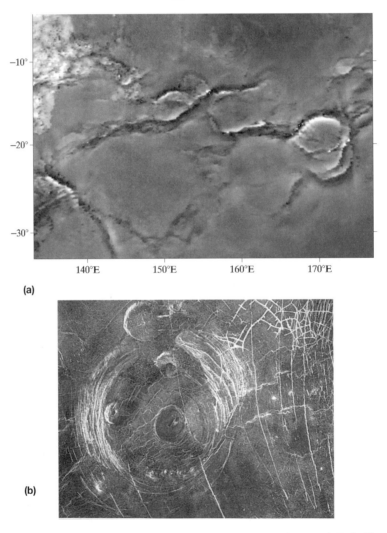

(a)

(b)

Fig. 8.8 Radar images of tectonic features on Venus. (a) Diana and Dali Chasmata. The frame is about 4200 km across. (MIT17/12/97, NASA, P. G. Ford and G. H. Petengill.) (b) A corona, about 150 km across, in Aino Planitia. (NASA/NSSDC P38340)

roughly circular plains, up to 600 km diameter but no more than about a kilometre high, and ringed by ridges and grooves. Some of them bear volcanic domes and display volcanic flows (Figure 8.8(b)). Their spatial relationships with other landforms indicate that they were created at different times. The tectonic interpretation is that each one is the combined effect of crustal uplift, volcanism, and sagging, in a prolonged and intricate sequence, not always with uplift first. A few might be highly degraded or deformed impact craters.

8.2.7 Tectonic and Volcanic Processes

In spite of the tectonic features just described, Venus lacks landforms that would suggest planet-wide plate tectonics. For example, long sinuous constructive margins like Earth's mid-oceanic ridges seem to be entirely absent. Moreover, the unimodal hypsometric plot indicates that there has been no widespread creation of a second kind of crust, and surface samples suggest that any second crust is at most rare.

If, as seems likely, plate tectonics on Venus is at most a local phenomenon affecting only a small fraction of the surface, this could be because the hot surface has made the lithosphere so plastic that widespread tectonic stresses are distributed, with the result that we get many lithospheric fractures that prevent the formation of a few large plates. On the other hand, it might be that the plate system is too rigid to be mobile because it lacks widespread continental-type plates, which on Earth might give the system a necessary lack of rigidity. An alternative or additional possibility is that the scarcity of water in the surface rocks will have raised their melting temperatures, so that a subducting plate would have to penetrate deep to melt (partially or wholly). There might be too much friction between a long length of plate and its surroundings for such deep penetration to occur.

❐ Plate tectonics is absent from some planetary bodies probably because they have thick lithospheres. Why is this an unlikely explanation for Venus?

Though increased melting temperatures, consequent upon dryness, would cause some lithospheric thickening, a thick lithosphere is unlikely because there are widespread tectonic features, and tectonism might still be occurring today. The high surface temperature also promotes a thin lithosphere. Estimates of lithospheric thickness of a few tens of kilometres are obtained from a close examination of how the surface has responded to stresses. However, with a hot surface and a general increase of temperature with depth the really rigid part might be only a few kilometres thick.

With no plate tectonics, how do we explain the tectonic and volcanic features, and how do we explain a possible global resurfacing about 300 Ma ago?

One explanation is that there is a system of convective hot plumes in the mantle, rather as seems to be the case underneath some of the plates on the Earth. Hot plumes on Venus could give rise to volcanoes that create some new crust, as in Figure 8.9. In this model these are the sites of the coronae. The sequence of events that create the coronae then include material being uplifted by a plume, volcanism, and sagging as the plume wanes. All stages in this proposed sequence are visible in different coronae. In some cases sagging seems to have preceded uplift. The tesserae

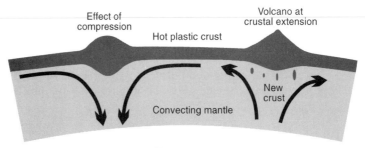

Fig. 8.9 Mantle convection and its effects in Venus

result from the *downgoing* part of a convective cycle (Figure 8.9). On this view the tesserae have been created by crustal compression. Some of the folded, faulted highlands might also have originated this way, such as Ishtar Terra (Figure 8.4). However, the topography of Ishtar Terra is also consistent with mantle uplift in this region. Such is the persistent puzzle that is Venus!

Support for the plume model comes from gravitational field measurements. Detailed gravitational analyses reveal that some topographic highs are in isostatic equilibrium due to crustal thickening, while others are probably supported dynamically. To the latter there is a contribution to the raised topography from isostatic adjustment resulting from the reduction in density consequent upon the greater temperatures in dynamically supported regions. These interpretations are in accord with the particular types of high terrain that fall into each gravitational category. Thus, many of the candidates for dynamic support, such as Beta Regio, show evidence that they are volcanic constructs that could have been created by a mantle plume that also provides dynamic support. By contrast, many of the candidates for isostatic support, such as Alpha Regio, are not obviously volcanic but look more like regions of crustal thickening such as could arise from crustal compression. The largest of these, Ishtar Terra, is so large that it might not simply be a thicker version of the surrounding crust, but also have lower density, in which case it is a rare example on Venus of crust analogous to continental crust on Earth.

On a global scale, one could interpret the mesolands as regions uplifted slightly by broad mantle upwellings, and the lowlands as regions above broad mantle downwellings.

The occurrence of a particularly large number of plumes about 300 Ma ago could possibly have generated enough magma at the surface to account for the widespread resurfacing that might have occurred around that time. An alternative explanation is that the lithosphere gradually thickens until plastic deformation within it decreases its viscosity to that of the underlying convective mantle. At this point there is global subduction of the lithosphere. A new lithosphere forms, and the cycle is repeated. However, there is as yet no general agreement on any one of these or any other explanations, even supposing such widespread resurfacing occurred.

8.2.8 Internal Energy Loss

You have seen that the interior of Venus today is probably not very different from that of the Earth. Therefore the interior energy sources have probably been similar in the two planets. In the case of the Earth, most of the internal energy has been escaping through plate recycling, but this does not seem to occur on Venus, and conduction through the lithosphere is probably weak. How then has Venus been losing its internal energy?

It is possible that mantle plumes, aided by delamination of the crust (Section 8.1.2) can get rid of much of the energy, in which case the coronae are important sites of heat loss. Subduction of lithosphere at certain chasmata could also make a contribution, because the lithosphere is cool and so would absorb internal energy from the interior. The episodic global subduction of the lithosphere that has been proposed to explain global resurfacing could result in enormous, if occasional heat loss. The new lithosphere then puts a thermal 'lid' on the interior, and the internal temperatures rise until the next subduction. However, as with many things to do with Venus, there is much uncertainty.

8.2.9 The Evolution of Venus

The interior evolution of Venus has probably been much the same as that of the Earth. The main differences are at the surface. No global system of plate tectonics has developed, perhaps because the high surface temperatures have made the lithosphere too plastic, or because the plate system is locked up through a scarcity of continental-type plates, or because the absence of water has raised the melting temperatures of the surface rocks. Instead, there might be a system of mantle plumes and downwellings that subduct little or no lithosphere but that account for many of the tectonic features. We therefore have predominantly one type of crust, a basaltic–rhyolitic crust derived by partial melting of a peridotite mantle, the crust gradually increasing in volume. Plumes, delamination, and subduction might account for most of the loss of internal energy.

Episodic global subduction of the lithosphere *might* occur, and this could account for the global resurfacing that might have happened about 300 Ma ago, and it is also another explanation for the loss of much of the internal energy. Any extensive resurfacing 300 Ma ago could alternatively have been caused by an outburst of mantle plumes at that time, that resulted in sufficiently copious production of basaltic lavas to obliterate almost all of the surface.

Volcanic activity might continue today, as might tectonic activity, and further resurfacing episodes are possible.

Question 8.2

What is a reasonable explanation for the absence of huge impact basins on Venus?

Question 8.3

Speculate on how the granites found on Venus by Venera 8 could have formed.

8.3 Io

Io is the closest of the four Galilean satellites to Jupiter. It is slightly larger than the Moon, and like the Moon it has a rocky surface and a rocky interior. Unlike the Moon it is volcanically active, and highly active at that.

❐ What are the main sources of energy to drive this activity?

The main source is tidal, with a supplement from radiogenic heating (Section 5.2.3).

The volcanic activity of Io was discovered on images returned during the flyby of Voyager 1 in 1979 and it came as a surprise to most planetary scientists. Voyager 1 revealed not only a surface free of impact craters and dominated by fresh-looking volcanic features, but also nine active volcanoes that made Io the most volcanically active body in the Solar System! Some of the volcanoes are visible in Plate 12.

Well before Voyager 1 it was known that Io had a reddish hue. Voyager 1 showed that this is due to an intricate pattern of red, orange, yellow, white, and black. Infrared spectrometry from the Galileo Orbiter indicates that the white areas are frosts of sulphur dioxide (SO_2). The other colours could be due to other sulphur oxides and to the various forms of sulphur itself. On the basis of cosmic relative abundances, Io could certainly have been born with sufficient sulphur for a surface layer several kilometres thick, and less than a few tens of metres would since have been lost to space. However, it seems likely that most of Io's sulphur is in an iron-rich core as iron sulphide (FeS), and that at the surface sulphur and its oxides are present only as a thin veneer that colours the surface. This veneer is presumed to have come from the volcanoes.

About 300 volcanic vents are known, randomly spread over the surface. Some are shield volcanoes up to a few hundred kilometres across and a few kilometres high, complete with summit calderas, and others are paterae, low-relief features with vents up to 200 km diameter. Some caldera walls are so high and steep that sulphur and its oxides would be insufficiently strong to constitute them. Silicates would be sufficiently strong, thus supporting the view that sulphur and its oxides are present only as a thin veneer. The vents switch on and off, and new ones appear, as indicated by changes in the three months between the Voyager 1 and 2 flybys, by subsequent observations from Earth of transient infrared hot spots, and by Galileo Orbiter observations in 1996–8. The estimated rate of volcanic resurfacing is 100 m depth per Ma, which explains the absence of impact craters.

The observed volcanic activity of Io includes volcanic plumes (Plate 12). Plumes require some gaseous content to create them, and vaporised sulphur or SO_2 seem likely, as opposed to water as on Earth. Lava flows are also observed. However, the SO_2 frost, caldera wall strength, Galileo Orbiter infrared measurements, and thermodynamic arguments all suggest that the dominant constituents of the plumes and of the lava are silicates, with sulphur and SO_2 as minor constituents. This is true of the lava too. Variations in the proportions of sulphur to silicate could explain the different types of flow and plume seen.

Though volcanoes are numerous, a second type of terrain covers most of Io's surface. These are flat, blotchy plains. Here and there erosion has exposed layers, each one a few hundred metres thick. A volcanic origin is presumed, involving lava flows and fallout from volcanic plumes.

The third type of terrain is mountains. These are dotted around the surface, though they constitute less than 2% of it. The largest are about 200 km across and 10 km high. Most of them do not seem to have a volcanic origin, and the plains sweep around the mountains in a way that shows the mountains to be older. They are too steep and too high to be made of sulphur and its oxides, and though they have a reddish tint, sulphur would be redder, and so they are presumed to be silicate.

Tectonic features on Io include faults attributed to tidal flexing of the lithosphere, but there is no sign of plate tectonics.

The estimated heat flow from the interior is a prodigious $10^{-9}\,\mathrm{W\,kg^{-1}}$, much greater than the $0.006 \times 10^{-9}\,\mathrm{W\,kg^{-1}}$ for the Earth. Indeed, this is several times more that the estimated input to Io from tidal heating! Much of the flow is in the radiation from broad regions at temperatures of about 300 K, which though less than the 1000 K or so of the infrared hot spots, is significantly greater than the 130 K mean surface temperature of Io. Cooling silicate lavas are a possible explanation of the warm regions, in which case Io might be in an atypically active volcanic phase, and the heat flow is then anomalously high.

The evolution of Io

Io presumably formed in a high-temperature environment near to protoJupiter. In this case it would have formed depleted in water and in more volatile substances. Subsequent radiogenic and tidal heating could soon have completed the devolatilisation. It seems certain that Io is differentiated into a silicate crust, a silicate mantle, and an iron core with iron sulphide, and that there is a thin lithosphere and convective asthenosphere (Figure 5.7, Section 5.2.2). The crust is probably rapidly recycled, being created volcanically and destroyed by melting at the base of the lithosphere.

Question 8.4

If Io were devoid of sulphur, what difference would this make to volcanism on Io, and to its surface?

8.4 Icy Surfaces: Europa, Enceladus, Triton

We end this chapter with a brief discussion of the three bodies with *icy* surfaces that seem to be volcanically or tectonically active today.

8.4.1 Europa

Europa is the second closest of the Galilean satellites to Jupiter, and was particularly well imaged by the Galileo Orbiter in 1996–7 (Plate 13). It is predominantly a rocky body, but is known from spectroscopy to be entirely covered in fairly pure water ice. The albedo in most areas is high, indicating that the ice has a clean surface. Areas of mottled terrain and dark plains might be where the surface has been darkened by ion implants from external bombardment, or by some non-icy component from interior and exterior sources. The existence of water ice is also indicated by Hubble Space Telescope UV spectra that have detected monatomic oxygen gas (O) in a thin atmosphere. This is very probably derived from water vapour through dissociation by solar UV. Models suggest that most of the oxygen is present as O_2, and that the atmospheric pressure is only $10^{-6}\,\mathrm{Pa}$.

The surface has almost no craters larger than a kilometre across. Therefore, the surface is nowhere older than about 200 Ma, and though viscous relaxation has played a role in this rejuvenation, there must also have been cryovolcanic resurfacing, perhaps continuing today. This will give a surface with the observed high albedo that will darken with age through ion implantation and dust infall. The surface is remarkably flat, with systems of dark ridges up to about 300 m high being the tallest features. These systems might be the result of thin cracks through which liquid water escapes to form a ridge on each side of the crack. The weight of the ridges produces further, parallel cracks, and the process repeats itself. Subsequent cracks can cross the older ones, to give the observed network of ridge systems. The ridges gradually disappear through viscous relaxation to leave smooth, dark bands. Ridges display faults, and there are areas where the surface has been broken up, liberating rafts of ice that moved away to be trapped by freezing water (Figure 8.10). Tidal stresses are probably responsible for disruption of the icy crust, though the patterns indicate that a shift in the rotation axis might also have contributed.

The smoothness of the surface and the ready access that liquid water seems to have to it indicates that the icy crust is only a few kilometres thick and that underneath it there is icy slush, perhaps mixed with rock, or more dramatically, widespread oceans of liquid water. The shell of water, ice and liquid, probably has an average thickness of roughly 100 km (Figure 5.7, Section 5.2.2). Throughout most of its depth the icy shell is plastic, and it thus constitutes an asthenosphere. Though tidal heating today is twenty or so times weaker in Europa than in Io, tidal heating coupled with radiogenic heat, perhaps with the aid of generally raised temperatures from solar radiation and residual primordial heat, could sustain an ocean or an icy slush. Heat from the underlying rocky mantle could have driven convection in the water, and this in turn could have created some of the cracks and ridges. Volcanism in the rocky mantle is possible, and this could vent material rich in organic compounds that

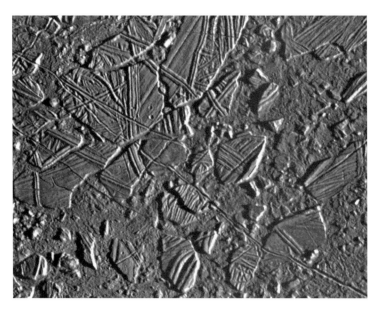

Fig. 8.10 The ridged and faulted surface of Europa, including rafts of ice frozen into new positions. The frame width is 42 km. (NASA/JPL P48526)

might darken some of the water that reaches the surface. It has even been speculated that life has originated in the oceans of Europa, and this is certainly worth investigation.

A less dramatic, less widely held view is that there is a solid icy asthenosphere, and it is partial melting in its upper reaches and at the base of the icy lithosphere that produces icy magma for cryovolcanism.

That Europa has a shell of water and Io does not was explained in Section 5.2.3. It is a result of the greater distance of Europa from protoJupiter.

Question 8.5

Describe the differences between the surfaces of the Galilean satellites that can be explained by the decrease in tidal heating that has occurred with increasing distance from Jupiter.

8.4.2 Enceladus

Enceladus is a medium-sized satellite of Saturn, with a mean density of about $1100 \, kg \, m^{-3}$.

❒ What does such a low density suggest about its interior?

The low density suggests a predominantly icy composition. The surface is certainly icy, and has an albedo close to 1, the highest in the Solar System. This suggests that the surface is very fresh. Its light-scattering properties are consistent with the texture of frost, perhaps the result of recent cryovolcanic eruptions on the body. At the low temperatures of Enceladus such eruptions would consist largely of water and ammonia. Evidence for eruptions is provided by Saturn's tenuous E ring (Figure 2.13, Section 2.3.2) which has a maximum density at the orbit of Enceladus. The particles in the E ring have estimated lifetimes of only about 10 000 years and thus need replenishing—ongoing cryovolcanism on Enceladus could be that source.

The high albedo veneer covers a richly varied landscape (Figure 8.11) with crater densities varying from almost zero to about that of the least cratered areas on the surfaces of the other satellites of Saturn. This indicates that even the oldest terrain postdates the heavy bombardment, and the youngest terrain could be very much younger, and certainly no older than 1000 Ma. Within each type of terrain the craters show various degrees of degradation, due in part to viscous relaxation. The most degraded areas might have been above interior hot spots analogous to mantle plumes in Venus.

There are various types of grooves and ridges, and these are concentrated in plains. Some of the grooves and ridges seem to be tensional features, others compressional, and yet others might be cryovolcanic fissures and vents.

If, as is probable, cryovolcanic and tectonic activity is occurring on Enceladus today, or subsided only in the relatively recent past, then the question of the heat source arises. The neighbouring satellites are similar in density; Mimas is about the same size as Enceladus, and Tethys is considerably larger, yet neither has signs of

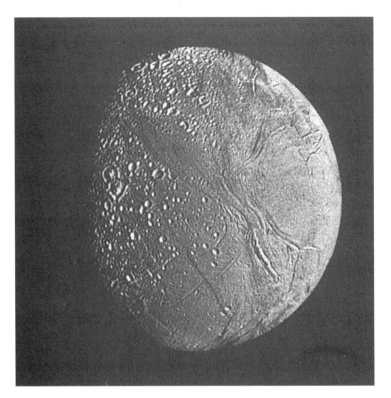

Fig. 8.11   Enceladus, from Voyager 2. (NASA/NSSDC P23956)

continuing or recent activity. This points to tidal heating for Enceladus, but not for its neighbours. Though the present eccentricity of the orbit of Enceladus is too small for sufficient tidal heating today, plausible changes in the eccentricity as a result of gravitational interactions between Enceladus and other satellites could have led to intermittent periods of greater tidal heating in the past, each lasting 10–100 Ma, with the last period ending relatively recently.

8.4.3 Triton

Much of what we know about Triton, by far the largest satellite of Neptune, is from the flyby of Voyager 2 in 1989, supplemented by Earth-based spectroscopy. Voyager 2 obtained high-resolution images of about two thirds of the sunlit hemisphere, with rather more of the southern than of the northern hemisphere being included. Plate 21 is a composite of these images.

Icy materials must comprise about a third of Triton's mass and 60% of its volume, probably concentrated into an icy mantle around a rocky core (Figure 5.7, Section 5.2.2). Though water ice is expected to be the dominant icy material, infrared spectroscopy reveals that at the surface nitrogen ice (N_2) dominates, and that CH_4 ice and CO ice are also present. These surface ices are presumably fairly thin veneers, derived from the mantle through their greater volatility than water. There are also patches of CO_2 ice, perhaps the result of chemical reactions triggered by impacts.

The surface ices, if pure, are colourless, and so the pinkish appearance of much of the surface (Plate 21) is presumably because cosmic rays and solar UV radiation have produced traces of complex organic compounds.

The surface is overlain by a thin atmosphere of nitrogen, plus a few per cent of methane and a trace of carbon monoxide. The presence of ices of nitrogen and methane at the surface is consistent with the atmospheric pressure of about 1.6 Pa and the 38 K mean surface temperature—at this temperature the combined pressures of nitrogen and methane if the gases are in equilibrium with their solids would be about 1.6 Pa. An indication that nitrogen and methane ices are no more than veneers is provided by topographic features such as 1 km high scarps—only water ice is strong enough at 38 K to support such topography.

There are three main types of terrain. High, smooth plains are presumed to be the result of a series of flows of icy lavas. At their edges they are seen to overlie the other two types of terrain.

❐ What does this imply about the relative ages of the three types?

The high, smooth plains must be the youngest, a conclusion borne out by the low density of impact craters on these plains. A similar type of surface is seen beyond the smooth plains, constituting the floors of four depressions 100 km or so across. A rugged area at the centre of each floor indicates a cryovolcanic origin for this infill, in which case the depressions could be calderas.

The second type of terrain is hummocky plains that border the smooth plains on their poleward side. Domes and a single ridge suggest volcanic creation of this type. It is the most cratered terrain on Triton, though only about as much as the lunar maria. No absolute ages can be deduced, but the hummocky plains, though still young among the surfaces of planetary bodies, are at least twice as old as the high, smooth plains.

Finally, there is cantaloupe terrain, which is unique to Triton. It gets its name from its resemblance to the skin of a cantaloupe (melon), and consists of a patchwork of approximately 10 km diameter dimples and bumps criss-crossed by grooves and troughs. Some of the grooves and troughs might be graben, and some have ridges running along their centre lines, which might be viscous extrusions. Some of the bumps might be extrusions of viscous ice. Fresh, long troughs cut all types of terrain, and might be recent graben. The high, smooth plains and hummocky terrain overlap cantaloupe terrain and so it must be the oldest type, yet there are few impact craters—this is a mystery.

In Plate 21 a polar cap is visible extending from the south pole towards the equator. It generally has a higher albedo than the rest of the surface, reaching 0.89 in its brighter parts (though even the darkest parts of the imaged surface have the rather high albedo of 0.62). Any north polar cap was hidden on the night side. The polar cap presumably consists of the same materials as the rest of the surface veneer. The cap's high albedo would then be the result of the seasonal deposition of pure ices, notably the more volatile nitrogen. This conclusion is supported by spectroscopic evidence that the atmospheric quantity of nitrogen varies seasonally, presumably as nitrogen frost is sublimed from one pole and deposited at the other.

Within the south polar cap four plumes have been identified in Voyager data. They rise straight up to an altitude of about 8 km, such high altitudes resulting from their buoyancy in the atmosphere. At 8 km the plumes are no longer buoyant and high speed winds then carry them sideways for 100 km or so. Dark streaks elsewhere in the polar cap indicate deposition from earlier plumes. The plumes are probably geysers from the sublimation of nitrogen below the surface, the nitrogen gas breaking through the overlying ice. The sublimation could be the result of heat flow from the interior, or from the solar heating of darkened ice beneath a metre or so surface layer of pure transparent ice.

Though the geysers are the only definite sign of ongoing activity on Triton, much of the surface is comparatively young, and models of the thermal evolution of Triton, in which there was powerful tidal heating early in its life, show that cryovolcanism could be occurring today as a result of the raised temperatures from primordial tidal heating plus radiogenic heat. The presence of nitrogen ice in the outer layers enables these heat sources to drive cryovolcanism because nitrogen is very volatile.

Question 8.6

Why is nitrogen cryovolcanism probably confined to Triton?

8.5 Summary

The bodies that are still subject to volcanic and tectonic activity comprise the Earth, Venus, and Io, which have rocky surfaces, and Europa, Enceladus, and Triton, which have icy surfaces. Rocky surfaces experience silicate volcanism and icy surfaces cryovolcanism. Impact craters are scarce as a result of resurfacing.

The Earth seems to be unique in having a global system of plate tectonics, in which the Earth's lithosphere is divided into a number of plates in relative motion. Oceanic crust is created at constructive plate margins from where the crust spreads, and it is destroyed at destructive plate margins where it is subducted into the mantle. At both types of margin there is a great deal of volcanic and seismic activity. Oceanic crust is derived by the partial melting of the mantle, and is basaltic–gabbroic in composition.

Continental crust is andesitic with a granitic–rhyolitic composition in its upper regions, and is derived from oceanic crust through a series of partial meltings at destructive margins and on the underside of continental crust. Continental crust is believed to be destroyed mainly through delamination. Plate motion is sustained by solid state convection in the asthenosphere, driven by the residual heat from primordial energy sources plus ongoing radiogenic heating. Plate recycling accounts for about 70% of the energy that reaches the crust, with mantle plumes and delamination accounting for most of the rest.

On Venus there is no widespread system of plate tectonics. It is unlikely that this is because of a thick lithosphere, but more likely to be because its lithospheric properties differ from those of the Earth. Instead of plate tectonics, there might be a system of mantle plumes and downwellings. Sample analyses indicate a predominantly basaltic crustal composition, in accord with the lack of an extensive system of plate tectonics. The predominance of one type of crust is also indicated by the unimodal hypsometric distribution. Volcanic activity has been widespread within the past few hundred million years, and little of the surface of Venus is older than about 300 Ma. This might be due to episodic global subduction of the lithosphere. Any extensive resurfacing 300 Ma ago could alternatively have been caused by an outburst of mantle plumes at that time, that resulted in sufficiently copious production of basaltic lavas to obliterate almost all of the surface. Plumes, delamination, and (limited) subduction might account for most of the loss of internal energy. Any episodic global subduction could make a major if intermittent contribution. Volcanic activity might continue today, as might tectonic activity, and further resurfacing episodes are possible.

The silicate volcanism on Io, and the cryovolcanism on Europa and Enceladus, depend largely on tidal heating. On Triton, the present cryovolcanic activity could be the result of the residual effects of primordial tidal heating plus radiogenic heating, the presence of the highly volatile nitrogen ice in the outer layers enabling these heat sources to be effective.

9 Atmospheres of Planets and Satellites: General Considerations

The atmospheres of the planetary bodies are richly diverse. This is apparent from Table 9.1, which lists some of the properties of the substantial atmospheres. As you can see, the list includes the main constituents of each atmosphere. This is specified as the **number fraction** of each constituent, that is, the number of molecules of the constituent divided by the number of molecules in the whole atmosphere. For convenience, I shall use that term 'molecule' to include constituents that are present as single atoms, such as argon in the Earth's atmosphere. In the case of the four giant planets there is no clear distinction between the atmosphere and the interior (Section 5.3), and so the number fractions are for the whole of the atmosphere above the altitude at which the pressure is 10^5 Pa, an altitude which is taken as the 'surface' of the giant.

The total quantity of an atmosphere can be expressed as the number of kilograms 'standing' on each square metre of the surface of the planetary body. This global mean value is called the **column mass** m_c, and these values are listed in Table 9.1. The total mass of an atmosphere is the column mass multiplied by the surface area of the body. The total is a very small fraction of the planetary mass — an atmosphere is a low-density veneer.

Table 9.1 also gives the magnitude of the atmospheric pressure p_s at the surface of the body. This is related to the column mass as follows:

$$p_s = m_c g_s \tag{9.1}$$

where g_s is the magnitude of the gravitational field at the surface (Section 4.1.2). To understand this relationship note that the right-hand side is the mass of atmosphere standing on unit area of surface, multiplied by the magnitude of the gravitational force per unit mass. The right-hand side is therefore the magnitude of the force exerted by the atmosphere per unit area of surface. A force per unit area is pressure,

Table 9.1 Some properties of the substantial planetary atmospheres

Planetary body	Major gases/number fractions	Escape speed[d] /km s^{-1}	Surface gravity[d] /m s^{-2}	Mean surface pressure/10^5 Pa	Column mass[e] /10^4 kg m^{-2}	Mean surface temperature/K
Venus	CO_2 0.96, N_2 0.035	10.4	8.88	91	102	735
Earth	N_2 0.77, O_2 0.21, Ar 0.009, H_2O 0.01	11.2	9.78	1.01	1.03	288
Mars	CO_2 0.95, N_2 0.027, Ar 0.016	5.0	3.72	0.0056	0.015	218
Titan[a]	N_2 0.82–0.99, CH_4 0.01–0.06, 0–0.06 Ar	2.64	1.35	1.5	11	94
Triton[b]	mainly N_2, some CH_4	1.45	0.78	1.6×10^{-5}	2.1×10^{-4}	38
Pluto[b]	mainly N_2, some CH_4 and CO	1.3	0.8	5.7×10^{-5}	7×10^{-5}	40
Jupiter[c]	H_2 0.863, He 0.135, CH_4 0.0018	59.5	24.7	1	0.405	165
Saturn[c]	H_2 0.963, He 0.033, CH_4 0.004	35.5	10.4	1	0.962	134
Uranus[c]	H_2 0.83, He 0.15, CH_4 0.02	21.3	8.8	1	1.14	78
Neptune[c]	H_2 0.79, He 0.18, CH_4 0.03	23.5	11.1	1	0.90	71

[a]The composition is for the middle atmosphere. There might be more methane in the lower atmosphere.
[b]Pluto's orbital eccentricity is large and its atmospheric properties vary as it goes around its orbit. The values are for the present, with Pluto not far from perihelion.
[c]The surface quantities for the giant planets are at the altitude where $p = 10^5$ Pa. The composition is for the total mass of atmosphere above 10^5 Pa.
[d]At the surface, at the equator and including the effects of rotation.
[e]Obtained by dividing the value in the fifth column by that in the fourth.

and thus the right-hand side is the magnitude of the surface pressure. Strictly, equation (9.1) is an approximation because g_s neglects the effect on p_s of rotation of the body, and the slight decrease of the gravitational field with altitude. In Table 9.1 the surface gravity is g_e, the acceleration of gravity that would be experienced by an object fixed at the equator, and so includes the slight effect of rotation.

Also included in Table 9.1 is the mean surface temperature T_s. In equilibrium, T_s is related to the mean surface pressure p_s and the mean surface density ρ_s via an equation of state (Section 4.4.3). Atmospheric gases are at relatively low densities. Therefore, they can be regarded as **perfect gases**, also called ideal gases, and we can then use the perfect gas equation of state. This equation applies locally, and is

$$p = \frac{\rho k T}{m_{av}} \qquad (9.2)$$

where k is a universal constant called Boltzmann's constant (Table 1.6), and m_{av} is the mean mass of the molecules in the atmosphere at the point in question. If we apply equation (9.2) at the surface then it allows us to calculate any one of ρ_s, T_s, and p_s, from the other two.

☐ In order to do so, what else needs to be known at the surface?

We also need to know m_{av} at the surface.

You can see from Table 9.1 that atmospheric compositions and column masses vary considerably from body to body. For the planetary bodies not listed, any atmospheres are far more tenuous than those included. Why atmospheres are so different is a topic for Chapters 10 and 11, where we examine individual atmospheres in detail. Before then, we will look at atmospheric processes in general. But first, we take a brief look at how atmospheres are studied.

9.1 Methods of Studying Atmospheres

One way of studying an atmosphere is from instruments within it, and the Earth's atmosphere has been widely studied in this way for centuries. For other atmospheres to be studied in this way it is necessary to fly a probe through the atmosphere, or to place a lander on the surface. Venus, Mars, Jupiter, and the extremely tenuous lunar atmosphere have had their atmospheres directly explored by probes or landers. But much of what we know about atmospheres comes from studying them remotely, either from the Earth's surface, from telescopes in Earth orbit, or from spacecraft that fly past the remote body or go into orbit around it.

Spectrometry is of great importance in remote studies of atmospheres. As a typical example, Figure 9.1 shows part of the infrared spectrum of Mars. This is the spectrum of the thermal radiation *emitted* by the Martian surface and then modified by passage through the Martian atmosphere. The atmosphere is cool enough so that it emits more feebly than the surface, and so the surface acts as a bright source of background illumination. Atmospheric molecules absorb the background at various wavelengths, and then emit the absorbed radiation at the same wavelengths, but

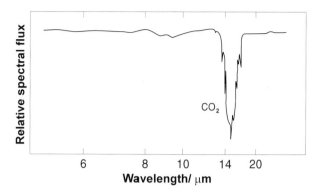

Fig. 9.1 Part of the infrared spectrum of Mars. Note that the smooth shape of the spectral flux from an ideal thermal source at 273 K has been subtracted. (Adapted from C. A. Beichman, JPL Publication 96-22 (1966))

equally in all directions, so the net effect is removal of some radiation from our line of sight. This results in a set of absorption lines, often narrower than the absorption lines produced by solids or liquids, and therefore readily distinguished from them. A particular molecule absorbs at a characteristic set of wavelengths, and therefore if absorptions at such wavelengths are detected, the presence of the molecule is inferred. The absorption line in Figure 9.1 lies at a wavelength known from laboratory studies to correspond to gaseous carbon dioxide and so it shows that carbon dioxide is present as a gas in the Martian atmosphere.

At the wavelengths in Figure 9.1 the solar radiation scattered from the Martian surface is much weaker than the radiation emitted by the surface, and it therefore makes a negligible contribution to the source on which the absorption lines are produced. At shorter wavelengths solar radiation dominates. The atmospheric constituents then make an imprint not only on the fraction scattered by the surface but also on the solar radiation on its way in to the surface.

Note that as well as a solid (or liquid) surface, clouds or deep atmospheric layers will do just as well as emitters or reflectors. Note also that, in addition to absorption lines, emission lines from atmospheres can also be observed in suitable circumstances, and their wavelengths also indicate composition.

So far we have used only the *wavelengths* of the spectral lines. There is clearly more to spectral lines than that: they have width, area, and shape; absorption lines have depth (Figure 9.1); emission lines have height. These characteristics can be used to infer three further properties of the atmosphere, namely, the number of molecules that produced the line(s), the temperature of the molecules, and their pressure. I will use the two absorption lines in Figure 9.2 to exemplify how these properties are extracted.

The number of molecules that produce a spectral line is the total number along our line of sight through the region in which the spectral line is produced. This total influences the *area* of the line, the greater the total, the greater the area. In Figure 9.2 the area of one of the lines has been shaded. The area of the other line extends beyond the confines of the figure. The area also depends on the average temperature

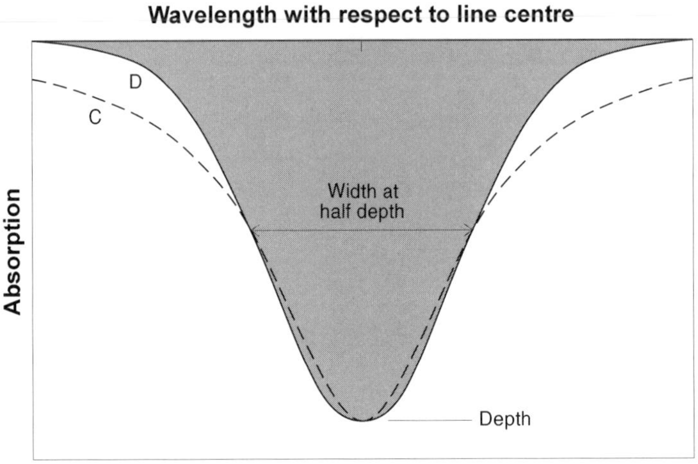

Wavelength with respect to line centre

Fig. 9.2 Two spectral absorption lines. The depth and width (at half depth) is the same in both cases, but the shape and area differ

along the line of sight. For example, consider a gas with a copious source of photons beyond it, i.e. a gas against a bright background. A particular spectral absorption line in the gas requires the molecule to be in a particular state before the photon is absorbed. The proportion in this state depends on the temperature of the gas. If the proportion is low, it is *as if* far fewer molecules are present, and the area of the line is consequently reduced.

Temperature also influences the *width* of the line (Figure 9.2). This is through the Doppler effect — the greater the temperature T, the greater the spread in the random velocities of the atoms or molecules, and therefore the greater the Doppler width. At a given temperature, the greater the mass m of the atom or molecule causing the line, the smaller the spread in the random velocities, and so the Doppler width is reduced. Overall, the Doppler width is proportional to $\sqrt{T/m}$. The spreading of the line due to this Doppler broadening gives the line a characteristic shape — line D in Figure 9.2.

❏ Will the Doppler effect change the wavelength of the *centre* of the line?

The directions of motion are random, so at any instant there are about half the particles moving away from us and about half towards us, regardless of the temperature. Therefore, the centre of the line remains at the same wavelength.

A spectral line can also be broadened due to collisions. If, during the absorption or emission of a photon, an atom or molecule collides with another, the photon energy will be changed slightly and hence so will its wavelength. The overall effect is that the width of the line is proportional to $p/\sqrt{T}$, where p is the pressure. This **collisional broadening** (also called pressure broadening) imparts a characteristic shape to a spectral line, different from that in Doppler broadening — line C in Figure 9.2. To show that the line width is proportional to $p/\sqrt{T}$ we need to use three fairly fundamental relationships. Do not worry if you are not familiar with them. They are:

that the line width is proportional to the frequency of collisions; that this frequency is proportional to $v\rho$, where v is the average speed of a molecule and ρ is the density; and that v is proportional to $\sqrt{T}$. Accepting these, you can see that the collision frequency and hence the line width is proportional to $\rho\sqrt{T}$. Using the perfect gas equation of state (equation (9.2)), we then obtain $p/\sqrt{T}$ for the factor to which line width is proportional.

Clearly a huge amount of information is embedded in a spectral line. Even more information is embedded in *sets* of lines. For example, the relative area of two different lines from the same region depends on temperature. The main point is that by examining the details of one or more spectral lines it is possible to establish the total number of molecules of the sort that produce the line(s), and the temperature and pressure of the molecules.

Different absorption lines can originate from different altitudes. Therefore by observing a variety of absorption lines it is possible to estimate altitude variations of composition, pressure, and temperature. Moreover, a *particular* line can be formed over a great range of altitudes, and this will influence the shape of the line. Therefore, a single line can also provide information about conditions at different altitudes. Variations with altitude can also be determined by observing how absorption lines change as the detector line of sight is swept across the disc of the planet. Near the centre of the disc we are looking straight through the atmosphere and the tenuous upper reaches have less effect than when we are looking near the edge of the disc.

Another way of investigating atmospheres is to use the radio transmission from an orbiting or flyby spacecraft when the craft passes behind (is occulted by) the planetary body as seen from the Earth. Then, to reach us, the radio waves have to pass through the atmosphere as in Figure 9.3, and they are refracted slightly. The angle of refraction can be measured and the mean refractive index along the path can then be calculated. During the occultation the path changes, which allows the refractive index versus altitude to be calculated. The refractive index depends on the mean density, the mean molecular mass, and on certain properties of the candidate molecules that are known from laboratory studies. Radio waves can penetrate clouds with particular ease, and so we can get information from depths that are obscured from us at other wavelengths. A similar technique at shorter wavelengths uses the occultation of a star by the planetary body.

Clouds can also be studied spectrometrically, though, as for all condensed matter, the spectral signatures that reveal composition are usually much less distinctive than for gases. As well as spectrometry, we can measure how the intensity of the solar radiation scattered by the cloud particles varies with the phase angle, i.e. the angle between the incoming solar radiation and our line of sight. This not only provides further information on composition, it also reveals particle shapes. Atmospheric circulation is revealed by following particular cloud features.

Question 9.1

Describe the effect on the spectral line in Figure 9.1 of

(a) A decrease in the number of carbon dioxide molecules in the Martian atmosphere
(b) An increase in the temperature of the Martian atmosphere
(c) An increase in the pressure of the Martian atmosphere.

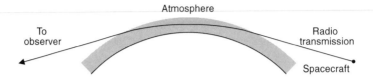

Fig. 9.3   The occultation method of studying planetary atmospheres

9.2 General Properties and Processes in Planetary Atmospheres

9.2.1 Energy Gains and Losses

In considering energy gains and losses, we need to include the gains and losses at the planetary body's surface. The giant planets have no surfaces in the sense used here, and so in relation to the giants the references to the surface in what follows are to be ignored. For bodies with well-defined surfaces but no atmospheres, the references to the atmosphere are to be ignored.

If we treat the atmosphere plus any surface as a single global entity, then energy can be gained in two major ways, namely, from solar radiation and by heat flow from the interior. There is only one major loss of energy — the loss to space of radiation emitted by the surface, by clouds, and by atmospheric gases. This will be at infrared wavelengths. Let W_{abs} denote the power absorbed from solar radiation, W_{int} the rate of heat flow from the interior, and L_{out} the radiant power emitted to space. A steady state requires

$$L_{out} = W_{abs} + W_{int} \tag{9.3}$$

Because of short-term variations in these three quantities, the steady state of interest is in the longer term, each of the three quantities then being averages over a period typically much greater than the orbital period. Table 9.2 gives the ratio L_{out}/W_{abs} for the planetary bodies with substantial atmospheres, and also for Mercury and the Moon, which are two of the many bodies that have tenuous or negligible atmospheres.

❐ What is the value of L_{out}/W_{abs} if there is negligible heat flow from the interior?

L_{out} then is equal to W_{abs}, and $L_{out}/W_{abs} = 1$. You can see that only for Jupiter, Saturn, and Neptune is this ratio substantially greater than 1. These infrared excesses were discussed in Section 5.3. For all other planetary bodies W_{int} is much less than W_{abs}.

W_{abs} is related to the flux density F_{solar} of solar radiation at the distance of the planetary body from the Sun via

$$W_{abs} = F_{solar} A_p (1 - a) \tag{9.4}$$

where A_p is the projected area of the surface (Figure 9.4), which for a spherical body of radius R is πR^2, and a is the **planetary albedo**, i.e. the fraction of the intercepted solar radiation that is reflected back to space by the surface and atmosphere. Table 9.2 lists values of a. F_{solar} is given by

Table 9.2 L_{out}/W_{abs}, a, and T_{eff} for some planetary bodies

Planetary body	L_{out}/W_{abs}[a]	a	T_{eff}/K
Venus	1.000	0.76	229
Earth	1.0002	0.30	see Q9.2
Mars	1.000	0.25	210
Titan	1.000	0.21	85
Triton	1.000	~0.85	~32
Pluto[b]	1.000	~0.3	~40
Jupiter	1.6(7)	0.44	120
Saturn	1.7(8)	0.46	89
Uranus	1.0(6)	0.56	53
Neptune	2.(52)	0.51	54
Mercury[c]	1.000	0.10	436
Moon[c]	1.000	0.11	271

[a]Values shown as 1.000 depart from being exactly 1 in the fourth or lower decimal places. This is also the case for the bodies not included here. Digits in brackets are uncertain.
[b]Pluto's distance from the Sun varies far more than those of the other planets, and it has a long orbital period. The values are for the late 1990s.
[c]Mercury and the Moon differ from the other bodies here in that they only have extremely tenuous atmospheres.

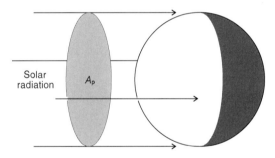

Fig. 9.4 The projected area of a planetary body

$$F_{solar} = \frac{L_\odot}{4\pi r^2} \quad (9.5)$$

where $L_\odot$ is the solar luminosity and r is the distance of the body from the Sun. Equation (9.5) assumes that solar radiation is uniformly distributed over all directions.

It is usual to express L_{out} as the **effective temperature** of the planetary body. This is defined in terms of the radiation from an ideal thermal source (Section 1.1.1). The power radiated by unit area of a such a source depends only on its absolute temperature T, and is given by σT^4 where σ is a universal constant called Stefan's constant (Table 1.6). For an ideal thermal source of surface area A the power radiated is therefore given by

$$L = A(\sigma T^4) \quad (9.6)$$

If A is the total area of a planetary body, and if L_{out} is the power radiated, then equation (9.6) *defines* T to be the effective temperature T_{eff} of the planetary body. Thus, *by definition* $L_{out} = A(\sigma T_{eff}^4)$, which can be rearranged as

$$T_{\text{eff}} = \left(\frac{L_{\text{out}}}{A\sigma}\right)^{1/4} \tag{9.7}$$

If W_{int} is negligible, then, for a spherical body (see Question (9.2))

$$T_{\text{eff}} = \left(\frac{F_{\text{solar}}(1-a)}{4\sigma}\right)^{1/4} \tag{9.8}$$

This is useful in that T_{eff} depends only on the solar flux density at the orbit of the planetary body and on its planetary albedo — the radius of the body has been eliminated. Table 9.2 gives values of T_{eff}.

However, a planetary body does not radiate in the manner .of an ideal thermal source, which raises the question of what T_{eff} means in terms of actual temperatures. For a body without an atmosphere the global mean surface temperature T_{s} will be approximately equal to T_{eff} if the surface is a very good absorber of radiation at the wavelengths of its emission. The less good it is, the greater will be T_{s} for a given T_{eff}. For a body with a substantial atmosphere, the radiation emitted to space comes from a range of altitudes in the atmosphere, as well as from the surface. In this case, T_{eff} is the temperature at some altitude above the surface, but this is not from where all the emission comes.

As well as total quantities, it is also important to consider the spectral distribution of the radiation. Figure 9.5(a) shows a typical spectrum of the radiation emitted to space by the Earth, and Figure 9.5(b) the spectrum of solar radiation. Note that whereas planetary emission is mainly at middle infrared wavelengths, solar radiation is mainly at visible and near-infrared wavelengths (see also Figure 4.13, Section 4.5.2). This is a consequence of the different source temperatures — the solar photosphere has a temperature around 5800 K, whereas the temperatures of planetary atmospheres and surfaces are much lower, typically a few hundred K. Because these two wavelength ranges are so separate, a surface or atmosphere can absorb a very different proportion of radiation in one range than in the other.

Question 9.2

Derive equation (9.8), and use it to calculate the missing value of T_{eff} in Table 9.2.

9.2.2 Pressure, Density, and Temperature versus Altitude

The temperatures and pressures in a planetary atmosphere depend on altitude. Here we shall concentrate on globally averaged altitude variations, deferring the horizontal ones to Section 9.2.6.

❐ If we know the temperature and pressure versus altitude, how can the density versus altitude be obtained?

The density versus altitude can be obtained from the perfect gas equation of state (equation (9.2)), provided we know at each altitude the mean mass of the molecules that constitute the atmosphere there.

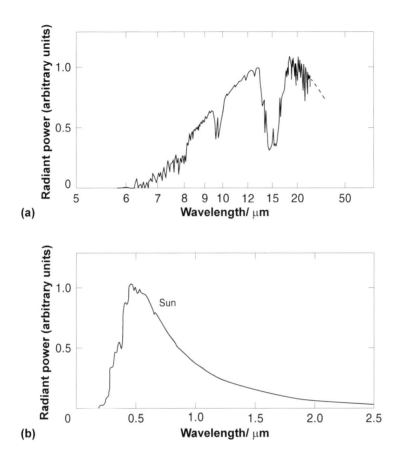

Fig. 9.5 (a) The Earth's emission spectrum (per unit frequency interval). (Adapted from V. G. Kunde *et al.*, JGR 79, 777 (1974).) (b) The solar spectrum (per unit wavelength interval)

Pressure and density versus altitude

Figure 9.6 shows pressure versus altitude in the Earth's atmosphere, which will serve to introduce the general ideas. How do we account for the rapid drop in pressure? On average over the globe, the atmosphere is neither expanding nor contracting, and so we can apply the hydrostatic equation introduced in Section 4.4.3. This gives the change in pressure δp for an increase δr in the distance from the centre of a spherically symmetrical body. A planetary atmosphere is a relatively thin layer on the surface and so in place of δr it is more sensible to use the increase δz in altitude z above the surface. If we again ignore planetary rotation, which has a relatively small effect on pressure, then

$$\delta p = -\frac{GM}{r^2} \rho \delta z \tag{9.9}$$

where r is distance from the centre of the body to a point in the atmosphere, M is the total mass within r, and ρ is the density at the altitude z. This equation shows that as altitude increases the pressure *must* decrease.

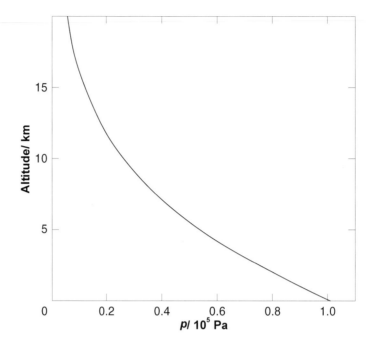

Fig. 9.6   Pressure versus altitude in the Earth's atmosphere

In equation (9.9) GM/r^2 is the magnitude g of the gravitational field at r (equation (4.6), Section 4.1.2), and with the atmosphere being a thin low mass layer we can use the surface gravitational field g_s everywhere within it. Thus, to a good approximation

$$\delta p = -g_s \rho \delta z \qquad (9.10)$$

Substituting for ρ from equation (9.2)

$$\delta p = -p[g_s m_{av}/(kT)]\delta z \qquad (9.11)$$

where T is the temperature at the altitude z. If the temperature is the same at all altitudes (isothermal atmosphere), and if m_{av} is also independent of altitude then it can be shown that equation (9.11) leads to

$$p = p_s \exp(-z/h) \qquad (9.12)$$

where p_s is the pressure at zero altitude and h is $kT/(g_s m_{av})$. The quantity h is called the isothermal scale height. For every increase in z of h the pressure falls by a factor of $\exp(1)$, which is 2.718 (to four significant figures). For the Earth h is about 8 km in the lower atmosphere — very much less than the Earth's radius.

Though atmospheres are not isothermal, and though m_{av} can change with altitude, equation (9.12) possesses the observed feature that the pressure falls rapidly with altitude. It follows from equation (9.2) that, for any realistic temperature variation, the density must also fall rapidly with altitude.

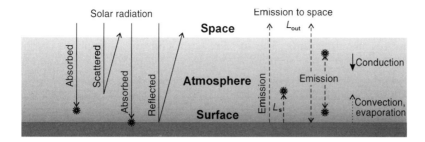

Fig. 9.7   Energy gains and losses in a planetary atmosphere

Temperature versus altitude

The temperature versus altitude depends on the various ways that energy is gained and lost at each level, and at the surface. These processes are summarised in Figure 9.7.

Consider first the surface of a body. It receives the solar radiation that has survived passage through the atmosphere. A fraction of this is absorbed by the surface, and the rest is reflected. The surface also receives radiation emitted by the atmosphere. Recall from Figure 9.5 the very different wavelength ranges of these radiation inputs, with solar radiation mainly at visible and near-infrared wavelengths (around $1\,\mu$m) and atmospheric radiation mainly at mid-infrared wavelengths (around $10\,\mu$m). The surface also receives heat from the interior, but (except for the giants) this is a sufficiently small fraction of the absorbed solar radiation that it can be ignored.

The surface loses energy by emitting its own radiation, again in the middle-infrared. It will also lose energy through the conduction of heat into the atmosphere in contact with the surface, followed by convection in the atmosphere. Convection was described in general terms in Section 4.5.2.

❏ What condition on the variation of temperature with position must be met for convection to occur?

For convection to occur, the temperature decrease with altitude must be at least as rapid as the adiabatic gradient. In the atmosphere convective plumes then rise, surrounded by denser, descending atmosphere that replaces the plume material. The convective plumes break up, and mix with the surrounding atmosphere, thereby transferring heat from the surface to the atmosphere. In planetary atmospheres the rate of temperature decrease with altitude is called the **lapse rate**, and so the adiabatic gradient is called the adiabatic lapse rate. If the actual lapse rate is less than the adiabatic value, there is no convection. If it is greater than the adiabatic value the rate of convection is so large that the adiabatic value, or something close to it, is quickly established.

For an atmosphere in which there is no condensation of any component in the rising plumes, the adiabatic lapse rate is given by

$$\Gamma_d = \frac{g_s}{c_p} \tag{9.13}$$

where c_p is the atmospheric specific heat at constant pressure, i.e. the quantity of heat that must be transferred (at constant pressure) to unit mass of the atmosphere to cause unit rise in temperature. The value of c_p depends on atmospheric composition. The only significant condensate in the Earth's atmosphere is water, and if water condensation is *not* occurring then the value of Γ_d is about $10\,\mathrm{K\,km^{-1}}$. The derivation of equation (9.13) is in standard texts on planetary atmospheres, though you can perhaps convince yourself that the equation should contain g_s and c_p, because g_s determines the rate of decrease of pressure with altitude (equation (9.10)), and c_p determines the temperature change of a parcel of atmosphere when it is in a new pressure environment.

The surface can also lose energy through the evaporation of liquids or solids (I'll use the term evaporation to include sublimation). It takes energy to convert a liquid or a solid into a gas, energy properly called enthalpy of evaporation, but more colloquially known as latent heat (Section 4.5). For this to be a significant energy loss the gas has to be removed so that further evaporation can take place. Convection provides one means of removal, by carrying the evaporated gas upwards.

In a convective column the temperature declines and might reach a value where the evaporant condenses in the atmosphere, to form liquid or solid droplets, which might constitute a cloud. The atmosphere is then said to be saturated with the condensable gas. When condensation occurs, energy is released, and so the energy lost by the surface is gained by the atmosphere. The energy released in condensation is another latent heat, and it has the effect of introducing additional thermodynamic properties of the atmosphere to the right-hand side of equation (9.13). For the Earth, if a region of the atmosphere is saturated with water vapour then the adiabatic lapse rate is about $5\,\mathrm{K\,km^{-1}}$.

Each level in the atmosphere where convection is occurring is thus heated not only directly by convection but also indirectly through the effect that convection has on promoting latent heat deposition. A level is also heated by absorbing a proportion of the solar radiation passing through it, and by absorbing some of the infrared radiation emitted by the other levels in the atmosphere and by the surface. Energy is lost by each level through its emission of (infrared) radiation. A further process that is effective in some circumstances is conduction. All these processes are shown in Figure 9.7. Together they determine the temperature at each atmospheric level.

A steady state

Though surface and atmospheric temperatures exhibit daily and seasonal changes, the average of the overall rate of gain at the surface is very nearly equal to the overall rate of loss if the average is taken over several orbits of the Sun. The same is true at each level in a planetary atmosphere. Therefore, the gains and losses in Figure 9.7 are nearly in balance, and so we are close to a **steady state**. It follows that the mean temperatures are very nearly constant.

The actual value of the mean temperature at the surface and at each atmospheric level depends on the way the various rates of energy gain and loss vary with temperature. Any change in any of the gains and losses, and the temperature is likely to change. For example, if the flux density of solar radiation F_{solar} were suddenly reduced, then the immediate effect at the surface would be a rate of gain of energy that was smaller than the rate of loss. This would result in surface cooling and a reduction in the rate of energy loss from the surface until a steady state was re-established at a new, lower surface temperature. Correct though this conclusion is, the details are *very* complicated, involving changes in energy transfer rates throughout the atmosphere, and between the atmosphere and surface.

Atmospheric domains

The lapse rate is one basis on which an atmosphere can be divided into different domains, and Figure 9.8 shows these domains in an idealised way. The **troposphere** is where the temperature decreases with increasing altitude, the lapse rate having a large positive value, in many cases equal or close to the adiabatic value. Indeed, the term 'troposphere' means 'turning sphere', which implies convection. The troposphere is where most clouds occur and is where weather changes can be rapid and large. In addition to the ubiquitous heat exchanges through the emission and absorption of infrared and solar radiation, a convecting troposphere also transports heat by convection, the result of the heating of the surface (and perhaps the lower troposphere) by the Sun, and in the case of the giant planets, mainly by heat from the interior. The troposphere extends up to the tropopause, where the lapse rate is sub-adiabatic. At the tropopause the atmosphere has become sufficiently thin that nearly all the infrared radiation emitted upwards by the atmosphere there escapes to space, and this is one of several factors that determine the altitude of the tropopause. The altitude varies with latitude, season, and weather.

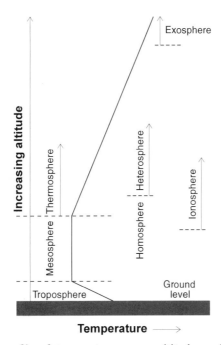

Fig. 9.8 A schematic profile of temperature versus altitude, and the major atmospheric domains

Above the tropopause there is the **mesosphere**, where the lapse rate is small. The mesosphere gains energy by the absorption of solar radiation — particularly at the UV wavelengths that lie just beyond visible wavelengths — and loses it by the emission of infrared radiation. There is no convection. The mesosphere ends at the mesopause. The **thermosphere** extends above the mesopause and is characterised by a *negative* lapse rate — the temperature *increases* with increasing altitude. It is heated by the absorption of solar ultraviolet radiation at much shorter UV wavelengths. Energy is lost through infrared emission and by the downward conduction of heat.

❏ Is there any convection in the thermosphere?

With a negative lapse rate there is no convection. The lapse rate is a complex outcome of the way the energy gains and losses vary with altitude.

You can see from Figure 9.8 that the atmosphere is also divided on bases *other* than the lapse rate. In the **homosphere**, atmospheric motions and frequent atomic and molecular collisions keep the atmosphere well mixed and therefore it has the same composition at all altitudes, except for any substances that condense (such as water vapour in the case of the Earth). Condensation depletes the local gas phase in the condensable substance, and the locality will be depleted as a whole if the condensed solid or liquid particles fall downwards (precipitate). In the **heterosphere** the mixing is weaker, with the result that molecules with greater mass are more concentrated in the lower heterosphere. In the **exosphere** the density is so low that collisions between molecules are sufficiently rare to allow the escape to space of a large fraction of any particles moving outwards with sufficient speed. Below the exosphere the density is higher, and so a particle is likely to collide before it escapes. In the **ionosphere** solar UV radiation has ionised a sufficient fraction of the molecules for the medium to display special properties, for example high electric conductivity.

Question 9.3

(a) Rewrite equation (9.12) in terms of density instead of pressure.
(b) In what domain in a planetary atmosphere should the actual decrease of pressure with altitude follow most closely the form of equation (9.12)? Give reasons.

Question 9.4

Describe the possible effects on a planetary atmosphere if the UV radiation from the Sun were to become much weaker.

9.2.3 Cloud Formation and Precipitation

A cloud consists of a thin dispersion of solid or liquid particles in a gas. Though a cloud is very effective at scattering light, the particles are separated by distances that are much greater than their diameter, and so the density of a cloud is not much greater than the density of the cloud-free atmosphere adjacent to the cloud.

Crucial to our understanding of cloud formation is the phase diagram of the substance that can condense to form the cloud particles. Take a region of the Earth's atmosphere as an example, where the condensate is water, and let's start with the region in a state where clouds cannot form. We thus have a mixture of water vapour and the other atmospheric gases, mainly nitrogen and oxygen. The atmospheric pressure is the sum of the pressures of the individual gases — oxygen exerts a pressure, so does nitrogen, so does water vapour, and so do each of the other constituents. Each gas in the atmosphere behaves much like a perfect gas, and for a mixture of perfect gases it is useful to note that $\rho = nm_{\text{ave}}$, and to rewrite equation (9.2) as

$$p = nkT \qquad (9.14)$$

where n is the number of molecules per unit volume — the number density (not the number fraction, which is global). For a mixture,

$$p = (n_1 + n_2 + n_3 + \ldots)kT \qquad (9.15)$$

where T is the temperature of the mixture, and n_i is the number density of the molecules of type i. The pressure contributed by a particular gas is called its **partial pressure**. For the ith constituent the partial pressure is

$$p_i = n_i kT \qquad (9.16)$$

In considering cloud formation we can usually use the phase diagram of the condensable substance, provided that we interpret the pressure as its partial pressure.

Figure 9.9 shows the phase diagram of water. Suppose the partial pressure of water vapour (gas) is at point A, and that the temperature at this point is that of the

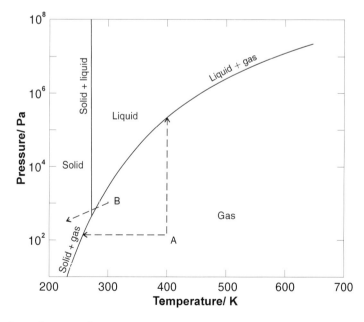

Fig. 9.9   Phase diagram of water. Note that water is unusual in that the solid+liquid line slopes to the left, though this is barely apparent on the scale here

atmosphere. Starting at A, consider an increase in partial pressure at constant temperature. This could be achieved by putting more water vapour into the atmosphere — equation (9.16) shows that the partial pressure is proportional to the number density. When the pressure reaches the 'Liquid + gas' line, liquid water droplets form. The number density of water molecules in the vapour is then fixed at the value corresponding to the partial pressure on this line. Any attempt to increase the number density will be obviated by further condensation.

Returning to A, consider a decrease in temperature at constant pressure.

❐ What sort of cloud particles form if the temperature is reduced sufficiently?

The 'Solid + gas' line is reached, so solid particles of water (ice) form. The formation of ice particles fixes the number density of water molecules in the vapour at the value corresponding to the temperature and partial pressure on this line. The 'Liquid + gas' and 'Solid + gas' lines together constitute the saturation line. The partial pressure on this line is called the **saturation vapour pressure** and the atmosphere there is said to be saturated with water vapour.

Droplet or ice-particle formation is delayed without the presence of other particles, such as dust, to act as condensation nuclei. Without such nuclei, and particularly if the decrease in temperature is rapid, the partial pressure will overshoot the saturation vapour pressure, and the atmosphere becomes supersaturated. This is not an equilibrium condition so ultimately the gas will condense, thus reducing the partial pressure to the saturation value — remember that a phase diagram, like an equation of state, is valid only for equilibrium conditions. It is also common for liquid droplets to appear initially, even if the temperature is low enough for ice to form. Such supercooled liquids are also in non-equilibrium, and will freeze.

Cloud formation is common in planetary atmospheres. Gas borne aloft by convection cools, and clouds form at a base altitude at which the saturation line of the condensate is reached. The curve in Figure 9.9 starting at B is one particular possibility in the Earth's troposphere. Clouds can also form above the cloud base if the conditions at higher altitudes lie on or to the left of the saturation line. Also, some cloud particles are carried upwards by vertical currents, even to the top of the convective column. Until condensation occurs, the partial pressure at all altitudes is a fixed proportion of the total pressure. Above the cloud base the gas phase is depleted in the condensable substance, and so p_i/p is less.

If cloud particles become sufficiently large, or if the upward currents are too slow, then the particles fall out of the cloud. If they do not wholly evaporate before they reach the surface the residue will reach the surface as rain, snow or hail — different forms of precipitation. Water vapour can also condense directly from the atmosphere on to the surface if the surface is sufficiently colder than the atmosphere. We then get dew, or frost. This return of water to the surface is the opposite physical process to the loss of water from the surface to the atmosphere through evaporation.

Question 9.5

Suppose that in some region of the Earth, the surface temperature is 293 K, and the lapse rate in the troposphere up to the cloud base is $7\,\text{K km}^{-1}$. If the partial pressure of water in the lower troposphere in this region is 100 Pa, use Figure 9.9 to estimate the altitude of the cloud base. Will the clouds consist of water droplets or ice crystals?

9.2.4 The Greenhouse Effect

In Section 9.2.1 you saw that a surface–atmosphere system loses energy to space in the form of radiation, largely at infrared wavelengths, at a rate given by the radiant power L_{out}. The radiant power L_s emitted by the surface alone is generally greater than L_{out}. The name given to this phenomenon is the **greenhouse effect** and it can be quantified by the ratio L_s/L_{out}. You saw that L_{out} can be represented by an effective temperature T_{eff}, which is an equivalent ideal thermal source temperature. L_s can likewise be represented by an effective temperature, but because the surface of a planetary body radiates approximately like an ideal thermal source, the effective temperature is not very different from the global mean surface temperature T_s. For this reason the magnitude of the greenhouse effect is more often quantified as $T_s - T_{eff}$. This difference is generally positive, so in some sense the greenhouse effect represents an enhanced surface temperature. (The greenhouse effect gets its name from the passive heating of greenhouses, though detailed calculations show that only a small fraction of the temperature rise in a greenhouse is associated with infrared radiation. Most of the rise is due to the trapping within the greenhouse of warmed air which cannot be convected away.)

It might come as a surprise to you that L_s can exceed L_{out} — at first sight it seems to imply a continuous build-up of energy in the atmosphere. However, this is not so. L_s and L_{out} are only two of the many energy flows involving the surface–atmosphere system (Figure 9.7). As long as the total rate of energy gain by the surface equals its total rate of loss then it will be in a steady state, and the same applies to the atmosphere and to the system as a whole. Any single rate of flow need not equal any other single rate — energy can flow at a much greater rate within a system than the rate at which it enters or leaves the system. In the case of L_s and L_{out}, the former originates from the surface, but the latter is a proportion of L_s plus some of the emission from throughout the atmosphere, notably in and just above the troposphere. The positive lapse rate in the troposphere then ensures that T_{eff} is less than T_s.

The greenhouse effect depends on the absorption by atmosphere constituents of the radiation that is emitted by the atmosphere and by the surface. If a constituent absorbs radiation then it must also emit radiation. Some of this emission will be absorbed by the surface, thus raising its temperature above what it would have been had the atmosphere been transparent to the radiation (and if all else was the same). If the atmosphere is non-absorbing then the surface would emit all of its radiation direct to space. Furthermore, there would then be no emission from the atmosphere. Therefore, L_s would equal L_{out} and the greenhouse effect would be zero. Emission from planetary surfaces and atmospheres is at mid-infrared wavelengths, and so it is the atmospheric absorptivity at such wavelengths that is important. Molecules consisting of a single atom do not absorb significantly at such wavelengths, and molecules with two atoms — diatomic molecules — are weak absorbers unless the atoms are of different elements. Molecules with more than two atoms are relatively strong absorbers. If a strong absorber is removed from an atmosphere then, if all other factors remain the same, T_s will fall, and the difference $T_s - T_{eff}$ will become less.

As well as this absorption–emission greenhouse effect, there is also a greenhouse effect from the scattering of mid-infrared radiation by cloud particles. Some of the radiation from the surface and lower troposphere will be scattered back to the surface by this means, resulting in further warming.

Question 9.6

(a) Suppose that the atmosphere of a planet consisted solely of single atoms of argon. Discuss whether the greenhouse effect would be significant.
(b) Explain why T_s would rise if a dry atmosphere had water vapour added to it.

9.2.5 Atmospheric Reservoirs, Gains, and Losses

The specific molecules in a planetary atmosphere are transient. They are removed to the surface by precipitation and also by chemical combination with surface materials. They are lost to space, and they can undergo chemical reactions within the atmosphere. The average time that a molecule of a particular constituent spends in a planetary atmosphere is called its **residence time**, and it is rather short, typically less than a few tens of days. Therefore, unless the constituent is replaced, it will quickly disappear from the atmosphere. The almost constant mass of the various constituents of planetary atmospheres shows that molecules are being replaced today almost exactly as rapidly as they are being removed.

Figure 9.10 illustrates this balance for a typical constituent. The constituent resides in a set of interconnected reservoirs, and the chemical form of the constituent need not be the same in all of these. The reservoirs are connected by the processes

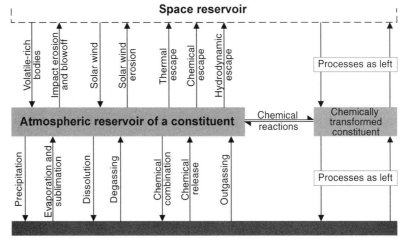

Fig. 9.10   The reservoirs of some atmospheric constituent, and the various types of process by which it can transferred from one reservoir to another

through which the constituent moves from one reservoir to another, and each process transfers the constituent at a particular rate. As long as the sum of the rates of loss from an reservoir is (very nearly) equal to the sum of the rates of gain, then the quantity of the constituent in that particular reservoir will be (very nearly) in a steady state. We will now examine the various processes involved.

Physical processes between the atmosphere and surface

You have already met precipitation as a physical process that removes a condensable constituent from the atmosphere, such as water from the Earth's atmosphere. Atmospheric gases are also removed through other physical processes, notably dissolution in liquids (such as the Earth's oceans) and, much more slowly, in solids. The reverse processes are evaporation (liquid to gas), sublimation (solid to gas), degassing (the reverse of dissolution). Note that for a steady state there is no requirement for these individual gains and losses to balance.

❏ Why not?

For a steady state, it is the sum of the rates of loss from a reservoir by *all* the various loss processes that must equal the sum of the rates of gain by *all* the various gain processes.

As well as evaporation and sublimation, a constituent can also enter the atmosphere through volcanism. This is called **outgassing**, and it can be accompanied by chemical changes not far below the surface so that the gases emitted are not the same as those at the source. The opposite process to outgassing is when an atmospheric constituent that has been physically trapped or chemically combined at the surface is subsequently buried by geological activity. Later, it might be outgassed again, though some proportion of outgassed material will never before have been part of the atmosphere — such gases are called juvenile.

Thermal escape

You saw in Section 9.2.2, that in the exosphere collisions between molecules are sufficiently rare to allow the escape to space of a large fraction of any molecules moving outwards with sufficient speed. The sufficient speed is called the **escape speed**, and it is given by

$$v_{\text{esc}} = \sqrt{\frac{2GM}{r}} \tag{9.17}$$

where G is the gravitational constant, M is the mass of the planetary body, and r is the distance from the centre of the body to the base of the exosphere. As you might expect, equation (9.17) shows that the escape speed increases as the mass of the planetary body increases, and decreases as the height of the exosphere increases.

The fraction of any constituent with sufficient speed to escape depends on the distribution of molecular speeds at the base of the exosphere. At this base collisions are just about frequent enough for the atmosphere to be treated as if it were in thermal equilibrium. In this case, the atmosphere has a well-defined temperature, and it determines the speed distribution for molecules of a given mass. The distribution is called the **Maxwell distribution** after the British physicist James Clerk Maxwell (1831–1879), who did much theoretical work in this area.

Figure 9.11 shows the Maxwell distribution for CO_2 at 400 K and 600 K, and for H_2 at 400 K.

❐ How does the distribution vary with molecular mass, and with temperature?

At a given temperature, the greater the mass of the molecule, the lower the speeds. For a given molecular mass, the higher the temperature, the greater the speeds. Note that a particular molecule changes its speed at every collision, and so a particular range of speeds in Figure 9.11 does *not* apply to a particular group of molecules, but to the fraction of the whole population that at any instant has speeds in that range. The speed at the peak of the distribution is called the most probable speed, and is given by

$$v_{mps} = \sqrt{\frac{2kT}{m}} \tag{9.18}$$

where k is Boltzmann's constant, T is the absolute temperature, and m is the mass of the molecule.

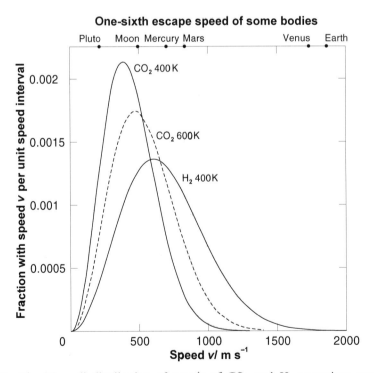

Fig. 9.11 The Maxwell distribution of speeds of CO_2 and H_2 at various temperatures, compared to one-sixth of the escape speed of some planetary bodies

In considering escape, T can be taken to be that at the base of the exosphere. The greater the value of T, the greater the value of v_{mps}, and the larger the fraction of a constituent that will have speeds in excess of the escape speed v_{esc}. The prominence of temperature in this escape process leads to its name, **thermal escape**. For a given exospheric temperature, the greater the mass of a molecule, the slower the rate of thermal escape, and this leads to discrimination between one type of molecule and another, between CO_2 and H_2 in the example in Figure 9.11. Thermal escape is thus mass-selective. Mass-selection extends to isotopes. The graphs in Figure 9.11 correspond to the isotopes ^1H, ^{12}C, ^{16}O, but had, for example, some hydrogen molecules included an atom of deuterium ^2H, to form $^2H^1H$ instead of $^1H_2$, then these molecules would have had a mass 3/2 times that of $^1H_2$, and a correspondingly slower rate of thermal escape.

Figure 9.11 includes $v_{esc}/6$ for some planetary bodies. If, throughout the 4600 Ma history of the Solar System, it has been the case that for some constituent $v_{mps} \leqslant v_{esc}/6$ then the accumulated fraction of the constituent lost by thermal escape will be small. Figure 9.12 shows more extensive data. Each planetary body is plotted at a point corresponding to $v_{esc}/6$ and the mean temperature at the base of its exosphere (which is at its surface if it has little or no atmosphere). This gives a broad indication of which constituents are more likely to have suffered extensive thermal escape.

Other physical processes between the atmosphere and space

Molecules in the exosphere can also be removed by the impact of charged particles from the magnetosphere or from the solar wind. If there is no magnetosphere then molecules that are ionised can also be removed by entrainment in the solar wind's magnetic field. Early in Solar System history, when the Sun was going through its T Tauri phase (Section 2.1.3), the greater UV flux must have generated ions in exospheres in great abundance, and the stronger wind must have been more effective in ejecting them. Therefore, the losses could have been considerable.

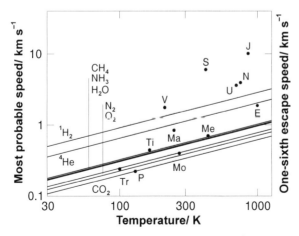

Fig. 9.12 The most probable speed v_{mps} versus temperature T for some major atmospheric gases, and associated data for planetary bodies

The opposite processes to losses to space are gains from volatile-rich bodies and the solar wind (Figure 9.10). These processes (as you will see) are thought to have been of great significance early in Solar System history, but the fluxes today are so low that they are negligible for all major atmospheric constituents in all substantial planetary atmospheres. Today, the major constituents can be replaced only from the surface.

Early in Solar System history, during the heavy bombardment, giant impacts must have been common, and instead of adding to the atmosphere they must have caused huge net losses. If the impactor had a radius greater than about the isothermal scale height of the atmosphere (Section 9.2.2) then pretty well the whole atmosphere would have been ejected, largely as a result of the shock-induced vibrations of the planet. This is called **blow-off**. Smaller impactors, less than the scale height of the atmosphere, would have caused more modest losses — partial ejection largely due to the explosion of hot vapour generated at the impact site. Once this **impact erosion** has started the scale height falls, making the atmosphere subject to erosion by smaller impactors, and these are more numerous. Considerable losses can consequently occur. Blow-off and impact erosion are *not* mass-selective.

Chemical processes

Chemical reactions can increase the kinetic energy of the products of the reaction, thus increasing their speed, particularly for products of lower molecular mass than the reactants. For example, a nitrogen molecule can be dissociated by a photon of solar UV radiation to give fast-moving nitrogen atoms

$$N_2(+UV \text{ photon}) \rightarrow N + N \qquad (9.19)$$

This can lead to escape to space, adding **chemical escape** to thermal escape.

❑ Where does the reaction need to take place for escape to be possible?

The reaction needs to take place in the exosphere.

For some constituents, chemical escape can be far more rapid than thermal escape, and like thermal escape, it is mass-selective, the greater the mass, the slower the rate of escape. Hydrogen (H or H_2) are the least massive molecules, and can therefore escape particularly rapidly through chemical and thermal means. If there is a large quantity of hydrogen escaping, then the high volume flow rate can entrain other molecules and carry them off in a process called **hydrodynamic escape**. It is thought to have been important early in Solar System history. Hydrodynamic escape is mass-selective, the greater the mass of the molecule, the less likely it is to be entrained, and the slower its rate of escape.

The involvement of the UV photon in reaction (9.19) makes this an example of a photochemical reaction. Such reactions are not only important in increasing loss rates. Thermospheres are in part warmed by the energy acquired through photochemical reactions, again involving solar UV photons. In the case of the Earth, the stratosphere is also warmed in this way, through the absorption of solar UV photons by ozone (O_3). This ozone is itself a product of photochemical reactions (more on this later). Indeed, there is a variety of important photochemical and other types of chemical reactions within planetary atmospheres that modify their compositions.

Chemical processes also lead to gains and losses at planetary surfaces. For example, if an oxygen-poor mineral is exposed on the surface, then atmospheric oxygen can combine with it in a process called **oxidation**. Gases can also be removed through adsorption onto grains, and through the formation of clathrates (Section 3.2.3), in which water ice entrains small molecules in its open, cage-like structure — molecules such as carbon dioxide (CO_2). Chemical processes can also *release* gases from the surface, such as in the weathering that occurs in chemical reactions between surface materials and water.

Question 9.7

State, with justification, which gain and loss rates are affected by each of the following factors:

(a) The mass of a planet (the radius of the exosphere being fixed)
(b) The temperature at the base of the exosphere
(c) The mass of the molecule of an atmospheric constituent
(d) The flux density of solar UV radiation.

9.2.6 Atmospheric Circulation

Atmospheric circulation is the combination of a variety of atmospheric motions on a global scale. The circulation distributes energy and materials throughout a planetary atmosphere, and is a crucial factor in regional climates. Some types of motions are important on only a single planetary body, and those are excluded from consideration in this general section.

Hadley circulation

Consider the case of a planet with an axial inclination considerably less than the 23.4° of the Earth. The solar radiation per unit area of its surface is then always greatest at the equator and least at the poles. This is because as latitude increases, the surface tilts more and more away from the Sun, and because a greater thickness of atmosphere has to be traversed. The surface at the equator thus becomes hotter than at the poles, and consequently warm atmosphere rises in the equatorial regions, moves polewards, and radiatively cools sufficiently to sink to the surface. The cycle is completed by the equatorward motion of the atmosphere at the surface, where it constitutes an equatorward wind. In the lower atmosphere of each hemisphere a huge convection cell is thus established, stretching from equator to pole, as in Figure 9.13. These convection cells transport energy away from the equator, and thus reduce the temperature change with latitude.

This kind of convection cell is called a **Hadley cell** after the British scientist George Hadley (1685–1768) who proposed the existence of such cells in the atmosphere of the Earth. They must not be confused with the smaller-scale convective plumes

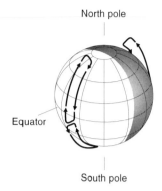

Fig. 9.13 A single Hadley cell per hemisphere on a slowly rotating terrestrial planet with a small axial inclination

described in Section 9.2.2, by which an adiabatic lapse rate can be established locally in a troposphere.

    You will see shortly that for the circulation in Figure 9.13 to occur it is also necessary for the planet to rotate no more than slowly. Among the planetary bodies with significant atmospheres, Venus rotates the most slowly, and it also has a small axial inclination. It is therefore reassuring to find that in its troposphere there is indeed a prominent single Hadley cell per hemisphere. The other bodies with significant atmospheres rotate much more rapidly, and so we need to consider the effect of this on Hadley circulation. This brings us to the Coriolis effect.

The Coriolis effect

Consider a parcel of atmosphere of mass m at the equator of a planet, and suppose that the rotation of the planet is carrying the parcel west to east at a speed v_e, the same as the equatorial surface speed. These speeds are with respect to an external observer. The parcel starts to move polewards. The motion could be at the top of a Hadley cell, though any other poleward motion will do. It moves to a latitude l, and gravity keeps it close to the surface, as in Figure 9.14(a). The parcel will increase its speed in the direction of rotation (not in the northerly direction) in accord with the principle of conservation of angular momentum. This principle emerges from Newton's laws of motion, and as applied to the parcel it states that the quantity mvr is constant, where v is its west-to-east speed, and r is its perpendicular distance from the axis of rotation, which has a value r_e at the equator (Figure 9.14(a)). The parcel mass m is constant, and thus as r decreases then v increases in proportion. At a latitude l the value of r is reduced to $r_e \cos(l)$, and therefore the west-to-east speed has increased to $v_e / \cos(l)$.

    We can now obtain an expression for the speed of the parcel *with respect to the surface* at a latitude l. With respect to the external observer, the surface at l is moving west to east at a speed $v_e \cos(l)$, and the parcel is moving west to east at a speed $v_e / \cos(l)$. Therefore, the speed of the parcel relative to the surface is given by

$$v_{rel} = \frac{v_e}{\cos(l)} - v_e \cos(l) \tag{9.20}$$

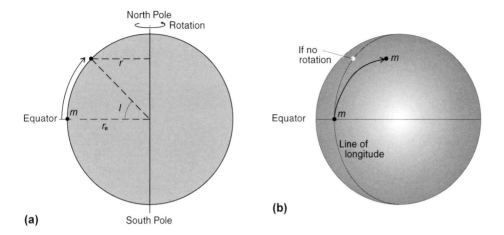

North Pole

Equator

South Pole

(a)

If no rotation

Equator

Line of longitude

(b)

Fig. 9.14 The Coriolis effect at a planetary surface

With $0 < l < 90°$ if follows that $0 < \cos(l) < 1$, and therefore v_{rel} is greater than zero. The speed relative to the surface is west to east. The parcel started with zero west to east relative speed, so in moving northwards it has picked up relative speed in the direction of rotation. This would also be the case for a poleward moving parcel in the other hemisphere. A possible path of the parcel across a planetary surface is shown in Figure 9.14(b).

❏ What happens to the parcel when it travels to the equator at the bottom of a Hadley cell, if it starts with v_{rel} as given by equation (9.20)?

The parcel will lose the relative speed it has acquired — as the distance from the rotation axis increases, the west-to-east speed of the surface increases, and the parcel loses speed to keep its angular momentum constant. In equation (9.20) we need to put $l = 0$ (the equator), and we see that then $v_{rel} = 0$. In reality the motion of the parcel is moderated in various ways, for example by friction between the atmosphere and the surface.

The west-to-east speeds of the atmosphere in Figure 9.14 are an example of the general case of a body (here a parcel of atmosphere) *viewed from a rotating frame of reference* (here the surface of a planetary body) having a tendency to accelerate in a direction that is perpendicular both to the motion of the body (here polewards along the Earth's surface) and to the axis of rotation. It will actually accelerate in this direction if it is free to do so. This tendency is called the **Coriolis effect** after the French physicist Gaspard Gustave de Coriolis (1792–1843).

The Coriolis effect has a profound influence on atmospheric circulation. From equation (9.20) you can see that it is proportional to the equatorial surface speed v_e and is therefore greater, the more rapid the rotation. One effect on a rapidly rotating planet, such as the Earth, is that the Hadley cells in Figure 9.13 are disrupted and do not reach as far as the poles.

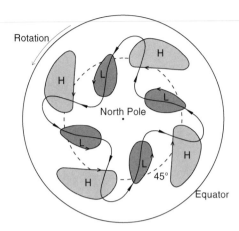

Fig. 9.15 Mid-latitude atmospheric waves

Other circulation patterns

Rotation can have a further important effect. Figure 9.15 shows mid-latitude circulation on a rapidly rotating planet. The rotation has resulted in a wave-like pattern of motion in the middle troposphere, which at the surface is correlated with the wind circulations labelled L for 'cyclones' and H for 'anticyclones' that are familiar from terrestrial weather maps. From such weather maps you will know that cyclones (low-pressure areas) and anticyclones (high-pressure areas) come and go, and so the details of the pattern in Figure 9.15 are not fixed. When atmospheric waves such as those in Figure 9.15 are present at mid-latitudes, they are the dominant mechanism of the poleward transport of energy. Many other sorts of atmospheric wave exist.

At a more local scale there are winds that arise from local temperature differences, such as occur between land and sea on Earth. Winds, however caused, can be deflected by large-scale topography into stable patterns called stationary eddies.

Question 9.8

(a) On Earth there is a Hadley cell extending from the equator to about 30° N and another extending from the equator to about 30° S. Calculate the east-west speed picked up by a parcel of atmosphere in travelling from the equator to a latitude of 30°, and state the direction of motion in the east–west direction in each hemisphere.

(b) Speculate on the effect on this Hadley circulation were the Earth to rotate much more slowly.

9.2.7 Climate

We all have some idea of what is meant by climate on Earth. It is a regional phenomenon — the climate of the coast of Kenya is unlike the climate of the Loire Valley in France — and it is a global phenomenon. Regionally or globally, the main factors that specify climate are temperature, precipitation, cloud cover and wind speed. These are the same as those that specify weather, the difference being that whereas weather can change on a daily or hourly basis, with climate we take a longer-term view. One aspect of **climate** is the average of each of the various factors over a sequence of years, usually 30 for the Earth. These averages do not change much from one 30-year sequence to the next, whereas large changes are common on the shorter time scales of weather.

In addition to the 30-year average, we also specify the variability. For example, as well as noting that the 30-year mean temperature for some region is 8 °C, we could add that in three years out of four the annual mean temperature is between 6 °C and 10 °C. We can also specify climate, including its variability, month by month, or season by season.

All planetary bodies have climate in some sense. In all cases climate, global or regional, depends on a huge number of factors. For example, global climate depends on the energy gains and losses for the surface and atmosphere as a whole, on energy exchanges within the atmosphere, and on the exchanges between the atmosphere and the surface. The greenhouse effect depends on atmospheric composition. Further factors include surface topography, and, at least in the case of the Earth, the behaviour of the oceans. Among many further significant factors is the rate of rotation of the planet, which determines the influence of the Coriolis effect on atmospheric circulation. So complex are climate systems that many aspects of planetary climates are poorly understood. Climate change is just as poorly understood, though there is ample evidence for large climate changes on several planets, including the Earth. Some particular cases are outlined in the next chapter.

9.3 Summary

A few atmospheres have been studied by means of instruments within them. Any atmosphere can be studied remotely, and spectral absorption lines produced within an atmosphere have been used to establish the composition, pressure, and temperature at various altitudes. Refraction of spacecraft radio waves and of starlight during occultation are additional sources of information.

For the surface–atmosphere system as a whole, solar radiation and heat flow from the interior are the two major gains of energy, with interior heat flow being significant only for the giant planets. Energy loss is through the emission of radiation to space. This loss is often expressed as an effective temperature.

Atmospheric pressure decreases rapidly as altitude increases. The variation of temperature with altitude is more complicated, and depends on the energy gains and losses at each altitude, as summarised in Figure 9.7. On the basis of the way the temperature changes with altitude, an atmosphere is divided into a troposphere, mesosphere, and thermosphere. On other bases different subdivisions are made. These subdivisions are summarised in Figure 9.8.

Clouds can form if the partial pressure of the condensable substance exceeds its saturation vapour pressure, though cloud formation is delayed without the presence of condensation nuclei. Ultimately the liquid droplets or ice crystals that form the cloud will precipitate out of it.

The absorption by the atmosphere of radiation emitted by the atmosphere and by any surface, raises the surface temperature. This is the greenhouse effect. It depends on atmospheric constituents that absorb mid-infrared radiation.

Atmospheric constituents are gained and lost by a variety of physical and chemical processes operating between a number of reservoirs, as summarised in Figure 9.10.

Atmospheric circulation is the combination of a variety of atmospheric motions. Motions that occur in several planetary atmospheres include Hadley circulation, Coriolis winds, and atmospheric waves with their associated cyclones and anti-cyclones.

Global or regional climates on a planetary body are specified in the main by the medium-term averages of temperature, precipitation, cloud cover, and wind speed, along with their variability. Climate, and climate change depend on a huge number of factors, and are not well understood.

Table 9.1 reveals the rich diversity of the substantial planetary atmospheres.

10 Atmospheres of Rocky and Icy–Rocky Bodies

In this chapter, attention is focused on the substantial planetary atmospheres, except for the giant planets — their atmospheres are the subject of Chapter 11. This means that the majority of planetary satellites are excluded from detailed consideration, and also the planet Mercury. From the rocky bodies only the Earth, Venus, and Mars are included, and from the icy–rocky bodies only Titan, Triton, and Pluto. Tables 9.1 and 9.2 list the main properties of the atmospheres to be considered.

10.1 The Atmosphere of the Earth Today

The Earth's atmosphere is, of course, the most familiar of all. From Table 9.1 you can see that the column mass is modest, but that it is the only substantial atmosphere in which oxygen is more than a minor constituent. This is not the only way in which our atmosphere is atypical.

10.1.1 Vertical Structure, Heating and Cooling

Figure 10.1 shows the globally averaged vertical structure of the Earth's atmosphere up to the top of the homosphere (Section 9.2.2). Only the lowest part of the thermosphere is included — it extends beyond the base of the exosphere, which is at about 500 km, the temperature rising as altitude increases. The thermosphere is heated mainly through the absorption of solar photons with wavelengths shorter than 91 nm — the extreme ultraviolet (EUV). The main absorbers are oxygen atoms that are themselves largely created with the aid of solar UV radiation in the range 100–200 nm. This radiation photodissociates oxygen molecules

$$O_2(+UV\ photon) \longrightarrow O + O \tag{10.1}$$

At times of high solar activity the EUV flux is greater, and consequently the thermospheric temperatures are higher, particularly above 120 km, and can reach

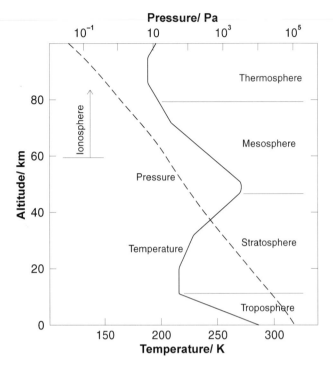

Fig. 10.1 The globally averaged vertical structure of the Earth's atmosphere, up to the top of the homosphere. The angular profile is an averaging artefact

2000 K or more at 500 km. The heating at each level in the thermosphere is balanced by cooling, mainly through the downward conduction of heat, but also by radiation from O and NO.

If you compare Figure 10.1 with the generic Figure 9.8 (Section 9.2.2) you will see that between the thermosphere and the troposphere there is a temperature bulge, a feature unique to the Earth. The upper part of the bulge is the mesosphere, and though the temperature decreases with altitude this is insufficient to promote much convection, heat losses and gains at any level being mainly through the absorption and emission of infrared radiation by CO_2. The lower part of the bulge, in which the temperature increases with altitude, is called the **stratosphere**. There is negligible convection here. The temperature bulge results largely from the absorption by ozone (O_3) of solar UV radiation in the wavelength range 200–300 nm. Heating is balanced by cooling, mainly through infrared radiation emitted by CO_2 and H_2O. The ozone is produced from molecular oxygen O_2 by a sequence of two chemical reactions that starts with the photodissociation of O_2 (reaction (10.1)) and finishes with

$$O + O_2 + M \longrightarrow O_3 + M \qquad (10.2)$$

where M is any atom or molecule involved simultaneously in this collision of O and O_2. Ozone production is significant in the altitude range 10–80 km.

As well as creating ozone, solar UV photons also destroy it. It is also destroyed by reactions with a variety of atoms and molecules. The quantity of ozone at any altitude depends on the local rates of production and destruction. At high altitudes the atmosphere is thin, and three-body collisions as in reaction (10.2) are rare, so there is little ozone. At low altitudes there is little ozone because much of the UV radiation that creates it has been absorbed at higher altitudes. The optimum is at about 25 km, so this is where the quantity of ozone is greatest.

Ozone is important to life on Earth because it screens the surface from most of the solar UV radiation that is damaging to life. It is therefore a matter of concern that substances are being released by human activity that are reducing the quantity of ozone. The reduction is particularly severe over the poles and mid-latitudes, where so-called holes in the ozone layer have appeared. The main substances that destroy ozone are nitrogen oxides released by high-flying aircraft, and chlorine from the dissociation of chlorofluorocarbons (CFCs).

The troposphere contains about 75% of the mass of the Earth's atmosphere. It is convective throughout most of its volume most of the time.

❏ What does this imply about the lapse rate in the troposphere?

It thus normally has an adiabatic lapse rate, for the reasons discussed in Section 9.2.2. The heating at any level in the troposphere is mainly through the absorption of infrared radiation from the ground and from other atmospheric levels, plus the latent heat of condensation of water. This water is made available through the convection in the troposphere, the convection itself also contributing to the heating. In the lower troposphere absorption of solar infrared radiation by water vapour makes a further heating contribution. The troposphere cools mainly through the emission of infrared radiation.

Most clouds are in the troposphere. These consist of liquid water droplets, or of flakes or particles of water ice. Some icy clouds occur in the lower stratosphere. At any one time approximately 50% of the Earth's surface is covered in clouds, and this raises the Earth's albedo considerably (Table 9.2).

The global mean surface temperature T_s of the Earth is 288 K, whereas the effective temperature T_{eff} is 255 K.

❏ What is the reason for the difference?

This is caused by the greenhouse effect (Section 9.2.4). It is mainly due to water vapour, a minor constituent, and to carbon dioxide, which is an even more minor constituent with a number fraction of only 0.000 36. Water vapour has the largest effect. Figure 10.2 shows the contributions of these two gases to the mid-infrared absorptivity of the Earth's atmosphere.

Question 10.1

Explain why convection and condensation play no role in heat transfer in the Earth's stratosphere.

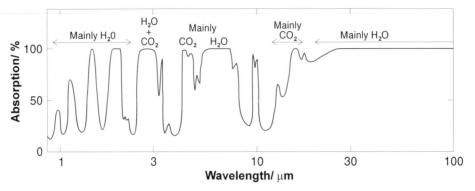

Fig. 10.2   The absorptivity of the Earth's atmosphere versus wavelength

10.1.2 Atmospheric Reservoirs, Gains, and Losses

Figure 9.10 (Section 9.2.5) summarises the various gains and losses that can affect the quantity of an atmospheric constituent. For the Earth today a major source is outgassing through volcanic activity. Most of the gases are recycled atmospheric constituents that have been buried by geological activity, though a small proportion of volcanic emissions might be juvenile. A major loss of atmospheric constituents is through the formation of rocks. Thermal escape and chemical escape occur, but at very low rates.

❐ Why are the thermal escape rates of the major atmospheric constituents oxygen (O_2) and nitrogen (N_2) so low?

Figure 9.12 (Section 9.2.5) shows that for both of these gases, at the present temperature at the base of the Earth's exosphere, v_{mps} is much less than $v_{esc}/6$, and so these gases must be escaping at very low rates.

The generic nature of Figure 9.10 obscures two components of the Earth's surface that have played a major role in determining the quantities of various constituents of the Earth's atmosphere. These components seem to be unique in the Solar System today. One of them is the oceans, and the other is the biosphere. The **biosphere** is the assemblage of all living things and their remains. It is a thin veneer covering most of the land, and it also occupies the oceans, particularly the surface layers. One of several important effects of the oceans and the biosphere is the dominant role they play in determining the atmospheric quantities of oxygen and carbon dioxide, as we can see if we examine the carbon cycle.

The carbon cycle

The main features of the present-day carbon cycle are shown in Figure 10.3. The main reservoirs of carbon are shown as a set of rectangles. Though the predominant repository of carbon in the Earth's atmosphere is carbon dioxide (CO_2—with a number fraction of 0.000 36), other chemical forms predominate in other reservoirs, and therefore it is the mass of *carbon* in each reservoir that is shown. The arrows

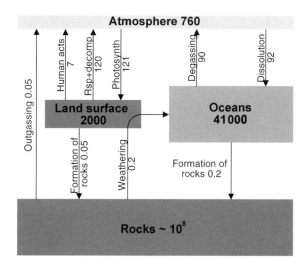

Fig. 10.3 The main features of the Earth's present-day carbon cycle. The size of each reservoir is given in 10^{12} kg of carbon, and the rates of transfer in 10^{12} kg of carbon per year

represent transfers between reservoirs, only the main ones being included. Each arrow is labelled with the rate of transfer of carbon between the two reservoirs connected by the arrow. The greatest rates involve the biosphere.

Consider first the arrow labelled 'photosynth' connecting the atmosphere to the land surface. This is **photosynthesis**, the process of building body tissue from carbon dioxide and water, and it is performed by green plants and blue–green bacteria with the aid of photons of solar radiation at visible and near-UV wavelengths. A long sequence of chemical and photochemical reactions can be represented as

$$n\text{CO}_2 + n\text{H}_2\text{O}(+\text{photons}) \longrightarrow (\text{CH}_2\text{O})_n + n\text{O}_2 \tag{10.3}$$

where n can have a variety of values. $(\text{CH}_2\text{O})_n$ are carbohydrates, and are the basis of further tissue building. The carbon dioxide is taken from the atmosphere, the water is taken from the soil, and the oxygen is released into the atmosphere. Most other organisms build body tissue by consuming other organisms. The carbon in tissue derived directly or indirectly from photosynthesis is called organic carbon.

Respiration and decomposition ('Rsp + decomp' in Figure 10.3) have the reverse effect of photosynthesis. Respiration is performed by most organisms, and in higher animals it is called breathing. In respiration, oxygen is taken in and converts some tissue to carbon dioxide and water. The chemical energy released by this process enables the organism to live and breed. After death, the dead tissue might be consumed, but if not it can decompose, and this too takes up atmospheric oxygen and releases water and carbon dioxide. Not all death results in the reversal of photosynthesis. Most dead tissue on land is in the form of leaf fall. A small proportion of this becomes buried, and over many millions of years it is transformed into the organic carbon components of rocks. Particularly rich examples are the fossil fuel deposits, of coal, oil, and natural gas. This transformation is included in 'Formation of rocks' in Figure 10.3.

Photosynthesis also occurs in the oceans — *inside* the ocean reservoir in Figure 10.3 — where it is performed by blue–green bacteria, particularly in the upper 100 m or so, where solar radiation can penetrate. As on land, the carbon involved in photosynthesis is in the form CO_2, in this case dissolved in the oceans. However, about 99% of the carbon in the oceans is in the form of the bicarbonate ion HCO_3^- and the carbonate ion CO_3^{2-}. Ocean organisms make use of this to form shells of calcium carbonate, $CaCO_3$. The carbon in these shells is called inorganic carbon to distinguish it from the organic carbon that is derived from photosynthesis.

Respiration and decomposition also occur in the oceans, where, as on land, a small proportion of the organic carbon escapes, in this case by settling to the ocean floors as a component of the sediments that can later form rocks. In addition, inorganic carbon also settles into the sediments, and this has led to the formation of huge deposits of sedimentary carbonate rocks, such as limestone and chalk, and their metamorphic products such as marble. Consequently, rocks are by far the largest reservoir of carbon on Earth (Figure 10.3). Carbonate rocks are exposed by geological processes and are weathered, particularly by water. The resulting dissolved (bi)carbonate ions return to the oceans. Volcanic activity returns CO_2 to the atmosphere.

The atmospheric quantity of carbon dioxide is determined by the accumulated effect of all the various rates of transfer in Figure 10.3. The quantity will be in a steady state if the sum of the rates of transfer into the atmosphere equals the sum of the rates of transfer out.

❐ Are these two sums equal in Figure 10.3?

The rate of gain is slightly greater than the rate of loss. Consequently the atmospheric quantity must be rising, and Figure 10.4 shows the observational evidence that it is. In fact, it was this sort of evidence that showed that there must be an inequality. The imbalance in recent decades has probably been the result of human activities, notably the burning of fossil fuels and forest clearance ('human acts' in Figure 10.3). The effect of these activities is to transfer some of the carbon in the rocks and on the land into the atmosphere, as CO_2. The rate of rise in Figure 10.4 is only about half of the rate of

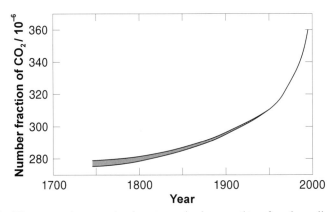

Fig. 10.4   The recent increase in the atmospheric quantity of carbon dioxide

release by human activities. This is because the rate of dissolution in the oceans and the rate of photosynthesis have increased in partial compensation.

The main concern about the increase in atmospheric CO_2 is that it is an important greenhouse gas (Figure 10.2) and therefore the Earth's surface temperature will rise if nothing else changes in compensation. There is good evidence that the Earth's global mean surface temperature has risen by about 0.7 °C in the last 100 years, and at least some of this is the result of the increase in atmospheric CO_2.

Oxygen and nitrogen

Photosynthesis is the main source of the oxygen O_2 in the Earth's atmosphere. A much smaller contribution comes from the photodissociation of water vapour. Therefore, if respiration and decay completely reversed the effect of photosynthesis, then the number of molecules of oxygen in the atmosphere and oceans would be nearly the same as the number of carbon atoms in the biosphere. In fact there is *far* more oxygen than this in the atmosphere, and an even greater quantity is dissolved in the oceans (mainly combined with sulphur, in the ion SO_4^{2-}). This is a consequence of the formation of the organic carbon component of rocks. From carbon stored in this way a huge amount of oxygen has been released from CO_2.

And yet most of the oxygen corresponding to the organic carbon in the rocks is not in the atmosphere and oceans. Over the ages most of the O_2 liberated from CO_2 has combined with oxygen-poor volcanic gases, and with freshly exposed oxygen-poor minerals in crustal rocks. This includes crustal rocks that have become (more) oxygen-poor as a result of oxygen ions supplied to the ocean in the formation of carbonates. As a result, the total mass of oxygen released through the formation of the organic carbon component of rocks is split today between the atmosphere, oceans, and surface rocks in the approximate ratios 1:2:40.

The biosphere and the oceans also influence the nitrogen content of the atmosphere. As a result, the nitrogen is now roughly equally split between the atmosphere, where it is present as N_2, and sedimentary rocks (and their metamorphic products), where it is stored mainly as nitrates (NO_3 group(s) in the mineral) and nitrites (NO_2 group(s) in the mineral). There is only a small amount of nitrogen in the biosphere and it is obtained from the atmosphere. Without the biosphere and oceans the atmospheric mass of nitrogen (as N_2) would be nearly double the present value. The quantity stored in nitrates and nitrites would be correspondingly small, sustained by electrical storms in the atmosphere. Without the biosphere but with the oceans nearly all the nitrogen would be in the oceans, in the form of the nitrate ion NO_3^-.

Question 10.2

Suppose that the Earth's biosphere is entirely destroyed, and that sufficient time passes for the Earth's atmosphere to reach a new steady-state composition.

(a) Explain what this new composition might be in terms of the major overall constituents.
(b) Why is it not possible for you to estimate the change in the Earth's mean surface temperature?

(c) Why would the solar UV radiation levels at the Earth's surface be greater?

10.1.3 Atmospheric Circulation

The atmospheric circulation in the Earth's troposphere, averaged over time, is shown in Figure 10.5. There are three cells per hemisphere. The ones nearest the equator are Hadley cells (Section 9.2.6). They extend to about latitudes 30° north and south, and thus cover about half of the Earth's surface. They are stabilised in latitude against seasonal changes by the slow thermal response of the oceans. Disruption by wave-like disturbances at mid-latitudes prevents them reaching higher latitudes. The polar cells are less prominent and less permanent. The mid-latitude cells are also impermanent, and are artefacts of averaging around the Earth various complex patterns, including waves. They are *not* Hadley cells — for example, the circulation is in the *opposite* direction to that conforming to the decrease of temperature with latitude. At mid-latitudes there is the system of cyclones and anticyclones shown in Figure 9.15. The Coriolis effect modifies the circulation within each cell by deflecting the north–south flow in the east–west direction, to give a diagonal flow. The resulting winds at the surface are shown schematically in Figure 10.5. In the tropics the winds are called the trade winds, and were utilised by sailing ships.

Wind directions are modified by friction between the atmosphere and the surface, and by other disturbances such as mountain ranges and continents. They are considerably modified by the oceans. The oceans also have circulation patterns, strongly influenced by the configuration of the ocean basins. Oceans and the atmosphere interact, and so the circulation of the one influences the other. This is a unique feature of the Earth.

10.1.4 Ice Ages

Perhaps the most dramatic aspect of the Earth's climate is the occurrence of ice ages. An **ice age** is characterised by a cooling of mid-latitude and polar regions to the point where polar ice caps, and any deposits of ice elsewhere, spread to give glacial

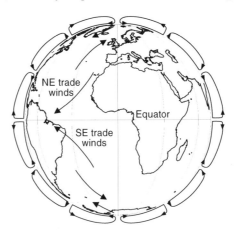

Fig. 10.5 Circulation in the Earth's troposphere. The vertical scale is greatly exaggerated

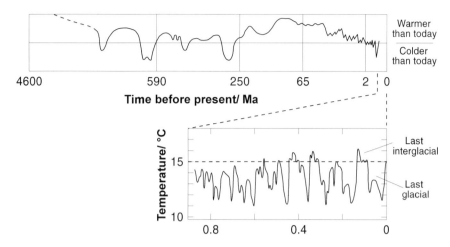

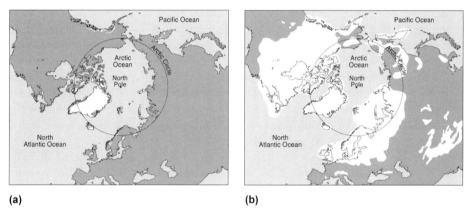

Fig. 10.6 The Earth's global mean surface temperature. The time axis is linear, but with changes of scale. (Adapted from L. A. Frakes, *Climates Throughout Geological Time*, Elsevier, 1979)

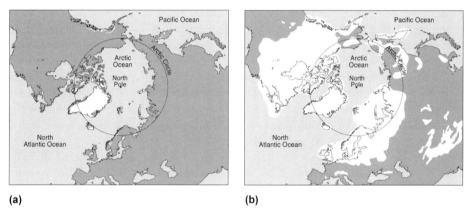

Fig. 10.7 (a) The present extent (white) in the northern hemisphere of permanent ice. (b) The maximum extent at the height of the last glacial period in the current ice age, about 18 000 years ago. (Adapted from B. S. John, *The Ice Age*, Collins, 1977)

conditions down to mid-latitudes. Within an ice age there can be short-lived warmings to give interglacial conditions in which the ice sheets retreat. Figure 10.6 shows the Earth's global mean surface temperature over the past 3000 Ma, inferred from geological and biospheric evidence. There was possibly an ice age around 2300 Ma ago, and another at 1000–600 Ma. There have been others since. We are presently in an ice age that started about 2.4 Ma ago, though in an interglacial period that has so far lasted about 10 000 years. The present extent of the permanent ice is shown in Figure 10.7(a), and the extent at the height of the last glacial period, 18 000 years ago, is shown in Figure 10.7(b).

The transition from glacial to interglacial conditions can be rapid, occupying as little as a century or so. It is thus possible that the current interglacial period could be over in 100 years' time, with permanent ice spreading again to mid-latitudes. However, any incipient glacial conditions could be delayed by human activities.

❐ How could human activities delay the onset of a glacial period?

A delay could result from the burning of fossil fuels and forest clearance. These activities are contributing to an increase in atmospheric carbon dioxide (Figure 10.4), and consequently to an increase in the greenhouse effect.

The causes of ice ages and of interglacials within ice ages have been much disputed, and the matter is not yet clear. There are several factors that have certainly had some influence on climate. The primary source of energy for the Earth's climate system is the Sun, and it is known that the luminosity of the Sun varies by up to about 0.5% on a time scale of centuries. On a longer time scale, stretching back to the origin of the Solar System 4600 Ma ago, there has been a gradual increase in luminosity from about 70% of its present value. Presumably this increase has not been smooth, though the details are unknown.

Further factors are changes in the Earth's orbit and rotation axis. The former arise largely from the gravitational field of Venus (which is nearby) and of Jupiter (which is massive), and the latter from the gravitational fields of the Sun and the Moon acting on the slightly non-spherical mass distribution of the Earth. The important orbital changes are in the eccentricity and in the direction of perihelion (precession of perihelion); the important changes in the rotation axis are in the inclination and in its direction (axial precession — Section 1.5.1). The changes are approximately periodic, the eccentricity ranging from approximately 0 to 0.06 with a predominant period of 93 000 years, and the axial inclination ranging from 22.0° to 24.5° with a predominant period of 41 000 years. Axial and perihelion precession combines to change the season in which perihelion occurs. The Earth's rotation axis precesses with a period of 25 800 years, and so the vernal equinox moves around the orbit with this period. Earth's perihelion precesses (in the opposite direction) with a period of 111 000 years, and so the vernal equinox moves around the orbit with respect to perihelion with a period of 21 000 years (you have to add the inverse of the periods).

None of these orbital and axial changes appreciably alters the mean distance of the Earth from the Sun, but they do alter the distribution of solar radiation across the globe. For example, averaged over a year the polar regions receive more solar radiation at 24° inclination than they do at 22°. The orbital and axial changes also affect seasonal patterns.

Another factor that influences climate is changes in the latitude distribution of the continents, and in the shape of the ocean basins.

❐ What is the cause of these changes?

The cause is plate tectonics (Section 8.1.2), and the time scale is $1–10^3$ Ma. Yet another factor is atmospheric composition. This has changed on time scales ranging from a year (a major volcanic eruption) to thousands of Ma — more on this in Sections 10.5 and 10.6.

The operation of one or more of these factors might not in itself be sufficient to cause the recorded climate changes. They can, however, act as triggers. For example, if the solar luminosity rises very slightly there will be little immediate effect on climate, but there will be some reduction in the ice cover of the Earth. This reduces the planetary albedo and so the mean global temperature rises slightly. This leads to a further retreat of the ice, and so on. This is an example of positive feedback — an initial change in temperature causes a further change in the same direction. The rise will not go on for ever. Ultimately if nothing else halts the feedback, then the total loss of all ice will. There is also negative feedback where an initial change in temperature leads to a *reduction* in the change. Negative feedback has a stabilizing influence, whereas positive feedback has a destabilizing influence. To explain ice ages, glacials, and interglacials, all the above factors plus feedback are probably necessary, different factors predominating on different time scales.

As well as ice ages there are less dramatic variations in climate, many of them driven by small variations in solar luminosity. There are also short-lived dramatic variations. The extinction of the dinosaurs and of many other species 65 Ma ago probably owed much to short-term cooling due to an asteroid impact, as outlined in Section 3.1.2.

Question 10.3

Present a qualitative argument to show that the solar flux density (power per unit area) on the Earth's surface at the poles, averaged over a year, is less when the Earth's axial inclination is zero than when it has its present value. (Assume no change in cloud cover.)

10.2 The Atmosphere of Mars Today

From Table 9.1 you can see that the atmosphere of Mars is much less substantial than that of the Earth.

❑ On a square metre of Martian surface, by what factor is the mass of the Martian atmosphere less than that of the Earth?

This factor is the ratio of the column masses:

$$\frac{1.03 \times 10^4}{0.015 \times 10^1} = 69$$

As well as a much smaller column mass, you can also see that the composition is very different. Whereas the Earth's atmosphere is dominated by nitrogen and oxygen, the Martian atmosphere is dominated by carbon dioxide.

10.2.1 Vertical Structure, Heating and Cooling

Figure 10.8 shows the globally averaged vertical structure of the Martian atmosphere when its dust content is low. In the kilometre or so near the ground there are huge

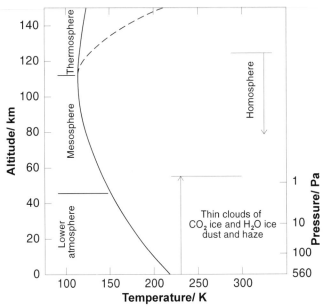

Fig. 10.8 The globally averaged vertical structure of the Martian atmosphere when its dust content is low. The two curves in the thermosphere correspond to extremes of low and high solar activity

diurnal variations, with temperatures at night plunging on average by roughly 100 °C below those shown in Figure 10.8. A major difference from the Earth (Figure 10.1) is that there is no mesospheric bulge of temperature. This is a direct consequence of the near-absence of ozone, itself a consequence of the very small quantity of oxygen. The lapse rate in most of the troposphere is close to the dry adiabatic value.

❐ What does the 'dry' mean?

This is the lapse rate when there is no condensation. In fact, water vapour does condense, and thin clouds of water ice can form. These are rather like cirrus, and not the billowy sort. But the number fraction of water vapour is only 0.0002, and so the latent heat release is slight. A more serious disturbance of the adiabatic lapse rate would be the condensation of carbon dioxide, but its low condensation temperature restricts its condensation to the coldest regions of the atmosphere. Some Martian clouds are visible in Figure 7.11 (Section 7.3.4).

In spite of the thin atmosphere, there is so much fine dust at the Martian surface that surface winds (up to the order of $10 \, \text{m s}^{-1}$) often raise enough of it into the lower atmosphere to change the lapse rate — absorption of solar radiation by the dust makes the lapse rate smaller. There is then little convection and it is more appropriate to refer to the 'lower atmosphere' rather than to the 'troposphere'. The dust consists of micron-sized red particles, that give a reddish tint to the Martian sky (Plate 9). Occasionally, global dust storms occur.

The Martian global mean surface temperature is considerably lower than that of the Earth, 218 K instead of 288 K. Though Mars has a lower planetary albedo than the Earth, it is at a greater distance from the Sun, and has a greenhouse effect of only about 8 °C, compared to about 33 °C for the Earth. The smaller greenhouse effect is largely a consequence of the much lower water vapour content on Mars — the column mass of CO_2 is actually *greater* than it is in the Earth's atmosphere.

10.2.2 Atmospheric Reservoirs, Gains, and Losses

The dryness of the Martian atmosphere is the result of the low temperatures — the atmosphere is in fact close to saturation. It is thus to be expected that atmospheric water vapour is in equilibrium with surface deposits. There is certainly one such deposit, and it is substantial — the northern permanent polar cap — and there is ample evidence for other surface deposits too. This evidence includes the water-related surface features described in Section 7.3.6, and the observations at the surface outlined in Section 7.3.7. The Martian meteorites contain rather modest proportions of water, though it is not possible to infer from this anything very useful about the water content of the regions from where they came. There is a wide range of estimates of the present quantity of water at and near the surface of Mars. These are usually given in terms of the uniform depth to which the water would cover the entire planet if the water were all liquid. For the Earth the total in the (near) surface and atmospheric reservoirs corresponds to a depth of 2700 m, whereas for Mars the estimates range from several tens of metres to a few kilometres. Even the smallest estimates are far larger than the equivalent of 10^{-5} m presently in the Martian atmosphere.

The major constituent of the Martian atmosphere, carbon dioxide, also condenses onto the surface, notably at the coldest parts. You saw in Section 7.3.5 that the seasonal caps at both poles on Mars consist largely of carbon dioxide ice, and that this is also the case for the permanent cap at the south pole (perhaps underlain by water ice). Carbon dioxide frosts also occur. This is unlike the major constituents of the Earth's atmosphere, where it never gets cold enough for oxygen and nitrogen to condense. Additionally, carbon dioxide can be locked up in water ice as a clathrate, and adsorbed onto clay minerals. It can also be present in carbonates, and though Earth-based spectrometry of atmospheric dust has provided inconclusive evidence for carbonates, all Martian meteorites contain Martian carbonates. However, unless Mars was warmer and wetter in the distant past, to the extent that oceans or large open bodies of liquid water existed, then extensive carbonate deposits are unlikely to have formed. Support for this view is the atmospheric ratio of CO_2 to N_2. This ratio is far higher than in the Earth's atmosphere (Table 9.1). The Martian value would be lower if, as on Earth, most of the CO_2 had gone to form carbonates. Estimates of the present quantity of carbon dioxide in some form in Martian (near) surface reservoirs range from the order of 10^2 times the amount in the atmosphere, to far greater quantities.

Atmospheric carbon dioxide is photodissociated at a high rate by solar UV radiation — the atmosphere is thin and almost devoid of ozone, so UV radiation can do its damage through the whole depth of the atmosphere. If there were no regeneration of carbon dioxide then it would be almost entirely photodissociated in 3000 years! That this has not happened is the result of various chemical reactions in the lower atmosphere that put the molecule back together again. Of crucial importance in these reactions is the trace of atmospheric water.

10.2.3 Atmospheric Circulation

Figure 10.9 shows three important modes of atmospheric circulation in the lower atmosphere of Mars. The contrast with the Earth is striking (Figure 10.5). With a rotation rate and axial inclination similar to the Earth, a similar type of circulation might be expected. However, the absence of oceans on Mars leads to a rapid change of surface temperature in response to seasonal changes in insolation. Therefore, in the summer the hottest surface is not at the equator but at the subsolar point. As a consequence there is a single Hadley cell stretching from the subsolar latitude across the equator to the less heated hemisphere (Figure 10.9(a)). In this hemisphere at mid-latitudes there are frequent cyclones and anticyclones. Around the equinoxes the Hadley cell breaks down.

In addition to the Hadley cell there is **condensation flow** (Figure 10.9(b)). This arises from the condensation of carbon dioxide at high latitudes in the winter hemisphere, notably at the seasonal polar cap. The associated reduction in pressure draws the atmosphere towards this region.

❐ Why is condensation flow unimportant in the Earth's atmosphere?

Condensation of a major atmospheric constituent does not occur on the Earth. Another component of the circulation is a **thermal tide** (Figure 10.9(c)). This is the result of the rapid cooling of the Martian atmosphere at night, leading to low pressure and a consequent flow from the warmer and consequently higher-pressure daytime hemisphere. Thermal tides occur on the Earth, but the thin atmosphere of Mars, dominated by the efficient infrared radiator carbon dioxide, leads to far greater night-time falls in tropospheric temperatures, typically by 100 °C. This leads to far stronger flow on Mars. Among other thermal effects there is a correlation with topography, with downhill drainage of cold air at night, and uphill ascent during the day.

10.2.4 Climate Change

In Section 7.3.6 it was noted that many water-related features on Mars indicate a warmer, wetter climate in the distant past. This will be considered in Section 10.6.3. You also saw (Section 7.3.5) that the polar deposits suggest an intricate series of more recent climate changes. It is thought that changes in the axial inclination of

(a)   (b)   (c)

Fig. 10.9 Atmospheric circulation in the lower atmosphere of Mars. (a) Hadley cell. (b) Condensation flow. (c) Thermal flow. (Adapted by permission from Figure 5 (p. 97) of *The New Solar System*, J. K. Beatty and A. Chaikin (eds), Sky Publishing, 1990)

Mars have been particularly important in the climatic changes reflected in the polar deposits.

Calculations of these changes extend to about 10 Ma ago, and show that the axial inclination has ranged from 13° to 42° — far greater than the 22° to 24.5° range for the Earth. This is because the Earth's axis is stabilised by the Moon. The Martian changes are quasi-periodic, with an underlying period of 0.12 Ma, amplitude modulated with a period of 2 Ma. At low axial inclinations the total annual solar radiation in equatorial regions is much greater than in polar regions, but the totals become more equal as the inclination increases (Question 10.3). The greater high-latitude temperatures at large inclinations have the effect of reducing the quantities of carbon dioxide and water in the polar reservoirs, and increasing those in the equatorial and atmospheric reservoirs. The net effect is an increase in atmospheric mass, and this is expected to increase the frequency of dust storms, which increases the dust content of the atmosphere and promotes deposition at the poles. A competing effect is the depletion of the polar reservoirs and a possible increase in wind scouring there. Thus, as the axial inclination varies with its two characteristic periods, there is a complicated interplay of erosion, deposition, and the dust/water ratio in the atmosphere.

Additionally, there is a 51 000-year cycle in the season at which perihelion occurs — the combined effect of axial and perihelion precession, and it is likely that the dust/water ratio in the atmosphere has also been changed by volcanic eruptions and by impacts. These effects would have been particularly significant on shorter time scales. Combined with the changes in axial inclination, it is plausible that they have led to the patterns of layering, deposition, and erosion seen in the polar deposits. Swings in climate across the globe are a necessary accompaniment to the changes reflected in the polar deposits.

Question 10.4

(a) Starting at midsummer in the southern hemisphere on Mars, describe the changes in the location of the Hadley cell as Mars goes around its orbit.
(b) Describe how the location of the Hadley cell at midsummer in the southern hemisphere changes as the axial inclination of Mars varies.

10.3 The Atmosphere of Venus Today

For a planet that has a mass, size and internal structure not very different from those of the Earth, the atmosphere of Venus is astonishingly different from ours. Like the Martian atmosphere it is dominated by carbon dioxide (Table 9.1), but the column mass is $102 \times 10^4\,\mathrm{kg\,m^{-2}}$ compared to $0.015 \times 10^4\,\mathrm{kg\,m^{-2}}$ for Mars and $1.03 \times 10^4\,\mathrm{kg\,m^{-2}}$ for the Earth.

10.3.1 Vertical Structure, Heating and Cooling

Figure 10.10 shows the globally averaged vertical structure of the Venusian atmosphere. Like the Martian atmosphere it has no mesospheric bulge of temperature.

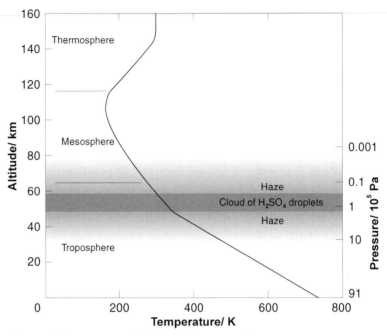

Fig. 10.10   The vertical structure of the Venusian atmosphere

❏ Why is this?

Again, this is because there is no strong absorber of solar radiation concentrated at these levels. The lapse rate is invariably adiabatic in the lower 35 km or so of the troposphere but from 35 km up to the cloud base at about 48 km it can be sub-adiabatic. This is because at these higher altitudes the atmosphere is more transparent to infrared wavelengths, and so can more readily exchange heat radiatively.

The main cloud deck extends from about 48–58 km, completely shrouding the planet all the time. The clouds consist primarily of droplets of sulphuric acid 1–10 μm in diameter. Above this deck there is a haze of 1–3 μm droplets, and below the deck a haze of 1–2 μm droplets with a sharp cut-off at 31 km. The haze droplets are also dominated by sulphuric acid. It seems that the droplets are formed at around 80 km altitude, the sulphuric acid (H_2SO_4) being produced by photochemical reactions involving minor atmospheric constituents including SO_2, the sulphur initially entering the atmosphere in the form of sulphur vapour as a minor component of volcanic gases. Once formed, the droplets descend to the upper tropopause where they are kept aloft by convection to form the main cloud deck. The droplets leak downwards, particularly those that grow large, and as they descend below the cloud deck they gradually evaporate in the ever-increasing temperatures, with none surviving below 31 km.

Below 31 km the atmosphere is very clear, and the temperature continues to increase to reach an astonishing mean value of 735 K at ground level. The effective temperature is 229 K, and so the magnitude of the greenhouse effect is an astonishing 500 °C or so. The greenhouse effect on Venus is sustained by the combined effects of carbon dioxide, water vapour, sulphur dioxide, and sulphuric acid droplets. These substances, plus the huge atmospheric pressure that broadens the spectral lines, make the atmosphere strongly absorbing over a wide range of mid-infrared wavelengths.

10.3.2 Atmospheric Reservoirs, Gains, and Losses

The atmosphere is very probably the main reservoir of carbon on Venus today. This is unlike the Earth where most of the carbon is in organic carbon deposits and in sedimentary carbonate rocks. The absence of oceans on Venus means that the carbonate generation rate is extremely low, whereas volcanism and high surface temperatures will gradually have destroyed any ancient sedimentary carbonate deposits and liberated CO_2 into the atmosphere. Support for this conclusion is the approximately similar mass of carbon in the Venusian atmosphere to that in all of the Earth's reservoirs. This is to be expected from their proximity in the Solar System.

A similar argument applies to nitrogen. The quantity in Venus's atmosphere is somewhat greater than that in the Earth's atmosphere, but the Earth also has nitrate and nitrite deposits, and when these are included, the two quantities are not so very different.

❏ Why are extensive deposits of nitrates and nitrites on Venus unlikely?

Nitrate and nitrite formation is promoted on Earth by the oceans and biosphere. It is possible that there were short-lived oceans early in Venusian history, but it is very unlikely that there was ever a biosphere. The widespread view is that there are no extensive deposits of nitrates and nitrites on Venus, and that consequently most of Venus's nitrogen is in the atmosphere.

With the global total quantities of carbon dioxide and nitrogen on Venus being similar to those on the Earth, it might be expected that this would also be the case for water. This is not so. The number fraction in the Venusian atmosphere is about 4×10^{-5}, and it is likely that the surface is far too hot for much water to be present, even if were bound to minerals. The quantity of water deep in the crust and in the mantle is unknown, but is unlikely to exceed terrestrial quantities. Excluding potential deep reservoirs in both planets, if all the Venusian water is in the atmosphere then Venus has about 10^{-5} times the water known on Earth! However, as for Mars, the trace of atmospheric water plays a role in reversing the effects of the photodissociation of CO_2.

The present low water content in the Venusian atmosphere could represent a balance between the gains from slow outgassing plus the occasional impact of comets and asteroids, and a slow rate of loss by photodissociation. Quite why there is so little water is a topic for Section 10.5.

10.3.3 Atmospheric Circulation

On such a slowly rotating planet with a small axial inclination, a single Hadley cell extending from equator to pole in each hemisphere is to be expected (Figure 9.13). There is evidence that such cells exists in the lower atmosphere. However, above these cells there are others, and the reasons for this are not well understood. There is also a thermal tide (Figure 10.9(c)), and though this is not as significant a component of the circulation as it is on Mars, the mass of the Venusian atmosphere is so great that the thermal tide could have influenced the rotation of the planet. Also, the associated mass redistribution leads to an extra torque exerted by the tidal force of the Sun, and this too influences planetary rotation. The net accumulated effect of the atmosphere on the rotation rate of Venus is unclear, though some astronomers have speculated that the atmosphere is largely responsible for the present slow retrograde rotation of Venus.

The Hadley circulation is apparent in the $5–10\,\mathrm{m\,s^{-1}}$ winds from equator to pole at the cloud tops. At the surface, due to friction between the atmosphere and the surface, the pole to equator flow is less than $1\,\mathrm{m\,s^{-1}}$, but transfers heat sufficiently rapidly to reduce equator-to-pole temperature differences to only about $10\,^\circ\mathrm{C}$. Surface friction also causes a low-altitude east–west wind speed of about $1\,\mathrm{m\,s^{-1}}$. This east-to-west drift increases with altitude becoming about $100\,\mathrm{m\,s^{-1}}$ at the cloud tops, particularly at high latitudes. These high speeds can be explained as a result of the upward transfer of angular momentum from the lower atmosphere, though the details are complicated. Figure 10.11 shows the combined effect of the north–south and east–west drifts at the cloud tops. This image is at ultraviolet wavelengths to enhance cloud features — the dark places are *not* breaks in the cloud.

Question 10.5

If *no* solar radiation penetrated the clouds of Venus, why would the lapse rate below the clouds be close to zero?

Fig. 10.11   Atmospheric circulation at cloud top level on Venus. (NASA/NSSDC 79HC222)

10.4 Volatile Inventories for the Terrestrial Planets

You have seen that the volatile substances that constitute a planet's atmosphere might also be present at or under the surface of the planet, perhaps in a different chemical form. In considering the origin of atmospheres we must consider the origin of volatiles totalled over all these possible reservoirs, and an essential first step is to attempt a volatile inventory for the present day. This inventory is then a basis for estimating the inventory at earlier times.

Figure 10.12 shows the *present* volatile inventories of water, carbon, and nitrogen for Venus, the Earth, and Mars. The quantities are the **global mass fractions** of CO_2, N_2, and H_2O, i.e. the total mass of each substance divided by the total mass of the planet. The quantities in other forms, such as carbon in carbonates, nitrogen in nitrates, water as hydroxyl (OH), are included by adjusting them to the masses of CO_2, N_2, and H_2O that these other forms would yield.

❏ Why do the major volatiles consist of hydrogen, oxygen, carbon, and nitrogen?

These are cosmically abundant elements, and they form relatively volatile compounds. (Note that volatiles are *not* the major repositories of oxygen. Most of a planet's oxygen is present in silicates and metal oxides. By contrast, only a small fraction of the carbon and nitrogen is *not* in the volatile reservoirs.)

The atmospheric quantities in Figure 10.12 are well known, but you have seen that the quantities at and beneath the surfaces are far less certain. Consequently, the global mass fractions shown in Figure 10.12 are lower limits, and in many cases the actual quantities could be very much greater. This is true even for the Earth — the

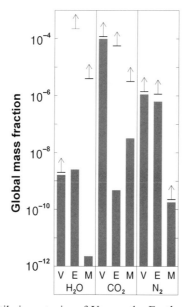

Fig. 10.12 Present volatile inventories of Venus, the Earth, and Mars. The bars are for the atmospheres, and the lines with arrows are the lower limits for the global mass fractions

quantities in Figure 10.12 are for the surface and crust. For example, there is also water in the upper mantle, but estimates vary from 60 ppm to over 200 ppm by mass, and even a trace in the greater volume of the lower mantle could add hugely to the inventory.

This is an unpropitious start! Nevertheless, Figure 10.12 indicates that the present volatile endowments of Venus, the Earth, and Mars, are very different. But what about the past? Were the inventories more similar then?

Tracing volatile inventories into the past

Any attempt to trace volatile inventories into the past must include a consideration of losses to space. There is strong evidence that Mars has lost to space a considerable proportion of its nitrogen. The evidence is in the atmospheric isotope ratio $n(^{14}N)/n(^{15}N)$, where $n(^{14}N)$ is the number density of atoms of the common isotope ^{14}N in the atmosphere, and $n(^{15}N)$ is the number density of ^{15}N atoms. Each number density includes the isotope present as single atoms and contained in the molecule N_2. In the Earth's atmosphere today $n(^{14}N)/n(^{15}N) = 272$, whereas in the Martian atmosphere it is about 170. The accepted explanation is the greater rate at which the less massive isotope has escaped from Mars, thus leading to its relative depletion there.

Nitrogen has been lost from Mars mainly by chemical escape, for example by the photochemical reaction

$$N_2(+UV \text{ photon}) \longrightarrow N + N \tag{10.4}$$

where the UV photon is provided by solar radiation. This reaction is only one of several, but in all of them the resulting nitrogen atoms are boosted to sufficient speeds to escape at a far greater rate than they do by thermal escape. The lighter isotope is lost at a faster rate because above the homopause the atmosphere is no longer well mixed, and the number of lighter N_2 molecules decreases less rapidly with altitude than does the number of heavier molecules. This leads to enrichment of the lighter isotope in the exosphere, from where chemical escape occurs. If the Earth and Mars started out with the same $n(^{14}N)/n(^{15}N)$ ratio, then it can be shown that the present difference in the ratios indicates that Mars has lost to space the order of 10^2–10^3 times its present atmospheric content of nitrogen.

❒ Why has the Earth lost little nitrogen through chemical escape?

Chemical escape of nitrogen from Earth has been slight because of the Earth's higher escape speed (Table 9.1).

Escape to space can also go some way towards explaining the dryness of Venus. The explanation starts with photodissociation of water in the upper atmosphere. The net effect of a series of reactions is

$$2H_2O(+UV \text{ photon}) \longrightarrow 2H_2 + O_2 \tag{10.5}$$

yielding hydrogen and oxygen molecules as gaseous components of the upper atmosphere. H_2 thermally escapes at only a low rate from Venus today because the base of the exosphere is cold, but in the past the upper atmosphere could have been warmer. Also, there was then a greater solar UV flux and a more copious solar wind, thus enhancing the rate of escape. Furthermore, though the exosphere is too cool today for thermal escape, loss of hydrogen might still be occurring through the action of electric and magnetic fields in the solar wind. These fields could accelerate hydrogen ions in the exosphere to escape speeds, the ions having been formed by photodissociation. It is thus conceivable that large quantities of hydrogen have been lost to space.

We must also get rid of Venus's oxygen—there would have been far more produced than the very small upper limit that we have for the atmosphere today.

❐ Why is thermal escape inadequate?

The mass of the O_2 molecule is much greater than that of H_2, so the molecular speeds of O_2 would have been much lower than those of H_2, and the rate of thermal escape correspondingly small (Figure 9.12, Section 9.2.5). It is possible that the oxygen has combined with surface materials, and some evidence for this is provided by the lander Venera 13, which found highly oxidised surface rocks.

The series of reactions summarised by equation (10.5) also occurs on Earth, but most terrestrial water is in the oceans and in other surface reservoirs, where it is protected from dissociation. Consequently, the Earth has lost little of its initial endowment of water. It was the hot troposphere of Venus and the consequent lack of precipitation that allowed water to reach high altitudes, thus exposing it to rapid photodissociation.

Support for this water loss mechanism from Venus is provided by the present-day isotope ratios $n(^2H)/n(^1H)$ for Venus and the Earth. The lower mass of ^1H enables its molecules and ions to escape to space at a greater rate than molecules and ions that include ^2H, and therefore on both planets $n(^2H)/n(^1H)$ tends to increase with the passage of time. Atmospheric and surface reservoirs exchange water, and with most terrestrial water residing in the surface reservoirs, protected from dissociation, this tends to maintain the terrestrial $n(^2H)/n(^1H)$ ratio at its initial value. On Venus there have been no oceans for a long time, if ever. Therefore, we would expect the $n(^2H)/n(^1H)$ ratio on Venus to be greater than on the Earth, and this is exactly what we find—about 120 times the terrestrial value of 1.6×10^{-4}.

With detailed models it is possible to estimate the amount of water that must have been lost to give such an enhancement. These models include the greater UV flux in the past. If the initial ratio of $n(^2H)/n(^1H)$ was much the same on both planets, and if the atmosphere of Venus has not been resupplied with water, then this loss process has got rid of the order of 10^2–10^3 times the mass of water presently in the atmosphere. This loss is equivalent to a global layer of liquid water of order 3–30 m.

The hydrogen isotope ratio $n(^2H)/n(^1H)$ on Mars also indicates loss to space—it is five times greater than that of the Earth.

The models that predict the losses to space make various assumptions, and therefore in tracing volatile inventories into the past there are further uncertainties to add to those about the present inventories. Moreover, processes that were concentrated early in Solar System history, such as impact erosion, blow-off, and hydrodynamic escape (Section 9.2.5) probably removed huge quantities of volatiles.

Therefore, we must now start at the earliest times and try to work forward, under the constraint that there has to be an acceptable match with what we can establish by working back.

Question 10.6

If Venus has lost to space 10^2–10^3 times the mass of water presently in its atmosphere, what would its global mass fraction of water have been before this loss, and how does this compare with the global mass fraction of water on Earth?

10.5 The Origin and Early Evolution of Terrestrial Atmospheres

Important evidence about the earliest times is provided by the inert gases.

10.5.1 Inert Gas Evidence

The complete suite of inert gases is helium (He), neon (Ne), argon (Ar), krypton (Kr), xenon (Xe), and radon (Ra). Recall that they are called inert because they are chemically unreactive, and they are called gases because at all but very high pressures or very low temperatures they exist in the gaseous phase — they are extremely volatile.

❑ Would you expect argon gas to be Ar or Ar_2?

In accord with their chemical unreactivity, the inert gases are present in atomic form, rather than combined into molecules.

Of particular importance here are the inert gas isotopes that are stable (not radioactive), and that have not been produced by radioactive decay — such radiogenic isotopes will have had their atmospheric quantities increased over time in some uncertain manner. All of radon's isotopes are unstable, and the non-radiogenic isotopes of helium are excluded because the atoms are light enough for helium to escape from the terrestrial planets. By contrast, a high and similar proportion of the other isotopes will have been retained. Moreover, the unreactivity and extreme volatility of the inert gases make it likely that nearly all of the initial endowment of each these isotopes is in the atmosphere. The global mass fractions of the atmospheric isotopes should therefore bear a primordial imprint. The non-radiogenic isotopes used most often in studies of planetary volatiles are ^{20}Ne, ^{36}Ar, ^{84}Kr, and ^{132}Xe.

Figure 10.13 shows the present-day global mass fractions of ^{20}Ne, ^{36}Ar, ^{84}Kr, and ^{132}Xe in the atmospheres of Venus, the Earth, and Mars. Also included are the values for the Sun, which is thought to have a composition similar to the nebula from which the planets formed. A striking feature of Figure 10.13 is that the graphs for the terrestrial planets have very *different* shapes from that for the Sun. Greater similarity between the Sun and the planets would be expected if a high proportion of today's planetary volatiles was captured from the solar nebula as a whole. This expectation

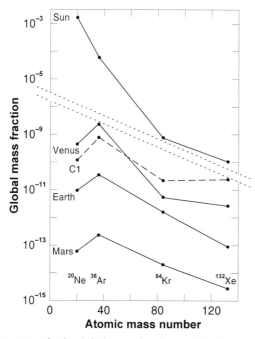

Fig. 10.13 The atmospheric global mass fractions of the important non-radiogenic inert gas isotopes, for Venus, the Earth, and Mars. The mass fractions for the Sun, and for the C1 carbonaceous chondrites are also shown

remains even allowing for subsequent mass-selective loss mechanisms such as thermal escape and hydrodynamic escape. Therefore, if the terrestrial planets ever did capture large quantities of volatiles from nebular gas and dust (mainly from the gas) then they have since been lost.

At a time estimated to be late in the formation of the terrestrial planets, the Sun very probably went through a T Tauri phase (Section 2.1.3). This was marked by a more copious solar wind and greater UV luminosity. It probably lasted about 10 Ma, long enough to scour the terrestrial planets of volatiles, partly through the ejection of upper atmosphere molecules by the impact of solar wind protons, and partly by the UV photodissociation of H_2 in the upper atmosphere, each H atom acquiring high non-thermal speeds. Some of these atoms escaped; others collided with other atoms and molecules, in some cases causing them to escape. The solar wind and the UV radiation also raised the temperature of the upper atmosphere, which increased the escape rate. If the abundance of hydrogen atoms became sufficiently high then their net outward flow would have caused hydrodynamic escape of other gases. The T Tauri phase also blew the remnant nebular gas out of the Solar System, and so after the T Tauri phase ended there was no gas left for recapture. Other loss mechanisms are described in Section 10.5.3.

The fifth graph in Figure 10.13 is for the C1 chondrites. These are a particularly primitive sub-class of the carbonaceous chondrite meteorites (Section 3.3.2), and are thought to have a broadly similar chemical composition to the planetesimals from which the terrestrial planets formed. Moreover, these inert gases would *not* have been lost to space when the planetesimals impacted the growing planet. In contrast to the Sun, the shape of the graph for the C1 chondrites is not very different from the shapes for the terrestrial planets. It is therefore an easy step to conclude that the terrestrial volatiles must have been delivered within the planetesimals. The C1 chondrites certainly contain the right sort of substances: they have water, mostly in hydrated minerals such as serpentine ($3MgO.2SiO_2.2H_2O$); and they have carbon and nitrogen, mostly within organic compounds. Furthermore, the non-volatile composition of comets is not very different from that of the C1 chondrites, and so if this broad similarity extends to the inert-gas profile then one cannot rule out icy–rocky planetesimals as important sources too.

Though the shapes of the graphs for the planets and the C1 chondrites are similar, they are not identical. Also, the graphs are separated by large differences in the values of the global mass fractions. We shall shortly explore these features, but first we need to consider more closely the acquisition of volatiles from planetesimals and other bodies during planet formation.

10.5.2 Volatile Acquisition during Planet Formation

The initial volatile endowment of the terrestrial planets is inextricably linked with the formation of the terrestrial planets as a whole. The solar nebular theory of their formation was outlined in Chapter 2, and we can pick up the story at the point where the inner Solar System was full of a swarm of planetesimals. The volatile content of the planetesimals at the distance where Mars will form was probably not much greater than that in the Venus region.

❐ Why should it have been any greater at all?

Some increase in volatile content with heliocentric distance is to be expected from the associated decline in nebular temperatures. Subsequently, with the growth of planetary embryos, close encounters between one embryo and another and between an embryo and a planetesimal made many orbits highly elliptical. Therefore, embryos and planetesimals began to criss-cross the terrestrial and asteroid zones, blurring any compositional differences.

By the time that Venus, the Earth, and Mars had grown to about 10% of the Earth's present mass ($0.1\ M_E$), the impacts of planetesimals were sufficiently violent for the planetesimals to be fully devolatilised. This resulted in massive atmospheres, mainly of H_2O, which is the most abundant volatile, but with CO_2, NH_3, SO_2 and other volatiles present also. Impact devolatilisation is expected to have been less overall for Mars because $0.1\ M_E$ is about the same as the planet's present mass. Nevertheless, the later arriving material would certainly have been devolatilised, and this would have created a considerable atmosphere on Mars too.

The appearance of massive water atmospheres led to large rises in surface temperature. This is partly because of the associated large greenhouse effect, and partly because of aerosols that not only added to the greenhouse effect but also absorbed solar radiation. With most of the kinetic energy of the impactors converted to heat at the hot surfaces, a magma ocean at about 1500 K was formed on Venus and the Earth (Sections 8.1, 8.2), but perhaps not on Mars. Mars might have lacked a magma ocean partly because the greater heliocentric distance would have reduced the solar heating of the surface, and partly because of lower impact speeds — the combined result of lower orbital speeds at the greater distance and the lower gravitational field of Mars. On Venus and the Earth, water dissolved in the magma oceans, stabilising the atmospheric mass at the order of 10^{21} kg. With a general decline in impacts the magma oceans solidified, giving up some of the water and other dissolved volatiles as they did so. In the case of the Earth most of the atmospheric water condensed to form oceans, but this might not have happened on Venus because of the greater solar heating closer to the Sun. If no magma ocean appeared on Mars, and if there was little subsequent geological activity that folded the surface into the mantle, then the mantle might be dry, with any condensed water confined near to the surface.

Whatever the earliest volatile inventories, they were subject early on to enormous modifications by various processes.

10.5.3 Massive Losses

Whereas the smaller planetesimals brought a net increase of the volatile endowment, larger planetesimals caused impact erosion, and the very largest and any left-over embryos would have caused blow-off (Section 9.2.5). Not all volatiles were equally affected. Water can condense or combine with surface materials more readily than the other common volatiles, and can thus obtain some protection. This can help to explain the dryness of Venus, where transfer to the surface was limited by the high surface temperatures resulting from the proximity to the Sun. The extent to which CO_2 was protected depends largely on whether any oceans of water appeared quickly enough for dissolution and carbonate formation to occur before extensive impact erosion.

Impact erosion can also explain the low global mass fractions of the inert gases in the Martian atmosphere (Figure 10.13) — the low mass of Mars and its proximity to the asteroid belt resulted in the loss of a much greater proportion of these gases. This was also the case for other volatiles, and can explain the probable scarcity of carbonates on Mars. It is also possible that impact erosion accounts for the discrepancy between the large quantities of water that seem to be needed to explain the Noachian fluvial features (always supposing that they were not caused by the cycling of a small quantity of water) and the far smaller quantities obtained from models that work back from the present $n(^2H)/n(^1H)$ ratio. These models invoke thermal escape, and therefore trace events back only to the end of impact erosion, and thus do not include processes that were important early in Solar System history but not since.

The other major early process is hydrodynamic escape (Section 9.2.5). Any iron in a magma ocean would react with water to generate hydrogen. The simplest reaction is

$$Fe + H_2O \longrightarrow FeO + H_2 \qquad (10.6)$$

where the water might be present as H_2O or chemically combined with minerals. Water in hydroxyl form undergoes an equivalent reaction. Hydrogen would have been generated in a similar way during core formation as liquid iron trickled through the mantle. The hydrogen built up in the atmosphere, and little water was left in the crust and mantle. Additional hydrogen in a water-rich atmosphere would come from the photodissociation of water. But regardless of how it was generated, the presence of large quantities of hydrogen in the atmosphere led at once to a large outflux of hydrogen to space.

❏ Why is this?

Hydrogen as H or H_2 has a low mass, so readily suffers thermal escape. Other molecules are entrained and so are lost too. Recall that the process is mass-selective and would have led to an increase in the $n(^2H)/n(^1H)$ ratio.

One entrained molecule was O_2 from the photodissociation of water, though models indicate that on Venus, Earth, and Mars most of the O_2 was retained, and was removed later by the oxidation of rocks. If there was sufficient CO, it too would have taken up a significant mass of O_2 in being oxidised to CO_2.

It seems likely that all the terrestrial planets suffered huge losses through hydrodynamic escape, blow-off, T Tauri activity, and impact erosion. Blow-off and T Tauri activity would have been most likely about 4600 Ma ago, and hydrodynamic escape before about 4200 Ma ago. Impact erosion would have been in step with the heavy bombardment, so would have been in steep decline about 3900 Ma ago. Almost entire atmospheres could have been removed, in which case the present atmospheres must largely be the result of subsequent outgassing, or of late veneers.

10.5.4 Late Veneers

With the growth of the giant planets, it is expected that icy planetesimals (comets) and other volatile-rich bodies were thrown across the terrestrial zone, and some were captured by the then (nearly) fully formed terrestrial planets, thus providing them with a late veneer. Even if these planets had been devoid of volatiles, this veneer could probably have provided them with considerable quantities. Subsequent geological activity would have buried much of this veneer into the mantle, except perhaps on Mars because of its lower level of activity. Today, the rate of acquisition of volatiles by this means is very low.

❏ Why did these bodies not cause impact erosion?

They were small bodies.

Evidence that at least some of the water on Mars came as a late veneer is provided by the Martian meteorites. Oxygen isotope ratios in some of the water are non-terrestrial. But the ratios are different from those in the oxygen in the silicate components. The water presumably came largely from the surface, whereas the silicates were derived from the mantle. It thus seems that the mantle and surface have had rather separate histories.

Further evidence that there was a late veneer and that it was due at least in part to icy planetesimals is provided by Figure 10.13. The C1 chondrite graph at its right-hand end is much flatter than the graphs for the terrestrial planets. This can be explained by volatile-rich bodies additional to bodies with a C1 chondrite composition, and icy–rocky planetesimals are a popular choice. Theories of their formation suggest a range of compositions. In the Jupiter region the volatile complement would have been dominated by water ice, plus small quantities of ices rich in carbon and nitrogen. As we go further out there were increasing quantities of ices of the more volatile substances, in accord with the decreasing temperatures. The gas-trapping efficiency also increases with decreasing temperature. The planetesimals from the furthest reaches, from beyond Neptune, would have had, as well as water, ices and trapped gases rich in nitrogen and carbon so that, overall, the elements C and N were in solar proportions. Of relevance to Figure 10.13 is that the inert gases would also have been present in solar proportions, except for an underabundance of helium and, to a lesser extent, of neon. A proportionate input from these icy–rocky planetesimals can give a match to the terrestrial graphs.

In order to account for many of the details of the volatile endowments of the terrestrial planets it is necessary to assume that each terrestrial planet received a particular mixture of the various bodies, the mix depending on the heliocentric distance of the planet. It is then possible to explain why Venus, the Earth, and Mars have global mass fractions of the inert gases that, on the whole, are smaller than their global mass fractions of H_2O, CO_2, and N_2. The various mixtures also explain certain differences among the three planets that relate to H_2O, CO_2, and N_2 alone. In particular, the three global mass fractions for each planet in Figure 10.12 can be turned into two ratios by dividing by the global mass fraction of water. We can be far more certain about the values of these ratios than we can of the individual global mass fractions, and it is the early values of these various ratios that can be explained. The details are beyond our scope.

Finally, note that icy–rocky planetesimals are hydrogen-rich bodies, and therefore they can give rise to chemical reactions that produce NH_3 and CH_4 in terrestrial atmospheres. These are powerful greenhouse gases and if they were abundant they would have been important in the evolution of the atmospheres, as you will see.

10.5.5 Outgassing

A late veneer does not preclude the possibility of continued outgassing, long after any magma ocean solidified and any core formation was complete. Indeed, outgassing must have been important on even the least geologically active of the three planets, Mars. Outgassing continues to this day on the Earth, probably on Venus, and possibly on Mars.

The mix of gases emerging from the interior depends on the mix of materials there and on the temperature. Any metallic iron present in the mantle, such as there would be before core formation, would remove oxygen from molecules, and thus convert much of the H_2O in the mantle to H_2 and much of the CO_2 to CO. The decomposition of any carbonates present would yield CO rather than CO_2. Without iron, at the sort of temperatures occurring in terrestrial planet mantles, the volatile-rich materials would outgas H_2O, CO_2, N_2, and traces of other volatiles.

Among these trace volatiles is ^{40}Ar from the radioactive decay of the unstable isotope of potassium, ^{40}K. Potassium is confined to rocky materials, and so if we know the atmospheric quantity of ^{40}Ar, and if we can estimate the amount of potassium in the interior of a planet, we can estimate the degree to which the planet has outgassed. We know the atmospheric quantities of ^{40}Ar for Venus, the Earth, and Mars, and a rough estimate of the potassium content has been made for the Earth, and also for Mars from studies of Martian meteorites. It seems that Mars is less outgassed than the Earth. This is in accord with its lower level of volcanic activity.

If indeed Mars is less outgassed than the Earth then this is a further factor in accounting for its lower global mass fractions of inert gases in Figure 10.13, and in accounting for any deficiencies in the near-surface quantities of other volatiles. However, we do not know the initial volatile inventory of the interior, and so we cannot use the degree of outgassing as a means of calculating the quantities of water and other volatiles that have appeared on the surface.

In spite of the many uncertainties, we nevertheless have rough estimates of the early volatile inventories of the terrestrial planets that, when inserted into plausible models of atmospheric evolution, do yield something like the present volatile inventories. These models are the subject of the next section.

Question 10.7

Describe the CO_2 data in Figure 10.12, and outline the processes that have determined the CO_2 endowment of the three planets. State to what extent the differences between the CO_2 endowments have been explained so far in this chapter.

10.6 The Later Evolution of Terrestrial Atmospheres

In considering the origin and early evolution of the terrestrial atmospheres we have been concerned largely with events that occurred between their origin 4600 Ma ago, and the end of the heavy bombardment at about 3900 Ma. We shall now be more concerned with events from about 4200 Ma onwards. By 4200 Ma any magma oceans had solidified, any blow-off and T Tauri activity had occurred, core formation, rapid outgassing and hydrodynamic escape were over. Impact erosion was continuing, but at a declining rate. Over this period the atmospheres of Venus, the Earth, and Mars, have taken dramatically different evolutionary paths. This has been against the backdrop of a gradual increase in solar luminosity as shown in Figure 10.14.

Some plausible scenarios are now presented. In considering each planet in turn some familiar material will be encountered as we blend the past into the present.

10.6.1 Venus

About 4200 Ma ago Venus probably had an atmosphere of carbon dioxide, water vapour, and nitrogen, perhaps overlying oceans of liquid water. The atmosphere was not as massive as it is today, but it contained sufficient water and carbon dioxide for a modest greenhouse effect. This, coupled with the proximity of Venus to the Sun, made the lower atmosphere warm. The warm temperatures sustained a high partial pressure of water, and the release of latent heat on condensation reduced the lapse rate to the point that atmospheric temperatures decreased rather slowly with height.

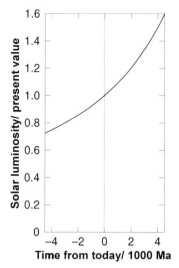

Fig. 10.14 The increase in solar luminosity from 4500 Ma ago until 4500 Ma into the future, according to recent models

Therefore, the number fraction of water was sustained to high altitudes, where it was subject to photodissociation and the consequent loss of hydrogen to space. The oxygen liberated has probably been lost through oxidation of the surface rocks, particularly as the surface temperature rose (Section 10.4).

Though water was being lost by photodissociation it was replenished by evaporation from surface reservoirs, thus maintaining its partial pressure. As the solar luminosity rose, the partial pressure of water in the lower atmosphere crept up the saturation curve. This increased the greenhouse effect which caused a further increase in partial pressure. This is an example of positive feedback, an inherently unstable feature that resulted in the complete evaporation of the oceans, perhaps by 3000 Ma ago, by which time the temperatures were high, though not as high as today, and the atmosphere contained a huge mass of water vapour.

Though most of this water was lost through photodissociation, the process slowed as the atmosphere dried out and the upper atmospheric temperatures consequently declined. Therefore, the last 10^6 Pa or so of water vapour must have been lost some other way. One possibility is incorporation into the crust, and subsequent burial in the mantle through geological processes. Volcanic activity can subsequently release such water, but in a steady state it is feasible that only a small proportion is ever in the atmosphere, which has therefore remained dry.

You have seen that photodissociation also explains the high ratio of $n(^2H)/n(^1H)$ in the atmosphere of Venus today.

❐ What is the explanation?

Photodissociation liberates hydrogen, and the lower mass of ^1H enables its molecules and ions to escape to space at a greater rate than molecules and ions that include ^2H. In addition, the earlier hydrodynamic escape of hydrogen also favoured the lighter isotope.

With the early loss of oceans, sedimentary carbonate formation was negligible, and so the CO_2 content of the atmosphere increased as volcanic activity released CO_2 from the crust and mantle, to form the massive atmosphere that we have today. This sustains the present temperatures through the greenhouse effect, enhanced by pressure broadening, and aided by the greenhouse effect of small quantities of other greenhouse gases.

There is an alternative scenario in which oceans *never* formed on Venus, but the end point is the same. This alternative is unlikely, if, as is widely believed, the devolatilisation of impactors and the late veneer produced large quantities of water over a relatively short time. If Venus did once have oceans of liquid water for hundreds of Ma then it is just possible that life began a short Venusian career. If so, there is just a chance that fossil remnants have survived to today.

10.6.2 The Earth

There is geological and biological evidence that the Earth has had oceans of water as far back as the evidence goes. The geological evidence is of metamorphosed sedimentary rocks 3800 Ma old, that must have been derived from ocean sediments. The biological evidence is the continuous presence of life since about 3850 Ma ago — only about 50 Ma after the end of the heavy bombardment. Though the details of the origin of life on Earth are still largely obscure, we do know that liquid water is essential for life in all its forms.

This stark contrast with Venus arises from the greater solar distance of the Earth. The atmosphere never became warm enough for water vapour to be abundant in the upper atmosphere — the upper troposphere has always been a cold trap where water has condensed and returned to the surface. Indeed, we have the opposite problem. The low luminosity of the youthful Sun (Figure 10.14) means that if the Earth's albedo and greenhouse effect were initially as they are at present, the oceans in that far-off time would have completely frozen over. The albedo would then have become so high that it would be difficult to unfreeze the oceans, even by today!

❐ How could different atmospheric composition have helped prevent global freezing?

If the Earth's greenhouse effect was sufficiently larger, the effect of reduced solar luminosity could have been offset. Something like 300–3000 times the present atmospheric mass of carbon dioxide is needed around 3800 Ma ago, with a slow decline rather well matched to the increasing solar luminosity, otherwise the temperatures would have become higher than we know they were. The total volatile inventory of carbon dioxide on Earth is at least 10^5 times that in the atmosphere, so the problem is one of controlling the atmospheric fraction. Could a carbon cycle accomplish this?

As soon as oceans appeared, they began to dissolve the carbon dioxide, and they reached a steady state with the atmosphere on the short time scale of the order of centuries. Carbonate formation in the oceans was slower, operating on a time scale of order 10^2 Ma. On this longer time scale a carbon cycle became established, but it was not in a steady state — the atmospheric CO_2 content must have declined as the outgassing rate decreased with the reduction in volcanic activity. It is therefore conceivable that there was always sufficient atmospheric CO_2 to prevent a global freeze but not so much as to cause temperatures to be too high. This rather fine tuning is aided by a negative feedback in which the net removal rate of CO_2 from the atmosphere decreases as the surface temperature decreases. This then allows outgassing to increase atmospheric CO_2 content, which increases the temperature. Such feedback can also help to melt surface ice, because ice cover also decreases the net CO_2 removal rate, and though it is unlikely that this effect of ice cover could have reversed a global freeze, it could have helped to prevent one.

Increased CO_2 content could also have helped warm the Earth in another, rather surprising way. The low temperatures high in the atmosphere when the Sun's luminosity was well below its present value could well have led to extensive CO_2 cloud cover. This would have increased the planetary albedo — just what we *don't* want. However, if the cloud particles were larger than a few micrometres they would have scattered infrared radiation emitted from below them and enhanced the greenhouse effect to an extent that more than compensated for the increased albedo.

Further assistance could have come from greenhouse gases that fill the gaps in the CO_2 absorption spectrum, such as NH_3 or CH_4. They can also reduce the CO_2 requirement. These gases are rapidly broken up by solar UV radiation so we need either a steady resupply or protection. The icy planetesimals that contributed to the late veneer could have provided a steady resupply. Moreover, CH_4 would be photodissociated in the upper atmosphere to yield a fine dust of solid hydrocarbons that could shield NH_3. Any assistance from such substances could have persisted until about 3800 Ma ago, and perhaps for a few hundred Ma later. The presence of NH_3 and CH_4 would have facilitated the synthesis of the huge organic molecules that life is based on, notably proteins, RNA, and DNA.

The occurrence of ice ages shows that the fine tuning that is needed to keep the Earth's surface temperature fairly constant as the solar luminosity rose has not been perfect. It is nevertheless impressive, and this has led to the suggestion by the British chemist James Ephraim Lovelock (1919–) that the biosphere (unconsciously) took an active part in the control. The cumulative mass of all organisms that have ever lived is at least an appreciable fraction of the Earth's total mass! Therefore, a significant effect is not out of the question, though this does not of itself indicate that the effect has been stabilising. The general notion of active biospheric control that tends to preserve optimum conditions for the biosphere is called the **Gaia hypothesis**. One of many ways in which it could operate is through the biogenic release of CH_4, and the release and uptake of CO_2.

The biosphere has certainly had a profound effect on atmospheric composition. You saw in Section 10.1.2 that it sustains almost the whole oxygen and nitrogen content of the atmosphere, and that is has promoted the removal of CO_2. When did the biosphere become active? We know that by about 3850 Ma ago simple single cells existed on Earth and that at about the same time an early form of photosynthesis was operating. By no later than 3400 Ma ago photosynthesis that generated oxygen existed, and this soon overtook the photodissociation of water as the main source of oxygen.

❐ Why has photodissociation always been a rather weak source of oxygen?

Most of the water has been in the oceans, beyond the reach of solar UV radiation. Let's focus on oxygen, and consider the way the atmospheric quantity has changed.

At first the oxygen from photosynthesis was almost entirely consumed by the oxidation of less than fully oxidised substances — rocks, volcanic gases, and biological materials. But ultimately there was sufficient oxygen for it to start to build up in the atmosphere. The record of atmospheric oxygen is revealed to some extent by geological and biological evidence. Rocks older than about 2200 Ma have sufficient pyrite (FeS_2) and uraninite (UO_2) to indicate that the oxygen partial pressure must have been less than about 20 Pa, otherwise oxidation would have largely destroyed them. A lower limit of about 10^{-6} Pa has been suggested by some scientists, on the basis of the abundance of Fe_3O_4, one of the more oxidised of the iron oxides. Photodissociation of water alone might have yielded a partial pressure of oxygen of around 5×10^{-4} Pa.

About 2000 Ma ago a new type of cell emerged upon which all of the higher forms of life is based. This is the eukaryotic cell, which has a nucleus and other internal structures, in contrast to the less structured prokaryotic cells that had earlier been the only kind. Even though all life was still in the sea (and single-celled) we can estimate the atmospheric oxygen that would have been needed to sustain the minimum oxygen content of water for eukaryotic cells to exist, and this corresponds to about 10^2 Pa of atmospheric oxygen. There are yet other geological and biological indicators leading to the inferred record of atmospheric oxygen in Figure 10.15. Note the large uncertainties! (It has even been argued by a few scientists that the mass of atmospheric O_2 has been within a factor of two of its present value throughout the last 4000 Ma, but this is very much a minority view.)

10.6.3 Mars

An important indication of atmospheric conditions early in Martian history is the evidence outlined in Section 7.3.6 that liquid water was stable at the surface during the Noachian epoch. At a time of lower solar luminosity this requires an enhanced greenhouse effect just as in the case of the Earth, but the requirements for Mars are

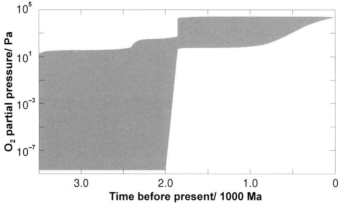

Fig. 10.15 The build-up of atmospheric oxygen on Earth. The grey shading shows the uncertainties. (Adapted from J. F. Kasting, *Science* **259**, 920 (1993))

more severe because of its greater distance from the Sun. Assuming a greenhouse effect not only from gaseous CO_2 but also from scattering by CO_2 cloud particles, a minimum atmospheric mass of CO_2 of roughly 100 times the present value is required—a column mass of about $1.5 \times 10^4 \, kg \, m^{-2}$, compared to a corresponding minimum requirement of about $0.1 \times 10^4 \, kg \, m^{-2}$ of CO_2 for the Earth at that time. This assumes that some water vapour was also present.

Such large quantities of atmospheric CO_2 might be asking a lot of the Martian carbon cycle, particularly if oceans of liquid water ever formed, though the greater weathering rate in the Noachian is consistent with greater atmospheric mass.

❐ How could sufficient warming have occurred with less CO_2?

The quantity (as on Earth) could have been less if the powerful greenhouse gases NH_3 and CH_4 were present. These substances could have been derived from icy planetesimals. Sufficient impacts could have been sustained to the end of the Noachian. Also, enhanced volcanic activity could have provided significant amounts.

But gradually the greenhouse gases became less abundant. Any NH_3 and CH_4 content was eliminated as these gases were destroyed by solar UV radiation faster than the declining rate of supply, and CO_2 was removed through impact erosion, adsorption into the regolith, and the formation of sedimentary carbonates. Removal by carbonates would have been particularly rapid if there were large open bodies of liquid water, though 'dry' formation of carbonates can occur if water vapour is present. Volcanic activity returned CO_2 to the atmosphere, but because of Mars's small size and consequent rapid interior cooling, volcanic activity declined steeply and early in Mars's history. With CO_2 removal still occurring, its atmospheric mass declined, and the clement conditions came to an end.

Since the end of the Noachian impact erosion has been slight, but CO_2 has continued to be removed from the atmosphere by regolith adsorption and by dry carbonate formation. The general decline in volcanic activity has thus presumably resulted in a gradual decrease in the mass of atmospheric CO_2.

What of the other volatiles? Since early in the Noachian the equivalent of a global depth of about 100 m of liquid water has been outgassed at most, and perhaps a lot less. A small fraction of this water has been lost by photodissociation and the escape of hydrogen, though most of it must have entered some surface or regolith reservoir. Chemical escape has deprived Mars of far more nitrogen than is presently in its atmosphere, and if there ever were open bodies of water on Mars then much of the initial endowment of nitrogen might now reside in nitrates.

Life on Mars?

Regardless of how Mars became as it is, it seems probable that during the first 1000 Ma or so after its formation, liquid water and the associated clement conditions existed on its surface. Life got going on Earth by about 50 Ma after the end of the heavy bombardment, so the question arises of whether life got going on Mars, and whether it survives today.

This latter question was put to the test in 1976 when Viking Landers 1 and 2 arrived and made surface observations for several years. Though three of the experiments were designed to detect life, none of them gave unequivocally positive results. Nor was there any evidence of organisms in the surface images. At the other extreme of scale, organic compounds (which in any case can have a non-biological origin) can be present in only very small quantities. Nevertheless it is just possible that in some warmer and perhaps wetter regions life clings on today. As to whether life once existed but is now extinct, many bioastronomers are optimistic, and therefore a search for Martian fossils is a high priority for future missions. One Martian meteorite, ALH84001, was once thought to display evidence of an early Martian biosphere, but most scientists now believe that the evidence is not there.

Question 10.8

Suppose that the solar luminosity increased considerably. Describe the possible consequences for the atmospheres of the Earth and Mars.

10.6.4 Mercury and the Moon

Mercury and the Moon have extremely tenuous atmospheres, with column masses of order $10^{-10}\,\mathrm{kg\,m^{-2}}$. The atmospheres are entirely exospheric. Thermal escape is largely to blame — see Figure 9.12 (Section 9.2.5), remembering that the surface temperatures of Mercury (Me) and the Moon (Mo) in daytime greatly exceed the mean values shown. Substantial proportions of such tenuous atmospheres are also lost through the UV ionisation of atoms and molecules, the ions then being carried off by the magnetic field in the solar wind. The solar wind is also a *source* of atmosphere, partly by supplying ions and atoms for capture, and partly by ejecting particles from the surface. UV radiation also ejects surface particles. Other sources are feeble outgassing, and the capture of volatile-rich bodies such as comets and carbonaceous chondrites. O, Na, and H are the dominant constituents of Mercury's atmosphere, (though the upper limit on Ar exceeds their measured values), and Ar and He for the Moon. In each case the atmospheric turnover is rapid.

If Mercury was initially well endowed with an atmosphere, and if this survived impact erosion, then Mercury's proximity to the Sun would have ensured that it developed a large greenhouse effect. Its low escape speed and the high UV radiation levels and intense solar winds so close to the Sun would soon have stripped it of all atmosphere and surface volatiles, leaving it with the tenuous atmosphere we see today. The surface of Mercury shows it to have been volcanically rather inactive, and this is because of its small size and rapid cooling. It might therefore be only partially outgassed, in which case juvenile volatiles might be appearing on its surface today — perhaps contributing to any ice deposits near the poles.

It is rather less likely that the Moon has an atmosphere awaiting outgassing.

❐ Why is this?

It is thought to have been created almost volatile-free by the manner of its birth (Section 5.2.1). Any late veneer was not retained in the face of the various loss processes, because of the Moon's low gravitational field.

10.7 Icy–Rocky Body Atmospheres

The only icy–rocky bodies with significant atmospheres are Titan, Triton, and Pluto. Among the icy–rocky bodies these have comparatively high escape speeds (Table 9.1), though this cannot be the only factor. The Galilean satellites have comparable escape speeds yet they all have negligible or extremely tenuous atmospheres.

❏ What is another factor in determining atmospheric retention?

Temperature must also be considered. Titan, Triton, and Pluto have low temperatures (Table 9.1), a result of the great distance of these bodies from the Sun. Escape is therefore at a low rate. However, a surface can be *too* cold, resulting in very low vapour pressures. For example, in the cases of Europa, Ganymede, and Callisto, the water ice that is so abundant at their surfaces is too involatile to have a significant vapour pressure even at their comparatively high surface temperatures. Therefore a further factor is the types of volatile material that have been available.

10.7.1 Titan

Titan, the Mercury-sized satellite of Saturn, has a remarkable atmosphere with a column mass eleven times that of the Earth. The mesosphere consists almost entirely of nitrogen (N_2), the remaining few per cent probably being mostly methane (CH_4). The vertical structure is shown in Figure 10.16.

A rich variety of carbon compounds exist as traces in Titan's atmosphere, particularly hydrocarbons (additional to methane). At the pressures and temperatures of the atmosphere many of these will condense to form solid or liquid particles, and it is these that constitute the haze layers in Figure 10.16. Typical particle sizes are 0.1–0.5 μm. The main haze layer is very deep and completely screens the surface from us at visible wavelengths (Plate 17). It is thought that a methane cloud deck lies beneath this layer. If this is so then the methane number density will be higher

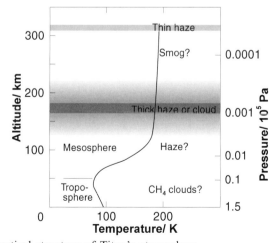

Fig. 10.16   The vertical structure of Titan's atmosphere

beneath the clouds than where it has been measured above the clouds, and the nitrogen number density will be correspondingly lower.

The global mean surface temperature of Titan is 94 K. There is a greenhouse effect due mainly to methane, with a magnitude of about 21 °C. This is partly offset by the haze which is transparent at mid-infrared wavelengths but absorbs solar radiation. Without this haze the surface temperature would be several degrees higher. Atmospheric circulation and the massive haze-laden atmosphere ensures very little temperature variation across the globe. The surface temperature is close to methane's triple point temperature (90.4 K) and so methane could exist as a solid, liquid, or gas at and near the surface, as shown by the methane phase diagram in Figure 10.17. If the methane number density at the surface is sufficiently high, then the partial pressure of methane will be on the saturation line in Figure 10.17. Any tendency for it to be higher causes condensation and precipitation, which maintains the partial pressure at the saturation value. In this case it is the surface temperature that determines the mass of methane in the atmosphere, just as in the case of water in the Earth's atmosphere.

A variety of other hydrocarbons presumably also precipitate onto the surface, conjuring a vision of a somewhat murky landscape of water ice and hydrocarbon ices, and hydrocarbon oceans, though observations indicate that if there are any oceans they are not widespread (Section 7.4.2).

10.7.2 Triton and Pluto

Triton, the large satellite of Neptune, has a tenuous atmosphere consisting largely of nitrogen with a few per cent methane and a trace of carbon monoxide (CO). The column mass is only $2.1 \, \mathrm{kg \, m^{-2}}$. Voyager 2 imaged a haze layer about 3 km thick centred at an altitude of about 3 km, through which the surface is readily seen. Nitrogen ice is present at the surface, plus methane, carbon monoxide, and carbon dioxide ices, and so the pressure and hence the mass of the atmosphere is determined

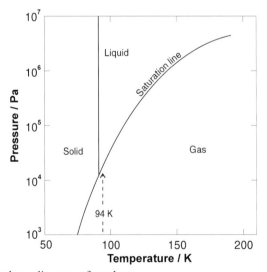

Fig. 10.17 The phase diagram of methane

by the mean surface temperature, which is 38 K. The surface temperature also determines the composition of the atmosphere, the vapour pressure of nitrogen in equilibrium with its ice being considerably greater than that of the other gases in equilibrium with their ices.

The high orbital inclination of Triton gives rise to huge seasonal changes over the 164-year orbital period of Neptune and so there is presumably a component of atmospheric circulation akin to the condensation flow on Mars.

Pluto is also known to have ices of methane and nitrogen at its surface, and also carbon monoxide ice. At the current mean surface temperature of 40 K there must be a predominantly nitrogen atmosphere with a trace of methane, much the same as on Triton. There is direct evidence for an atmosphere with this sort of composition. Pluto has an eccentric orbit and is currently not far from perihelion. At aphelion, in 2113 it will be about 1.7 times further from the Sun, with a correspondingly far colder surface and an atmosphere far more tenuous than it is now. As Pluto goes around its orbit the column mass is estimated to vary over a considerable range.

10.7.3 The Origin and Evolution of the Atmospheres of Icy–Rocky Bodies

Titan

When Voyager 1 in 1980 provided the first direct measurement of the composition of Titan's atmosphere, the predominance of nitrogen caused some surprise, though there were theoretical indications that nitrogen would be present. Our understanding of the composition is now better developed, and the following explanation is widely accepted.

There are probably two sources of the copious quantity of nitrogen in Titan's atmosphere. First, when Titan formed in the outer solar nebula, the temperature was so low that nitrogen gas (N_2) was trapped within the water ice that constituted a large fraction of the icy–rocky planetesimals that formed Titan.

❏ What kind of substance is this water ice–nitrogen combination?

This is a clathrate. Much of this nitrogen was released on accretion. The subsequent slight heating of Titan, tidal or radiogenic, has since released more.

Second, in the cold outer solar nebula, ammonia is expected to have been a significant constituent of the icy planetesimals. At the low surface temperature of Titan, ammonia has a very low vapour pressure which is why, if it is still present in the ices, it has not been detected in the atmosphere. However, via a series of chemical and photochemical reactions, ammonia vapour can be converted into N_2 and H_2. The H_2 will suffer thermal escape, which explains the small quantity in the atmosphere, but the nitrogen is retained. Operating over the 4600 Ma history of Titan, this can account for some of the nitrogen. It can only account for most or all of it if the surface of Titan was 50 °C or so warmer in the past. This condition is necessary to prevent the condensation of an intermediate product in the reaction sequence, hydrazine (HNNH). If hydrazine condenses then it cannot undergo a crucial photochemical reaction and the sequence halts. It is plausible that the surface was sufficiently warm to prevent condensation, through greater tidal or radiogenic heating in the past.

Overview of the icy–rocky body atmospheres

Among Pluto and the large satellites of the outer Solar System it is only Titan that has a massive atmosphere. Triton and Pluto have tenuous atmospheres, and the remaining large satellites have negligible atmospheres. The explanation starts with the sort of ices that could have condensed in the outer Solar System during its formation, and in the sort of gases that the ices could have trapped. At Jupiter's distance from the Sun the nebular temperature was so high that the icy planetesimals contained water ice as the only significant volatile, with only small quantities of carbon- and nitrogen-rich ices. Little gas would have been trapped. The present surface temperatures of Europa, Ganymede, and Callisto are not sufficiently high to generate a significant atmosphere from the relatively involatile water ice. Io probably formed too close to Jupiter to ever have had much water ice, but if it did, tidal heating would have driven it off.

From Saturn outwards the more volatile ices of methane and ammonia could condense in the planetesimals, and at the lower temperatures trapping of significant quantities of gases like N_2 and CO was possible. Titan is sufficiently warm for a massive atmosphere to have been derived from these ices. Further out, Triton and Pluto are too cold for such a massive atmosphere — the nitrogen and methane is largely condensed on the surface. Only small quantities of inert gases would have been trapped in the planetesimals, so it is unsurprising that these have not yet been detected.

In Section 10.5.2 you were reminded that planetesimals were not confined to the zone in which they formed, which raises the question of why the Galilean satellites did not acquire atmospheres similar to that of Titan, as a late veneer. A possible explanation is the high speeds acquired by planetesimals as they fell towards the Sun, made even higher as they subsequently fell towards Jupiter. The impacts would then have been too violent for volatiles to be retained. Moreover, if the Galileans already had atmospheres, this is one way in which they could have been removed — by impact erosion.

Question 10.9

Suppose that Titan and Triton swapped places today! Discuss whether the atmosphere of Titan would become like that of Triton, and vice versa.

10.8 Summary

Tables 9.1 and 9.2 and Figure 10.12 present the basic properties of the atmospheres discussed in this chapter.

The Earth's atmosphere today differs from that of Mars and Venus in that it consists largely of oxygen and nitrogen rather than carbon dioxide. This is a consequence of the action of the oceans and the biosphere. The abundance of oxygen has led to the unique feature of a stratospheric 'bulge' in temperature, caused by the absorption of solar UV radiation by ozone O_3, derived from O_2.

The greenhouse effect is small on Mars, because of its thin, dry atmosphere. It is larger on Earth, because of the greater quantity of atmospheric water vapour, supplemented by carbon dioxide (and traces of other greenhouse gases). On Venus the greenhouse effect is very large, sustained by a huge mass of carbon dioxide, with the assistance of far smaller quantities of water vapour, sulphur dioxide, and the sulphuric acid droplets that constitute the planet-wide cloud.

The circulation of the Earth's atmosphere is dominated in the tropics by one Hadley cell per hemisphere, by planetary waves at mid-latitudes, and by a cell in each polar region. The Coriolis effect limits the extension of the Hadley cells, and also deflects the winds. On slower rotating Venus there is just one Hadley cell per hemisphere in the lower troposphere, stretching from equator to pole. Mars (except near the equinoxes) has a single Hadley cell stretching from the subsolar latitude to the winter hemisphere. There is also condensation flow, and a thermal tide stronger than that on the Earth and Venus.

The present volatile inventories of the Earth, Mars and Venus show substantial differences. It is not possible to deduce their initial volatile inventories by adjusting the present inventories using loss processes that are still operating. This is because the effects of blow-off, the T Tauri phase of the Sun, impact erosion, and hydrodynamic escape, all of which are thought to have occurred early in Solar System history, are very uncertain and yet caused huge losses. A consistent picture is one in which there is rapid loss of volatiles from embryos and planetesimals during planet formation, followed by outgassing due to core formation. Much of these early atmospheres were then lost, and volatiles subsequently reached the surfaces through further outgassing and through collisions with a variety of volatile-rich planetesimals.

The three terrestrial atmospheres have evolved quite differently from each other largely because of their different distances from the Sun. On Venus the volatiles that have been retained are largely in the correspondingly massive atmosphere, except for water, most of which has been photodissociated, with loss of hydrogen to space and incorporation of oxygen into rocks. On the Earth, the oceans and the biosphere have 'locked' most of the carbon into carbonates, and about half the nitrogen into nitrates. Photosynthesis has resulted in a build-up of atmospheric oxygen over the last 2000 Ma. Mars probably has much of its volatiles in a variety of near-surface deposits, because of its greater distance from the Sun. Martian volatiles have also been lost to space, through thermal and chemical escape.

The increase in solar luminosity, and the greenhouse effect as determined by evolving atmospheric compositions, have been of enormous importance in the evolution of the volatile inventories on Venus, the Earth, and Mars, and consequently in their long-term climate changes too.

The smaller terrestrial bodies — Mercury and the Moon — are devoid of significant atmospheres because of their low escape speeds and proximity to the Sun.

Titan, Triton, and Pluto have atmospheres determined by the types of icy materials available to them when they formed, and their surface temperatures. At the large solar distances of these bodies, ices of methane and ammonia are expected in addition to the water ice that can also condense closer to the Sun. The water ice could also contain nitrogen, as a clathrate. The atmosphere of Titan is more massive than that of Triton and Pluto because Titan is closer to the Sun and hence warmer. Chemical reactions in the atmosphere of Titan could have enhanced the quantity of nitrogen.

Other large icy–rocky bodies, and Io, lack significant atmospheres because of the lack of volatile materials that would be gaseous at their surface temperatures.

11 Atmospheres of the Giant Planets

With Jupiter, Saturn, Uranus, and Neptune we come to four planets where, as Table 9.1 shows, the atmospheres are dominated by molecular hydrogen (H_2) and atomic helium (He), and not by water and substances rich in carbon and nitrogen. Moreover, for these giant planets the distinction between the atmosphere and the interior is blurred. This is apparent from the interior models discussed in Section 5.3. In Jupiter and Saturn the atmospheres blend seamlessly into the molecular hydrogen envelope, with no surface separating them. In Uranus and Neptune icy and rocky materials get more abundant with depth, but probably gradually, and again with no interface.

Because of cloud formation, but additionally because of the huge depth of the atmospheres, variations of composition with depth are to be expected. In Table 9.1 the compositions of the giant planet atmospheres are given for the *total* atmosphere above a specified depth — that at which the pressure is 10^5 Pa. These are number fractions. To specify the composition at a *particular* depth, one could use the local number density of each constituent (Section 9.2.3), but in the case of the hydrogen-dominated giants it is more usual the specify the number of molecules of each constituent in any local volume divided by the number of hydrogen molecules (H_2) in the same volume. This is the **mixing ratio** of the constituent, in this case with respect to H_2.

Table 11.1 gives the mixing ratios for the giant's atmospheres at depths of a few times 10^5 Pa. They have been obtained by a variety of remote techniques, including the use of microwave emission from beneath cloud and haze that obscures optical radiation. In the case of Jupiter, the Galileo probe in December 1995 provided direct analysis. Just how representative the data in Table 11.1 are of the atmospheres as a whole will emerge throughout this chapter.

Though there are similarities between the four atmospheres, there are also differences, and consequently the atmospheres can be grouped into two pairs, Jupiter and Saturn forming one pair, Uranus and Neptune the other. The atmospheres within each pair resemble each other more closely than they resemble the atmospheres in the other pair.

11.1 The Atmospheres of Jupiter and Saturn Today

Plates 11 and 16 show the highly coloured, richly structured cloud tops of Jupiter and Saturn, at once strikingly different from the planetary bodies that we have so far

Table 11.1 The atmospheric composition of the giant planets, given as mixing ratios with respect to H_2

Species[a]	Jupiter	Saturn	Uranus	Neptune
H_2	1	1	1	1
He	0.156	0.03	0.18	0.23
CH_4	2.1×10^{-3}	4×10^{-3}	0.03	0.03
NH_3[b]	2.5×10^{-4}	2×10^{-4}	—	—
H_2O[b]	$< 3.7 \times 10^{-4}$	—	—	—

[a]Only abundant species are shown, as measured in the upper molecular hydrogen envelopes at a depth of a few times 10^5 Pa.
[b]A dash indicates that no useful upper limit is available.

considered. The patterns are more distinct on Jupiter partly because the overlying haze there is thinner (because of the higher temperatures), and partly because on Saturn the lower gravity has caused the upper cloud deck to be spread over a greater range of altitudes, and it is consequently less sharply defined.

11.1.1 Vertical Structure

Figure 11.1 shows the vertical structure of the atmospheres of Jupiter and Saturn, as deduced from a variety of measurements by Earth-based telescopes and by spacecraft, involving spectra and images at many wavelengths from UV to radio. In the case of Jupiter much of the data at pressures greater than 10^5 Pa have come from the Galileo probe and Orbiter.

For both planets the lapse rate up to the tropopause is close to the adiabatic value for the mix of gases present, and so the atmosphere is probably convective. The energy source is mainly heat welling up from the interior, supplemented among the

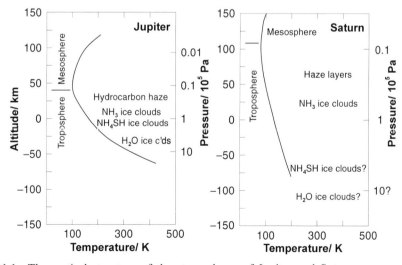

Fig. 11.1 The vertical structure of the atmospheres of Jupiter and Saturn

clouds by the absorption of solar radiation. In the mesospheres and thermospheres there is no convection, and the lapse rate is determined by radiative transfer between the different altitudes, by radiation to space, and by the absorption of solar radiation. The generally lower temperatures for Saturn (at a given pressure) are due to the greater distance of Saturn from the Sun and the lower heat flux from its interior. The Jovian exosphere is particularly hot, perhaps because of atmospheric waves travelling upwards, or bombardment by magnetospheric electrons, or a reduction in methane content — methane promotes radiative cooling.

For each planet, three layers of cloud are shown in Figure 11.1, plus some layers of haze. In each case the uppermost cloud layer consists of ammonia ice particles that form cirrus-like sheets. At lower altitudes there is a layer of ammonium hydrosulphide (NH_4SH), perhaps mixed with ammonium hydroxide (NH_4OH), and lower still a layer of water is shown. In all three layers the cloud particles are solid. Cloud formation was outlined in Section 9.2.3.

❐ Try to recall the essential features of the process.

As altitude increases the partial pressure of each atmospheric gas decreases, and in the troposphere the temperature decreases too. If, for any constituent, the partial pressure and temperature meet the saturation line in the phase diagram of the constituent then the constituent can condense, either as ice crystals or as liquid droplets, depending on the temperature. The lowest altitude at which this occurs will be at the base of the clouds. The higher the partial pressure of a constituent in the atmosphere below the cloud base, the lower the altitude of the cloud base will be.

The uppermost cloud layer is, of course, directly observable but, particularly for Jupiter, it is not unbroken, and in polar regions it is largely absent. Nevertheless, much of the general picture in Figure 11.1 was for a long time based more on modelling than on measurement. It was measurements from the Galileo Orbiter at infrared wavelengths that played a large part in confirming the general picture, and in adding detail, such as the altitudes of the various deeper cloud layers shown in Figure 11.1.

The Galileo probe held the promise of the direct sampling of cloud particles at accurately measured altitudes, and indeed it returned data down to a pressure of 24×10^5 Pa. But it went through an almost cloud-free hole! Nevertheless, it detected a patchy cloud layer with a base at about 0.6×10^5 Pa and a very tenuous layer with a base at about 1.5×10^5 Pa. An increase in opacity below 9×10^5 Pa might have been the thin upper reaches of a third cloud layer. The compositions of the particles in these layers was not measured, but the highest level cloud is presumed to be the NH_3 cloud. The very tenuous layer could be the NH_4SH cloud — the probe mass spectrometer detected sufficient NH_3 and sulphur compounds in the gas in this region to make NH_4SH particles. If there was a third layer then it might have been H_2O.

In 1994, before the Galileo spacecraft reached Jupiter, the fragmented comet Shoemaker–Levy 9 plunged into the Jovian atmosphere. Unfortunately, less has been learned than had been hoped, in large part because it is difficult to distinguish between comet material and Jovian material. However, gases were observed leaving Jupiter at greater than the escape speed, so the impact did teach us something about impact erosion.

Saturn is not as well observed, and so the compositions of the clouds and the altitudes of the cloud bases have been inferred from the measured temperatures and pressures at each altitude, and from atmospheric composition. The measurements are incomplete, particularly of composition, and so they are supplemented by reasonable assumptions.

11.1.2 Composition

Hydrogen and helium

Table 11.1 gives the mixing ratios at depths of a few times 10^5 Pa, which is well into the homospheres. Hydrogen is readily detected, but even though helium is a substantial constituent it has not been easy to establish its quantity. This is because its strong spectral signatures are at UV wavelengths. The Earth's atmosphere is opaque at such wavelengths, and in any case the UV signatures of helium are obscured by the UV signatures of molecular hydrogen (H_2). Spacecraft have detected helium spectrally in the atmospheres of Jupiter and Saturn, but only above the homopause, where the mixing ratio is not the same as in the homosphere. Then, in 1995 the helium mixing ratio for Jupiter obtained by direct sampling of the homosphere by the Galileo probe, and it is this value that is given in Table 11.1. Until then, the quantities of helium in the homosphere had been obtained only by indirect techniques, and this is still the case for Saturn and the other giants.

In one indirect technique the total pressure of the atmosphere is obtained from the collisional broadening (Section 9.1) of the infrared spectral lines of H_2. It is assumed that, except for H_2, the atmosphere contains little else except atomic helium (He). Thus, the partial pressure of He is assumed to be the difference between the measured total pressure and the measured partial pressure contributed by H_2 (and any other measured substances, notably CH_4). In another indirect technique, data from radio occultation has been combined with infrared measurements to obtain the mean molecular mass of the atmosphere. On the assumption that the atmosphere is dominated by H_2 and He, the mixing ratio of He is then calculated. Because both techniques use infrared and radio wavelengths at which the giant planet atmospheres are relatively transparent, the helium content has been determined to some way below the uppermost cloud layer, to where the pressure is several times 10^5 Pa.

Hydrogen and helium do not condense in the giant atmospheres, nor are they significantly depleted by chemical reactions. Therefore the values in Table 11.1 apply to great depths. In the case of Jupiter and Saturn the values probably apply throughout the molecular hydrogen envelope down to the metallic hydrogen interface, deep in the interior. There is, however, evidence that the helium mixing ratios in Table 11.1 are lower now than in the past because of the downward-settling of helium in the *metallic* hydrogen mantle — this depletes the upper mantle in helium which in turn depletes the envelope. This phenomenon and the evidence for it was described in Section 5.3.1, where models of the evolution of Jupiter and Saturn suggest that the depletion should be greater in Saturn.

❐ Do the values in Table 11.1 bear this out?

If Saturn and Jupiter started out with similar mixing ratios of helium, then today we would expect the mixing ratio in the atmosphere of Saturn to be less, and this is exactly what we find, as Table 11.1 shows.

Other substances

With the other atmospheric constituents we have to beware of the possibility of strong compositional variations with altitude. That the values in Table 11.1 apply at a level well into the homosphere is no guarantee against this.

❒ Why not?

In a homosphere, by definition, there is sufficient mixing to prevent *gases* from separating. This does not rule out separation through condensation. Condensation will lead to the formation of cloud and haze. In the troposphere convection will carry the cloud particles above the cloud base, but this will only partially offset the depletion. Moreover the *gas* phase will still be heavily depleted, and it is the gas phase that many compositional detection methods sense. Another cause of altitude variation is chemical reactions within the atmosphere. Above the clouds it is photochemical reactions driven by solar UV radiation that are important, giving rise to layers of photochemical smogs consisting largely of hydrocarbons that are much scarcer lower down. The ozone layer in the Earth's atmosphere is another example of a photochemical product.

Table 11.1 shows that in the upper molecular hydrogen envelopes the most abundant atmospheric constituents after H_2 and He are (in order) CH_4, NH_3, H_2O. To what extent do the mixing ratios of these compounds represent the whole molecular hydrogen envelope?

Methane (CH_4, the main repository of carbon) does not condense anywhere in the atmospheres of Jupiter and Saturn, nor is it significantly depleted by photochemical reactions. There is, however, a significant chemical conversion very deep in the molecular hydrogen envelope, where the temperatures exceed about 1200 K. Models show that a proportion of the methane is converted to carbon monoxide via the reaction

$$CH_4 + H_2O \longrightarrow CO + 3H_2 \tag{11.1}$$

Evidence that this occurs is the detected trace of CO, presumably brought up by convection. But even if a significant fraction of methane is present as CO, the C/H_2 ratio for the whole envelope must be almost exactly equal to the measured methane fraction in its upper reaches. This is partly because it makes no difference whether the C is in a CH_4 molecule or in a CO molecule, and partly because the quantity of H_2 produced by reaction (11.1) is very small. What *would* make a difference would be the production deep in the envelope of significant quantities of a *condensable* carbon compound that was then confined there. There is no evidence, theoretical or observational, that this has happened.

The dominant repository of nitrogen in a hydrogen-rich atmosphere at the pressures and temperatures of the observed homospheres is NH_3, and the altitude variation in its mixing ratio has been followed down to several times 10^5 Pa. In Figure 11.1 you can see that in Jupiter such pressures occur beneath the NH_3 and NH_4SH clouds, and it is this deep measurement that is given in Table 11.1. (Note that this is the one value that is *not* from the Galileo probe — the probe gave only an upper limit, and this was greater than previous measurements.) Therefore, it is thought that the N/H_2 fraction in Table 11.1 is typical of the molecular hydrogen envelope of Jupiter. The position with Saturn is more marginal — the measurements might not reach the supposed NH_4SH cloud level.

For each planet the variation of the NH_3 mixing ratio with depth provides additional information on the composition of the cloud layers. The ratio increases by several orders of magnitude as we move down through the uppermost cloud deck, lending support to the view that this deck consists of NH_3 crystals. For Jupiter there is also a clear increase in the NH_3 fraction as we traverse deeper altitudes, and this is in accord with the Galileo Orbiter findings that a second cloud deck exists, and that it consists of particles of NH_4SH, presumably with some NH_4OH. Chemical models predict the formation of such substances in the giant atmospheres by the reactions

$$NH_3 + H_2S \longrightarrow NH_4SH \qquad (11.2)$$
$$NH_3 + H_2O \longrightarrow NH_4OH \qquad (11.3)$$

the amounts of the products depending on the partial pressures of the reactants, and on the temperature. The models predict that NH_4SH is more abundant than NH_4OH and that at altitudes above about 1.5×10^5 Pa it is an important repository of sulphur. However H_2S (hydrogen sulphide) is the main repository of sulphur in the *gas phase* in a hydrogen-rich atmosphere — NH_4SH condenses. If NH_3 was initially much more abundant than H_2S then reaction (11.2) is a plausible explanation of the very small quantities of H_2S found in the Jovian atmosphere above the NH_4SH cloud level. As yet H_2S is undetected in Saturn's atmosphere.

Before the Galileo probe measured the composition of the Jovian atmosphere, H_2O (the oxygen repository) had been detected only above the water cloud deck. Unsurprisingly the mixing ratios are very small, well below the sensitivity of the Galileo probe. But the probe failed to detect water anywhere, even though it traversed the altitude range at which water clouds were expected. This expectation rested on the assumption that the water mixing ratio in the envelope corresponds to the O/H ratio in the Sun. In fact, the probe upper limit corresponds to a value about five times less. Is this really the upper limit for the whole envelope? Probably not. The probe passed through a relatively cloud-free region called a hot spot, so named because the lack of cloud allows us to see deep into the atmosphere. The sparseness of cloud is probably because such spots are where the atmosphere is descending — more on this shortly. Therefore the Jovian O/H ratio could exceed the probe upper limit, though whether the Jovian O/H ratio is the same as that of the Sun is a topic for Section 11.3.

Question 11.1

From Table 9.1, use the number fractions for Jupiter to calculate the average mixing ratios above 10^5 Pa. Compare your values to those in Table 11.1. How would you expect the values for NH_3 (not included in Table 9.1) to compare?

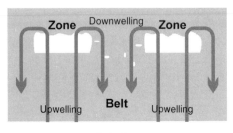

Interior

Fig. 11.2 An explanation of the belts and zones of Jupiter

11.1.3 Circulation

The atmospheric circulation of the two planets is revealed by the cloud patterns (Plates 11 and 16). Though the details of the patterns change on a time scale of days, the largest-scale features have changed little in their broad appearance over the three centuries or so of observations. This indicates a stable system of atmospheric circulation. The most widespread large-scale feature is the banding parallel to the equator. The bright bands are called zones and the dark ones are called belts. Figure 11.2 illustrates a widely accepted model of the banding on Jupiter, based on temperature measurements and on vertical cloud motions.

The zones are marked by high, cool clouds freshly formed near the tops of convective columns, and therefore consisting of clean crystals (presumably of NH_3). The belts are where the atmosphere is sinking to complete the convective cycle. They are freer of cloud because the condensates have been frozen out as cloud particles in the zones, from where a substantial fraction of the particles precipitate (presumably), rather than get carried into the belts. The sparseness of the cloud in the belts exposes deeper-lying, warmer regions that are more richly coloured, and they are darker at visible wavelengths because they are poorer reflectors of sunlight. This convective model of the banding also explains the presence at Jovian cloud top altitudes of CO and phosphine (PH_3). These are present only as trace gases, but far above the quantities that should exist in chemical equilibrium at the low temperatures at these altitudes.

❐ What does this indicate?

This indicates that they have been borne aloft from deep below where the higher temperatures have created them.

On Saturn the correlation between zones and ascending atmosphere and belts and descending atmosphere is weak, and so the convective model in Figure 11.2 might not apply.

For both planets we still have to account for the zones and belts forming bands rather than some other geometrical form, and for the orientation of these bands parallel to the equator. A clue is supplied by the winds (as revealed by cloud motion). They are predominantly parallel to the equator with speeds typically as in Figure

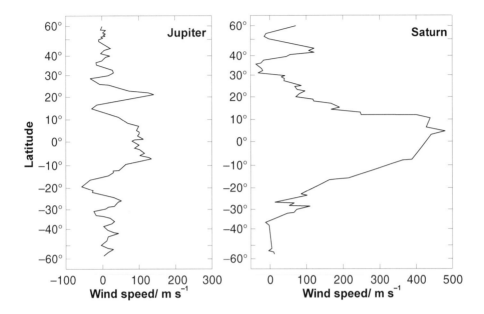

Fig. 11.3 Winds on Jupiter and Saturn. Positive values are winds from west to east

11.3, where positive values are west to east, and negative are east to west. The speeds are with respect to the rotation of the deep interior, which is deduced from the periodicity of radio emissions from charged particles orbiting under the influence of the planets' magnetic fields. The details won't concern us, but the result is that the rotation period of the deep interior of Jupiter is 0.4135 days, and 0.4440 days for Saturn. The rotation periods at the cloud tops are somewhat smaller than these values at all but high latitudes.

Broadly, there is a correlation between the wind speed and the banding, with changes in wind speed as we go from belt to zone or zone to belt. The rapid rotation of Jupiter and Saturn gives a strong Coriolis effect. This tends to deflect the north–south motion at the top of the convection cycle into an east–west motion, and this must be a factor in creating the east–west winds and the associated bands. Whether it is sufficient is unknown.

Another, perhaps dominant circulation pattern is illustrated in Figure 11.4. It extends throughout the molecular hydrogen envelope. Models of rapidly rotating fluid spheres with adiabatic temperature gradients give rise to such concentric cylinders of fluid. If this pattern exists in Jupiter and Saturn then the bands could be the tops of the cylinders. The difference between the thicknesses of the molecular hydrogen envelopes in Jupiter and Saturn could then help account for the differences in the wind patterns between these two giant planets.

Whatever the cause of the banding, the atmospheres of Jupiter and Saturn differ from those of the planetary bodies we met earlier in that solar radiation is the main atmospheric energy source only in and above the upper troposphere.

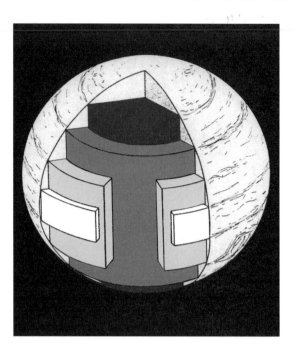

Fig. 11.4 A possible circulation pattern in the molecular hydrogen envelopes of Jupiter and Saturn. (Adapted by permission from Figure 14 (p. 148) of *The New Solar System*, J. K. Beatty and A. Chaikin (eds), Sky Publishing, 1990)

Deeper down there is little or no solar radiation, though there is a copious flow of energy outwards from the hot interiors. The importance of this flow might be apparent in the Galileo probe measurements of wind speeds, namely, 80–100 m s^{-1} at the top of the NH_3 cloud, increasing to about 170 m s^{-1} 35 km deeper, and then remaining at about this value for as far as the probe reached, about 100 km below the cloud top, to where (over most of Jupiter) little solar radiation penetrates. If the winds are powered only by the absorption of solar radiation, the speeds might be expected to have decreased considerably as the probe sank. That they did not indicates that the outward flow of internal energy might be the main power source of these winds. If so, this supports the concentric cylinder model of the banding.

Further insight into the nature of the circulation in Jupiter and Saturn is provided by the temperatures in the upper troposphere—these vary only slightly with latitude. The decrease of solar irradiance with latitude tends to produce a temperature decrease from equator to pole. That the measured decrease is so slight must be largely the result of equator-to-pole circulation, and perhaps partly because of heat flow from the interior, which is expected to be independent of latitude.

Ovals and spots

As well as belts and zones, Jupiter and Saturn display ovals of various sizes. Cloud motions within and around the ovals show them to be vortices, also called eddies—

regions of rotating atmosphere rather like cyclones on Earth. The atmosphere is probably rising at the core and descending in a spiral outside the core. Vortices are concentrated at the boundaries between the bands. This is where wind shear is greatest, and consequently the vortices often develop into wavy streaks. The vortices probably derive their rotational energy from the motion of the rising currents of atmosphere. The largest vortex, the Great Red Spot on Jupiter (Plate 11, lower right) has a length twice that of the Earth's diameter, and it has existed for at least 100 years, probably for over 300 years. Its longevity is a result of its size. From when it became sufficiently large, its rotational energy has been so huge that it has survived encounters with smaller features. Longevity also owes something to the lack of a solid surface to dissipate rotational energy. The Great Red Spot rotates once every six days and extends both above and below the surrounding clouds. Three white vortices south of the Great Red Spot are each about the diameter of Mars, and are at least 50 years old. Smaller vortices on both planets have been seen to come and go. Generally, vortices have lifetimes that decrease as their size decreases, the smallest seen having lifetimes of only a few days.

Other patchy or wispy features are not vortices. In some cases they are glimpses through the upper cloud layer. These include irregular dark brown spots and irregular blue-grey/purple spots, though at about 10^3 km across they are much larger that the word 'spot' implies. The dark-brown spots are thought to be glimpses of a lower cloud layer, whereas the blue–grey/purple spots are the hot spots — so warm that they are thought to be cloud-free areas, giving us a view limited only by the depth to which sunlight can penetrate, a limit imposed by the scattering of sunlight from atoms and molecules — this is **Rayleigh scattering** (and it also accounts for the blue skies on Earth). Clearly the underlying clouds are patchy, in accord with the Galileo probe findings.

❐ What kind of spot do you think this probe descended through?

The probe descended through a blue–grey/purple spot.

Saturn has fewer features disturbing the banding, and nothing on the scale of the Great Red Spot. Instead of really large spots there are meandering waves. There is little understanding of the reasons for these differences.

11.1.4 Coloration

It remains to account for the colours of Jupiter and Saturn. The substances that are thought to dominate each cloud layer — NH_3, NH_4SH, NH_4OH, and H_2O — are all colourless. Clouds consisting of these substances would therefore appear white. White-ish features are confined to high altitudes and are presumably freshly condensed ammonia particles almost free of colouring agents. Other features have presumably had time to become coloured. It needs only a trace of a coloured substance to cause the intensity of the observed colours. One group of possible colouring agents, as on Io, is various forms of solid particles of sulphur, derived photochemically from sulphur compounds below the upper cloud deck. Some forms of solid sulphur are yellow, others are brown. Another possibility is phosphorus, solid particles of which can be yellow or red. Phosphine (PH_3) has been detected as a trace in both Jupiter and Saturn, though it is not known if the appropriate chemical reactions occur to form phosphorus particles. Yet other possibilities are various compounds that can be formed by chemical reactions involving some or all of CH_4, NH_3, and H_2S. But in spite of all these possibilities the actual colouring agents and the reasons for their spatial distribution are unknown.

Question 11.2

Suppose that the atmospheres of Jupiter and Saturn consisted *only* of hydrogen and helium.
(a) Discuss plausible differences this would make to: the clouds; the temperatures in the upper molecular envelope; the facility with which the atmospheric circulation could be investigated.
(b) State why the solar-driven component of the circulation would be changed.

11.2 The Atmospheres of Uranus and Neptune Today

Plates 19 and 20 show that Uranus and Neptune are blander in appearance than Jupiter and Saturn, and that Uranus is blander than Neptune. The predominant visual impression in Plates 19 and 20 is the result of the Rayleigh scattering of sunlight from a deep layer of gas, the bluish-green tint arising from absorption of red wavelengths by methane. The slightly greener tint of Uranus could be due to a deep-lying layer of methane cloud.

11.2.1 Vertical Structure

Figure 11.5 shows the vertical structures of the atmospheres of Uranus and Neptune, deduced from much the same variety of measurements used for Saturn. The observable atmospheres are generally colder at comparable pressures than those of

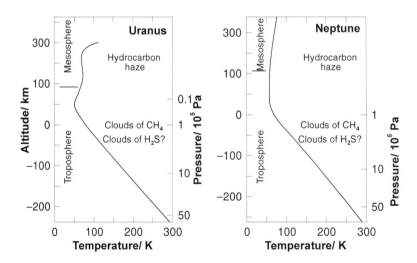

Fig. 11.5 The vertical structure of the atmosphere of Uranus and Neptune

Jupiter and Saturn, a consequence of the greater solar distance. Heating is by solar radiation down to about the 10×10^5 Pa level, which includes the upper troposphere. It is not known whether the lapse rates are as large as the adiabatic values. Below this level the increase of temperature with depth depends on internal heat sources. These greater depths have been explored by microwave observations that have revealed that whereas in Neptune the increase is probably at the adiabatic rate, in Uranus the lapse rate is generally a bit smaller. Therefore, the deep troposphere of Neptune is probably convecting, whereas that of Uranus is probably not, or only weakly so. One would therefore predict that there must be a far lower rate of escape of heat from the interior of Uranus than from the interior of Neptune. This is consistent with the barely detectable infrared excess from Uranus, and the greater excess from Neptune (Table 9.2). Models indicate that deeper into the interior the lapse rate is adiabatic in both planets.

In the upper troposphere there are rare flecks and streaks of cirrus-like clouds. These are white because they are at too shallow a depth to be seen though a tinted atmosphere. Higher still there is a thin haze of photochemical compounds, mainly hydrocarbons. The low temperatures result in extensive freezing out at low altitudes of many condensable atmospheric components, giving rise to deep-lying cloud decks. The composition of the visible and deep-lying clouds can only be inferred, but evidence from atmospheric composition, atmospheric temperatures, and the CH_4 phase diagram strongly suggests that the high-altitude white clouds are CH_4 ice, whereas any deep-lying cloud deck is probably CH_4 droplets, or, less likely, H_2S. If there is a deep-lying CH_4 cloud deck, then models suggest it might be underlain by H_2S clouds, perhaps underlain in turn by H_2O clouds.

11.2.2 Composition

Table 11.1 shows that, like Jupiter and Saturn, the atmospheres of Uranus and Neptune consist mainly of H_2 and He, though He is rather more abundant in Uranus and Neptune, particularly compared to Saturn. These different He/H_2 ratios have a ready explanation based on our models of the interiors. The interiors of Uranus and Neptune do not attain pressures sufficiently high for metallic hydrogen mantles to form. Therefore, there is no possibility of the downward concentration of helium arising from its insolubility in metallic hydrogen. Consequently the elemental He/H ratio measured in the observable atmospheres of Uranus and Neptune is the same as that in the whole planet. The interiors of Jupiter and Saturn *do* attain sufficient pressures for metallic hydrogen mantles to form, but whereas Jupiter is so hot that only slight segregation has occurred, the cooler interior of Saturn has led to considerable segregation. It is thus possible that the whole-planet He/H ratios are similar in all four giant planets. We shall return to this important point in Section 11.3.

Another difference is the supposed CH_4 condensation. This is likely, and therefore, in determining the CH_4 content it is important to measure the composition at depths where the temperature is sufficiently high to prevent condensation. The measurements in Table 11.1 are at such depths, and you can see that CH_4 is more abundant in Uranus and Neptune than in Jupiter and Saturn. This is another important point for Section 11.3. High in the atmospheres, Uranus and Neptune are greatly depleted in CH_4. This lends strong support to the view that there is a deep-lying CH_4 cloud deck.

H_2O has not been detected in either atmosphere. This is not surprising in view of the low temperatures.

❏ Why is it not surprising?

H_2O is less volatile than CH_4, so it would condense very deep in the atmospheres. Whether there is a very deep-lying H_2O cloud deck is unknown, but if there is, and if the particles are liquid then we can explain another observation—the scarcity of NH_3. This is no surprise in the upper troposphere, again because of the low temperatures, but microwave observations indicate that NH_3 depletion persists to greater depths. NH_3 is very soluble in liquid H_2O, so this scarcity could be the result of its dissolution in H_2O droplets. However, there are other possible explanations.

One is that the nitrogen/sulphur ratio is so low that NH_3 has largely disappeared into the formation of NH_4SH, which would form deep-lying clouds. Another possibility relates to the chemical reaction that produces NH_3 from N_2

$$N_2 + 3H_2 \longrightarrow 2NH_3 \tag{11.4}$$

This reaction is favoured in the low-temperature conditions in the atmospheres of Uranus and Neptune, but it requires a catalyst if it is to proceed at a high rate. If there are no suitable catalysts, and if the original form of nitrogen were N_2 rather than NH_3, then this could also help to explain the scarcity of ammonia.

In the upper troposphere of Neptune CO and HCN have been detected, but not in the case of Uranus. This might be another indication that convection is absent in some layers of the atmosphere of Uranus. In hydrogen-rich atmospheres at the low temperatures in the upper tropospheres, chemical reactions would quickly reduce CO and HCN to very small quantities unless they were being replenished at a sufficient rate from the deep interior. Convection can fulfil this role. In the case of CO, it is CO itself that would be brought up, whereas for HCN it is N_2, which then participates in reactions that yield HCN. The direct detection of N_2 is beyond present capabilities because of its weak (infrared) spectral lines.

11.2.3 Circulation

The circulation of the atmospheres, as for Jupiter and Saturn, is revealed largely by bands, and by the motion of cloud features. The speeds of motion are again with respect to the rotation period of the interior, and this period is again determined from the periodicity of radio emissions. It has been more difficult to establish the circulation of Uranus. This is because it displays fewer atmospheric features than Neptune (Plates 19 and 20). The rare, elusive white flecks of cloud in the upper troposphere of Neptune are even rarer and more elusive on Uranus. Though no main cloud decks are visible for either planet, a deep-lying haze layer on Neptune exhibits dark belts and dark spots. A large dark spot was discovered in the southern hemisphere by Voyager 2 in 1989, and named the Great Dark Spot (Plate 20). It

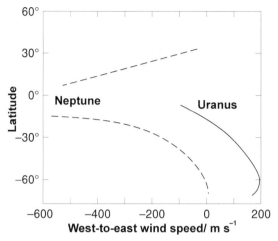

Fig. 11.6 Winds on Uranus and Neptune

covered about the same fraction of Neptune's surface as the Great Red Spot covers of Jupiter's surface, and also seems to have been a vortex. It had vanished when Neptune was examined by the Hubble Space Telescope a few years later, but another one had appeared in the northern hemisphere. The dark spots can be interpreted as holes in the haze layer, giving a view into deeper regions that appear darker because they scatter less sunlight. The deep haze layer of Uranus has very weak banding, and no spots have been seen.

With fewer haze and cloud features on Uranus—perhaps a result of the lower heat flow from the interior—the wind speed data are few and uncertain. Figure 11.6 shows a rather idealised picture for both planets. The equatorial value for Uranus is from radio occultation data. Though there are considerable differences in the details between Figure 11.6 and the corresponding Figure 11.3 for Jupiter and Saturn, all four planets exhibit circulation that is predominantly parallel to the equator.

❒ What is the cause?

This is the result of planetary rotation. One striking feature on Neptune is the large negative wind speeds—the atmosphere is predominantly rotating more slowly than the interior. This might reflect some deep-seated circulation, though at present the cause is largely a mystery. There is, however, a plausible explanation of Neptune's small temperature difference between poles and equator. As in the cases of Jupiter and Saturn, this could well be the combined effect of a high rate of heat flow from the interior, which is independent of latitude, and equator-to-pole circulation efficient enough to offset the greater equatorial insolation.

In the case of Uranus the large tilt of the rotation axis means that during the Voyager 2 flyby in 1986 the south pole was pointing almost directly towards the Sun and the north pole away from the Sun. Given the 84-year orbital period of Uranus this had been the situation for some years, and it remained much the case for several years afterwards. Thus, if solar heating is important in the circulation of Uranus it would be tempting to conclude that the circulation we have observed over the last decade or so is impermanent. This is not necessarily so. Averaged around Uranus's orbit the insolation is greater at the poles than at the equator, and the thermal response time of the troposphere is of the order as the orbital period. Consequently, a Hadley circulation transferring heat from the poles to the equator could be a fairly persistent feature in each hemisphere. It can be shown that such a flow is consistent with the variation of wind speed with latitude in Figure 11.6. However, measurements show that the pole is not much hotter than the equator. Therefore, there is an efficient redistribution of absorbed solar radiation across the whole planet. The precise mechanism is unknown.

Question 11.3

In Table 11.1, the mixing ratios of helium in the atmospheres of Uranus and Neptune are indirect estimates, based on the amounts needed to top up the atmospheres to their total masses. These estimates have assumed that the proportion of N_2 is negligible. In which of these two planets might this assumption fail, and what is the evidence? What would be the effect on the estimate of the helium mixing ratio?

11.3 The Origin of the Giant Planets — A Second Look

The atmospheres of the giant planets are so massive that they are not distinct from their interiors. The origin of the atmospheres is therefore inseparable from the origin of the giant planets as a whole.

In the solar nebular theory (Chapter 2), the planets form from a disc of gas and dust around the Sun. This disc has much the same composition as the young Sun, which is preserved today outside the core in which fusion reactions are occurring. The composition by mass is about 71% hydrogen, 27% helium, and about 2% 'heavy' elements (Section 2.2). The *gas* in the disc has much the same He/H mass ratio as the young Sun, but it is somewhat depleted in those heavy elements that form the icy and rocky compounds that make up the dust grains in the disc. Conversely, the dust grains are enriched in heavy elements, and depleted in hydrogen and helium. Helium is almost entirely absent from the grains, and only a small fraction of the hydrogen is present. This is in compounds such as H_2O, NH_3, and CH_4 at the distances that the giant planets formed from the Sun. In the kernel theory of the formation of a giant planet (Section 2.2.5), the dust accretes to form icy–rocky planetesimals and a number of these become assembled into a kernel of several Earth masses of rocky and icy materials, massive enough to capture nebular gas and other planetesimals. We thus have two distinct sources of giant material — nebular gas (with a trace of dust), and icy–rocky planetesimals. The planetesimal material tends to concentrate into a central core, and the nebular gas into a surrounding envelope.

Can we reconcile the data in Table 11.1 with this theory? To see whether we can it is convenient to convert the data into elemental mass ratios with respect to hydrogen. For example, the He/H mass ratio is the ratio of the mass of helium in a region to the mass of hydrogen in the same region. For elements such as N, C, and O that form compounds, the mass of the element is the sum over all the forms in which it has been

Table 11.2 Elemental mass ratios with respect to hydrogen in the giant planet molecular envelopes and in the young Sun

Ratio[a]	Jupiter	Saturn	Uranus	Neptune	Young Sun
He/H[b]	0.31	0.06	0.36	0.36[a]	0.39
C/H	1.3×10^{-2}	2×10^{-2}	1.8×10^{-1}	1.8×10^{-1}	4.3×10^{-3}
N/H[c]	1.8×10^{-3}	1×10^{-3}	—	—	1.5×10^{-3}
O/H[c]	$< 3.0 \times 10^{-3}$	—	—	—	1.5×10^{-2}

[a]Values for the giants are derived from Table 11.1, except that the He/H ratio for Neptune has been reduced on the assumption that the outer envelope contains a number fraction of N_2 of 0.003.
[b]The values for Uranus and Neptune are not significantly different from the Sun.
[c]A dash indicates that no useful upper limit is available.

measured. Table 11.2 gives the elemental mass ratios for C, N, O, and He for the four giant planets. Values for the young Sun have been added. Note that for Neptune, the He/H ratio has been adjusted on the assumption that there is a number fraction of N_2 of 0.003 in the outer envelope (Question 11.3). What do these ratios tell us?

The He/H ratio

The He/H mass ratio in the captured gas is expected to be much the same as in the young Sun.

❏ Why is this?

This is because helium has remained almost entirely in the gas, and only a small fraction of the hydrogen has been removed from the gas in the form of compounds in the planetesimals. This expectation is borne out in Uranus and Neptune within the uncertainties of the data for these planets. Jupiter has a slightly but significantly smaller ratio, and Saturn a considerable smaller ratio. Remembering that the data are for the outer envelopes of the giants, this is a further indication that there has been slight settling of helium in the metallic hydrogen mantle of Jupiter, and considerably more in that of Saturn. This is in accord with the expectation that the mantle of Jupiter is at a considerably higher temperature than that of Saturn (Section 5.3.1).

There seems to have been no settling in Uranus and Neptune, and this is in accord with the lower interior pressures and the consequent absence of metallic hydrogen.

The C/H ratio

The C/H ratios in Table 11.2 are thought to apply to the whole envelope — methane does not form clouds in Jupiter and Saturn, and the values for Uranus and Neptune are probably below the methane cloud base. You can see that for all four giant planets the C/H ratio exceeds the solar value. Moreover, the ratio increases from Jupiter, to Saturn, to Uranus and Neptune. How can these excesses be reconciled with the supposed formation of the envelopes from captured nebular gas that was at least mildly depleted in heavy elements?

The answer is probably that the planetesimal material becomes hot enough during and after planet formation to lose some of its more volatile constituents. These constituents thus enrich the surrounding envelope in certain heavy elements, including carbon. If all four giant planets have acquired about the same mass of planetesimals (of order 10 Earth masses), and if their planetesimal materials have all outgassed to roughly the same extent, then the C/H ratio in the envelope will be inversely related to the mass of the envelope — the greater the mass of the envelope, the greater the dilution of the outgassed carbon. It follows that there would then be an inverse relationship between the C/H ratio and the total mass of the planet.

❐ Is this what we find?

This is exactly the feature in Table 11.2. Further carbon enrichment in all four cases would come from the subsequent capture of volatile-rich bodies, notably icy–rocky planetesimals and comets.

In the alternative, protoplanet theory of giant planet formation (Section 2.2.5), all four giants are predicted to have an overall composition similar to the Sun. It is therefore very difficult to account for an excess of carbon in the atmospheres. Indeed, core separation of heavy elements would cause a depletion. Capture of planetesimals and comets could offset this depletion, though perhaps not completely, and moreover this might well not produce the strong differences between the giants. In any case, models of the solar nebula that should produce protoplanets seem to have difficulty in doing so.

Other mass ratios

The atmospheric $^2H/^1H$ ratio (not shown in Table 11.2) also increases from Jupiter to Saturn, to Uranus and Neptune. A good explanation of this increase starts in the interstellar medium (ISM), where the dust was formed that subsequently became a component of the solar nebula. The formation of most compounds in the ISM depends on reactions between ions and molecules. In ion–molecule reactions involving hydrogen, there is a tendency for the heavy isotope to become concentrated in condensable substances. For example, in HCN the ratio $^2HCN/^1HCN$ is greater than the overall $^2H/^1H$ ratio. In the solar nebula the dust is thereby enriched in ^2H (deuterium), and consequently the planetesimal material in likewise enriched. The remainder of the argument parallels the one just given for the C/H ratio — dilution on outgassing. The protoplanet theory has difficulty explaining the $^2H/^1H$ ratios too, and for similar reasons to its difficulty with the C/H ratios.

The N/H ratios for Jupiter and Saturn are roughly the same as the solar value, and so the N/C values are less than solar. This could be because nitrogen was less readily incorporated into icy–rocky planetesimals than carbon. This is possible because much of the carbon and nitrogen in the solar nebula could have been present as CO and N_2, rather than as CH_4 and NH_3, in which case the planetesimals would preferentially acquire carbon, CO being less volatile than N_2. Outgassing within the giants then leads to a less than solar N/C ratio. For Uranus and Neptune, NH_3 has not yet been detected, implying N/H considerably less that the solar value, which is difficult to reconcile with the outgassing model. However, you have seen that if the nitrogen is largely in the form N_2 it would not yet be detectable. It is also possible that NH_3 has been taken up in the formation of NH_4SH, or that it has dissolved in deep-seated water clouds.

We have no useful measurements of H_2O in giant planet atmospheres. The atmospheres above any H_2O clouds are expected to be very dry, but you have seen (Section 11.1.2) that the Galileo probe penetrated to depths below where water clouds lie elsewhere in Jupiter, and was only able to set an upper limit on O/H of 20% of the solar ratio. Though this might not be typical of Jupiter as a whole, it is possible that water, being less volatile than the repositories of carbon and nitrogen, outgassed from planetesimal material to a far smaller extent. It is also possible that different compounds have different solubilities in metallic hydrogen. About 85% of Jupiter's hydrogen is in the metallic phase, and so a modest excess solubility of water could lead to a large depletion in the molecular hydrogen envelope.

In spite of some outstanding puzzles, we can conclude with some confidence that the atmospheres of the giant planets are consistent with an origin of the giant in which a kernel of icy–rocky materials captured nebular gas (with a trace of dust) and other icy–rocky planetesimals. The composition of the envelopes was then modified by outgassing of the kernels, and to some extent by the subsequent capture of volatile-rich bodies.

Question 11.4

Neon (Ne), like helium, is a volatile inert gas. Discuss whether you would expect the Ne/H ratio to be the same in the giant planets as in the young Sun. Given that neon is soluble in helium, would you expect the Ne/H ratio measured in the outer envelope of each giant to be the same as for the planet as a whole?

11.4 The End

We have now completed our study of the Solar System as it is today, and of how it might have originated, and of how it might have evolved to its present condition. By way of valediction let's consider briefly what awaits it in the future.

The interiors of all planetary bodies will continue to lose energy, and for those bodies whose geological activity is dominated by interior sources of heat, the level of geological activity will decline. Atmospheric loss mechanisms will continue, and with the decline in outgassing that accompanies a decline in geological activity, a decrease in atmospheric mass will occur until the growing luminosity of the Sun (Figure 10.14, Section 10.6) begins to drive volatiles from surface repositories.

In the case of the Earth, for the next 1000 Ma or so, the increasing solar luminosity will make the Earth warmer, though negative feedback should stabilise the temperature to some extent. The feedback operates through a decrease in the size of the greenhouse effect — as the solar luminosity increases it is expected that (in a rather complicated way) the carbon cycle will reduce the atmospheric quantity of CO_2 (unless an advanced civilisation liberates a great mass of it). But the feedback will not be fully effective and so the surface temperature will rise, and this will put more water into the atmosphere. The consequent increase in the greenhouse effect now introduces a positive feedback. One scenario for 1500 Ma into the future has a global mean surface temperature of about 350 K, only 1 ppm of CO_2 in the atmosphere, and the onset of rapid water loss from the atmosphere through photodissociation and hydrogen escape. Further temperature rise liberates CO_2. The Earth is now firmly on the evolutionary trail followed by Venus early in Solar System history. As a result, perhaps by 2500 Ma from now, the Earth will have a Venus-like atmosphere and a corresponding high surface temperature.

At some time during the demise of the Earth as we know it, Mars might again become clement, as carbon dioxide gas and water vapour are released from surface and subsurface reservoirs.

The increasing solar luminosity in Figure 10.14 is the result of the continuing conversion of hydrogen to helium in the solar core (Section 1.1.3). An inert inner core of helium grows in size, surrounded by a shell in which hydrogen is still undergoing fusion. About 5000 Ma from now this shell will be so thin that the rate of energy release by fusion within it will be insufficient to support the interior. The core will then contract and, in a complicated process, the rest of the Sun will expand and cool, the surface taking on an orange tint. The huge increase in the surface area of the Sun will far outweigh the decrease in surface temperature, and so the luminosity will increase greatly. The Sun will reach beyond the present orbit of Venus, nearly as far as the Earth. However, the Sun loses mass in the form of an intense solar wind, so that the planetary orbits increase in size, in the case of the Earth to about 1.7 AU. This might spare the Earth from consumption, and perhaps Venus too, though not Mercury. On the other hand, the tidal interaction between the bloated Sun and the Earth will tend to reduce the size of the Earth's orbit, so it will be touch and go! In any case, the atmospheres of Venus, the Earth, and Mars, and all surface volatiles, will have escaped to space long before. The peak surface temperature of the Earth will be about 1600 K.

The Sun will have then made the transition to a red giant, where the gravitational energy released by the contraction of the helium core will have raised core temperatures to the point where helium undergoes fusion to form carbon, releasing energy at a sufficient rate to maintain a pressure gradient that stabilises the star. The red giant phase will last about 1000 Ma, and Pluto, and some of the satellites of Uranus and Neptune, might become suitable for life.

But this is a turbulent time, with luminosity variations and a copious solar wind. It comes to an end in a runaway instability in which the Sun will cast off a significant fraction of its mass, a supersonic blast of hot gas tearing though the Solar System on its way to forming a huge shell of gas called a planetary nebula (nothing to do with planets). The solar remnant will contract to become an Earth-sized body called a white dwarf. Though its surface will be far hotter than that of the Sun today, it will be so small that its luminosity will be feeble. Its gravitational field will also be much reduced, and it is likely that some planets will escape from the Solar System. These, and the planets that remain in solar orbit, having been cooked and blasted, will then freeze. Any intelligent beings that might have survived will probably have to journey to other stars to find somewhere to live. The existence of exoplanets will give them great expectations.

11.5 Summary

Tables 9.1 and 9.2 list basic properties of planetary atmospheres, and Tables 11.1 and 11.2 give further data for the giant atmospheres.

All four giants have atmospheres dominated by molecular hydrogen (H_2) and helium (He), with small quantities of less volatile substances that form clouds. There are traces of other (unknown) substances that colour the clouds. The clouds are banded parallel to the equator. The circulation revealed by the clouds is predominantly east–west, and a factor in this is the large Coriolis effect in these rapidly rotating planets that acts on convection patterns. For Jupiter and Saturn there is probably a major contribution from deep-seated circulation patterns in the molecular envelope, in the form of concentric cylinders.

Jupiter, Saturn, and Neptune have adiabatic lapse rates in their tropospheres, the result of heat welling up from the interior, supplemented by the absorption of solar radiation in the upper tropospheres. Uranus has a sub-adiabatic tropospheric lapse rate in the lower troposphere, consistent with the low rate of upwelling heat.

The composition of the giant planets' atmospheres can be explained by the capture by icy–rocky kernels of nebular gas during the formation of the Solar System, with subsequent modifications due to the downwards segregation of helium in the metallic hydrogen mantles of Jupiter and Saturn, the outgassing of the kernels, and the capture of volatile-rich bodies.

The Solar System will undergo huge changes when the Sun evolves into a red giant, and then into a white dwarf.

Question Answers and Comments

Note that comments are in brackets { }. These are not expected as part of your answer.

Chapter 1

Question 1.1

At the lower photospheric temperature in the past the spectrum was at longer wavelengths. Also, the lower luminosity in the past gave a smaller area under the spectrum. Since then, the spectrum has shifter to shorter wavelengths and the area under the spectrum has increased.

Question 1.2

Figures 1.4 and 1.5 together show that the medium-sized planets — the four terrestrial planets — are closest to the Sun, the largest, the Earth, being the third one out. Then comes the largest planet, Jupiter, then the somewhat smaller planet Saturn, and then the two smaller giant planets, Uranus and Neptune. Finally we come to Pluto, the smallest planet of all. The broad correlation of size with distance is therefore that the largest planets lie at intermediate distances, with smaller planets closer to, and further from, the Sun. {The reasons for this are discussed in Chapter 2.}

Question 1.3

The semimajor axis of Fortuna is obtained from equation (1.4):

$$\frac{a_F}{a_E} = \left(\frac{P_F}{P_E}\right)^{2/3}$$

where the subscript E denotes the Earth, and F denotes Fortuna. For the Earth, a_E is 1 AU and P_E is 1 year. The orbital period P_F of Fortuna is given in the question as 3.81 years. Therefore

$$a_F = (1\,\text{AU}) \times \left(\frac{3.81\ \text{years}}{1\ \text{year}}\right)^{2/3} = 2.44\,\text{AU}$$

Question 1.4

The perihelion distance of a planet from the Sun is $a(1-e)$ (Figure 1.8). Taking a and e from Table 1.1, the perihelion distances of the Earth and Venus are, respectively, 0.983 AU and 0.718 AU. Thus the Earth–Venus distance is 0.265 AU.

From the radar data, this distance is also $(3.00 \times 10^5\,\text{km s}^{-1}) \times (264\ \text{s}/2)$, which is 3.96×10^7 km. Therefore, $1\,\text{AU} = 3.96 \times 10^7\,\text{km}/0.265 = 1.5 \times 10^8$ km to two significant figures.

Question 1.5

(a) {One possible sketch is shown. You might have adopted a different viewpoint, but the relationship between the orbit of comet Kopff, the ecliptic plane, and ♈ should be the same. Also, the shape of the orbit should (roughly) reflect its eccentricity, with the Sun distinctly off-centre, and with some allowance for the oblique viewpoint.}

(b) Denoting the perihelion distance by q, Figure 1.8 shows that

$$q = a(1 - e)$$

and so

$$a = \frac{q}{1 - e}$$

Thus, putting in the values of q and e given in the question

$$a = \frac{1.58\,\text{AU}}{1 - 0.54} = 3.43\,\text{AU}$$

The aphelion distance is (Figure 1.8) $a(1+e)$, which is 5.28 AU. The orbital period P_K is obtained from Kepler's third law in the form in equation (1.4), rearranged as

$$P_K = P_E \left(\frac{a_K}{a_E}\right)^{3/2} = (1\ \text{year})\left(\frac{3.43\,\text{AU}}{1\,\text{AU}}\right)^{3/2} = 6.35\ \text{years}$$

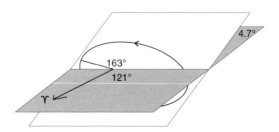

Fig. Q1.5(a)

This is 6 years and 0.35×365.24 days, which is 128 days. Thus, the first perihelion date in the twenty-first century is 6 years 128 days after 2 July 1996. This puts it in November 2002.

(c) If the orbits are in the same plane they will intersect, but they are not — the orbits have different inclinations. In this case they will intersect only if the two orbits have just the right relative sizes for the difference between the longitudes of the ascending node — given that the orbits do not intersect we must conclude that this condition is not met.

Question 1.6

Applying equation (1.6) to the Earth

$$M_\odot + m_E = \frac{4\pi^2 a_E^3}{GP_E^2}$$

$M_\odot \gg m_E$, so

$$M_\odot = \frac{4\pi^2 a_E^3}{GP_E^2} = \frac{4\pi^2 (1.496 \times 10^{11} \text{m})^3}{(6.672 \times 10^{-11} \text{N m}^2 \text{ kg}^{-2})(3.156 \times 10^7 \text{ s})^2}$$

Therefore

$$M_\odot = 1.989 \times 10^{30} \text{ kg}$$

Applying the same procedure to Jupiter

$$M_\odot \approx \frac{4\pi^2 (7.783 \times 10^{11} \text{ m})^3}{(6.672 \times 10^{-11} \text{ N m}^2 \text{ kg}^{-2})(3.743 \times 10^8 \text{ s})^2}$$

Therefore

$$M_\odot \approx 1.991 \times 10^{30} \text{ kg}$$

The values of a and P in Table 1.1 are only given to four significant figures, so these two values of the solar mass just about agree within this precision. However, we have actually calculated $M_\odot + m_{\text{planet}}$, and m_J is about $M_\odot/1000$, whereas M_E is only about $M_\odot/300\,000$. Therefore, the second value needs to be reduced by the factor $1/1000$, giving much better agreement. {The accurate value for $M_\odot$ is 1.9891×10^{30} kg.}

Question 1.7

(a) If the Sun rotated more rapidly then its departure from spherical symmetry would be greater. Therefore, the orbital elements of Venus would become more variable. {This would not apply to a planet in an orbit in the equatorial plane of the Sun. No planets orbit in this plane.}

(b) If the mass of Jupiter were doubled it would exert a greater gravitational force on the other planets. Again, the orbital elements of Venus would become more variable. {Also, because Jupiter would more greatly distort the other planets tidally, they would have a greater influence on Jupiter itself.}

(c) If the Sun entered a dense interstellar cloud the gravitational and non-gravitational forces on Venus would increase. Once more, the orbital elements of Venus would become more variable.

Question 1.8

Table 1.1 shows that the axial inclination of Venus is 177.4°, and so makes an angle of only 2.6° with the plane of its orbit — nearly ten times less than the case of the Earth. Therefore, seasonal variations would be small, unless its orbit were very eccentric. However, Table 1.1 shows that the orbital eccentricity is only 0.007, less than half that of the Earth. {The seasonal variations on Venus are indeed small.}

Question 1.9

If our calendar were based on the sidereal year, then, because of the precession of the equinoxes, the seasons would gradually move through the year. For example, 12 900 years from now, in the northern hemisphere spring would start in September, and the summer solstice would be in December.

Question 1.10

The orbit of Mars is moderately eccentric, whereas the Earth's orbit is only slightly eccentric (Table 1.1). Consequently, the opposition distance is more sensitive to the position of Mars in its orbit than to the position of the Earth in its orbit. It is thus particularly important to have Mars near perihelion when it is at opposition. From Figure 1.22 it is clear that when Mars is at perihelion, then for it to be in opposition the Earth has to be about two thirds of the way around its orbit from its own perihelion in early January. Therefore, the date of such an opposition is late August/early September. {In fact, late August.} Because the Earth is at aphelion in July, the closest oppositions will be slightly earlier. {In fact, in mid-August.}

Question 1.11

Eclipses can only occur when the Moon is at or near a node of its orbit when at the same time the nodes are on or near the line from the Earth to the Sun. If these nodes are *not* fixed with respect to the Earth's orbit around the Sun then the line-up will move through the months of the year, and consequently eclipses are not confined to particular months.

Question 2.1

(a) The greater the mass of a planet the larger the displacement of the star from the centre of mass of the system (equation (1.7)), and therefore the larger the orbit of the star around the centre of mass. This makes it easier to detect and measure the stellar orbital motion.

(b) For a stellar orbit of given size, the angular size of the orbit is greater the nearer the star, and therefore the easier the orbital motion is to detect and measure.

Question 2.2

The feature already apparent in the existence of circumstellar discs is feature 2 in Table 2.1 — that the orbits of the planets lie in almost the same plane {the plane of the circumstellar disc}, and that the Sun lies near the centre of this plane.

Question 2.3

(a) The average orbital angular momentum of a planet is given to a good approximation by

$$l_{orb} = mva$$

where m is the mass of the planet, v is its average orbital speed, and a is the semimajor axis of the orbit. The average speed is given by $2\pi a/P$ where P is the orbital period. Therefore

$$l_{orb} = m\left(\frac{2\pi a}{P}\right)a = m\left(\frac{2\pi a^2}{P}\right)$$

Using Kepler's third law

$$P = ka^{3/2}$$

we obtain

$$l_{orb} = m\left(\frac{2\pi a^2}{ka^{3/2}}\right)$$

Therefore

$$l_{orb} = \frac{2\pi}{k}ma^{1/2} \tag{2.2}$$

(b) We use equation (2.1) in the form in part (a)

$$l_{orb} = m\left(\frac{2\pi a^2}{P}\right)$$

with values for m, a, and P from Table 1.1. Thus, for the Earth

$$l_{orb} = 5.974 \times 10^{24}\,\text{kg}\left(\frac{2\pi(1.496 \times 10^{11}\,\text{m})^2}{3.156 \times 10^7\text{s}}\right) = 2.662 \times 10^{40}\,\text{kg}\,\text{m}^2\,\text{s}^{-1}$$

From equation (2.2)

$$l_{orb} = l_{orb}(\text{Earth}) \frac{ma^{1/2}}{m_E a_E^{1/2}}$$

So, from Table 1.1, for Jupiter

$$l_{orb} = (2.662 \times 10^{40} \text{ kg m}^2 \text{s}^{-1}) \left(\frac{1.8987 \times 10^{27} \text{ kg}}{5.974 \times 10^{24} \text{ kg}} \right) \sqrt{\frac{7.783 \times 10^{11} \text{ m}}{1.496 \times 10^{11} \text{ m}}}$$

$$= 1.930 \times 10^{43} \text{ kg m}^2 \text{ s}^{-1}$$

Similarly

$$l_{orb} = 2.513 \times 10^{42} \text{ kg m}^2 \text{ s}^{-1}$$

for Neptune.

Question 2.4

The ice line divides the dust into two zones, with water ice dominating beyond this boundary, and rocky materials within the boundary. The ice line is roughly where the present division between the terrestrial planets and the giant planets lies (Figure 2.6).

Question 2.5

(a) Assume that all the dust in the sheet in the distance range 0.8–1.2 AU forms planetesimals in that range, and that no planetesimals come from elsewhere. In this case the total mass M of the planetesimals is the area between these distances times the average column mass. To sufficient accuracy, from Figure 2.6

$$M = \pi((1.2 \text{ AU})^2 - (0.8 \text{ AU})^2) \times 10^2 \text{ kg m}^{-2}$$

Converting AU to metres (Table 1.6), we get

$$M = 6 \times 10^{24} \text{ kg}$$

This is similar to the Earth's mass of 5.974×10^{24} kg, and so this suggests that much of the dust sheet in the range 0.8–1.2 AU went to form the Earth. This is consistent with 0.8 AU being beyond the orbit of Venus, and 1.2 AU being within the orbit of Mars.

(b) The size range of planetesimals is given in Section 2.2.3 as 0.1–10 km across, so use a typical size of 1 km across. {A typical size of 0.5 km is acceptable, but 5 km is not — there will be only a small proportion this big.} Taking the typical size to be the diameter of a sphere, the typical volume V is

$$V = \tfrac{4}{3}\pi(500 \text{ m})^3 = 5 \times 10^8 \text{ m}^3$$

With a mean density given in the question as 2500 kg m^{-3}, the typical mass m is 1×10^{12} kg. {It is sufficient to work to one significant figure.} Thus the number of planetesimals in the range 0.8–1.2 AU is of order

$$\frac{M}{m} = \frac{6 \times 10^{24} \text{ kg}}{1 \times 10^{12} \text{ kg}} = 6 \times 10^{12} \text{ \{A lot!\}}$$

Question 2.6

The features in Table 2.1 that apply to the terrestrial planets are (in the numbering in the table) 2, 3, 6, 7, and 9. All these features can be explained by solar nebular theories. {The rotation of Venus is not discussed explicitly in Chapter 2. Its retrograde rotation is discussed briefly in Section 10.3.3.}

Question 2.7

If the protoSun went through its T Tauri phase much earlier than in Figure 2.12, then the growth of the giants would have been stunted. Jupiter and Saturn would have acquired less massive envelopes of hydrogen and helium, and Uranus and Neptune might have got no further than the kernel stage.

Question 2.8

The satellites of the giant planets in Table 1.2 (additional to Triton), that are likely candidates for a capture origin, are as follows:

Jupiter: Ananke, Carme, Pasiphae, Sinope
Saturn: Phoebe
Uranus: S/1997U1, S/1997U2
Neptune: (only Triton)

These all have orbital inclinations greater than 90°, so are in retrograde orbits. They are also in large orbits, well away from the giant planet.

Question 2.9

(a) The present ring system of Saturn is the outcome of the accumulated effects of various processes that replenish ring particles, and various processes that remove them. The time scale on which these processes operate is much shorter than the age of the Solar System. Therefore, if the replenishment rate were to fall substantially below the removal rate, then at some time in the future it is possible that the ring system could be much less extensive.
(b) By the same arguments as in part (a), it is possible that the ring system of Jupiter will be much more extensive in the future: this would require the replenishment rate to rise substantially above the removal rate.

Question 3.1

(a) The orbital inclinations of Pallas and Euphrosyne are unusually large for belt asteroids, the inclinations of belt asteroids being predominantly less than 20°. The eccentricity of Bamberga is considerably larger than the typical range of 0.1–0.2 for belt asteroids.

(b) To be at or near L_4 or L_5, the semimajor axis of the orbit must be similar to that of the planet (Figure 3.3) {and the eccentricity has to be similar}. Comparison of the semimajor axes in Table 1.3 with those in Table 1.1 shows that none of the asteroids is at or near any L_4 or L_5 points.

Question 3.2

It depends on how rapidly the number of asteroids increases with decreasing size. Consider the total mass in a certain range of size Δr centred on r_1. The total mass is the number of asteroids in the range times the average mass of an asteroid in the range. If Δr is centred on a smaller size r_2, the average mass decreases, and this can offset the greater number of asteroids in this new range, so that the total mass is considerably less than in the first case. {In Figure 5.4, unless the graph turns up unrealistically sharply at sizes below those shown, there is only a small fraction of the total mass in the asteroid belt awaiting discovery.} {You might have answered the question in a different way, perhaps without using symbols. Does your answer contain the essential idea, and does it express it succinctly?}

Question 3.3

(a) All the asteroids in Figure 3.6 are considerably smaller than the approximate 300 km radius above which gravitational forces make a body roughly spherical.
(b) A small spherical asteroid could have been born spherical, in which case it is not an unaltered collisional fragment, which would be non-spherical. Alternatively, regardless of its initial shape, it could have since suffered extensive erosion by dust impact, which tends to make asteroids spherical, and little by way of volatile loss, which tends to make them non-spherical.

Question 3.4

Among the five most populous asteroid classes — S, M, C, D and P — the reflectance spectrum of Eros (Figure 3.9) has a similar shape (up to a wavelength of 1.1 μm) to that of class S, with class D as the only other plausible possibility (Figure 3.7). Its albedo is medium, like class S, and not low, like class D, so it seems to be class S. {It is!}

Its surface is thus likely to be made of iron–nickel alloy mixed with appreciable proportions of silicates. It is more likely to be found in the inner region of the asteroid belt. {Eros is a member of the Amor group, with a semimajor axis of 1.46 AU. It is thus closer to the Sun than the asteroid belt, but could well have originated in the inner belt.}

Question 3.5

The semimajor axis is given by (Figure 1.8)

$$a = \frac{q}{1-e} = \frac{0.9766 \, \text{AU}}{1 - 0.9055} = 10.33 \, \text{AU}$$

The orbital period, from Kepler's third law (equation (1.3)) is

$$P = 10.33^{3/2} \text{ years} = 33.2 \text{ years}$$

It is thus a short-period comet, with rather a long period for the Jupiter family. It thus seems to be Halley-type, and this is confirmed by its large orbital inclination. {It's inclination is over $90°$, so it is in a retrograde orbit.}

Question 3.6

At 30 AU from the Sun there might be a coma consisting of the most volatile icy substances evaporated from the nucleus by the (feeble) heat of the Sun. Any such coma could well be too faint to be seen from the Earth. As it approaches the Sun the coma grows as less volatile materials, notably water, begin to evaporate, and the coma becomes visible. Evaporation is from vents in a residue of ice-depleted dust that covers the surface of the nucleus, and the venting gas carries dust into the coma. A huge hydrogen cloud forms, and also ion tails, dust tails, and perhaps other types of tail. The ion tail points away from the Sun; the dust tail only approximately so, and it is curved. As it recedes from the Sun the gases recondense on to the nucleus so the tails fade away and the coma also. There might be sporadic outbursts from reactivated vents that break through the dusty surface.

Question 3.7

The other property is the time spent near the Sun. For example, a comet in a low eccentricity orbit will spend most of its orbital period not far from its perihelion distance, whereas a comet in a high eccentricity orbit will dash through the inner Solar System, and therefore spend only a very short time near perihelion.

Question 3.8

The feature of the orbits of Halley-type comets that indicates an inner Oort cloud origin is their generally larger orbital inclinations than the other short-period comets. E–K belt comet orbits generally have low inclinations, and would tend to give rise to low inclination short-period comets.

Question 3.9

Most meteorites are never found because:

- They fall into the oceans (which cover most of the Earth's surface)
- They fall in remote parts of the continents
- They get buried or otherwise obscured before anyone passes by
- They are mistaken for terrestrial rocks.

Question 3.10

For the ratio of carbon to iron to be much the same as in the Sun, the meteorite would be of the most primitive kind, i.e. a carbonaceous chondrite. Such meteorites have relative abundances of all the elements much like those in the Sun, except for those elements that remained as gases, and so were not retained by the meteorites. Helium is one of these, and this is why in carbonaceous chondrites the ratio of helium to carbon is far smaller than in the Sun.

Question 3.11

If the meteorite came from the outer metre or so of a larger body then it will have been exposed before it was liberated, and so its exposure age would be greater than the time ago that it was liberated. However, its exposure age could not exceed its solidification age because cosmic ray tracks cannot survive in a liquid.

Question 3.12

To fully differentiate, an asteroid needs to have been larger than a few hundred kilometres across. There might never have been many of these. The mantle will lie beneath the crust, not readily exposed by collisions, except in the rare collisions with a comparably large asteroid. Furthermore, if the crust is not thin, then as it is not made of weak chondritic material, it is unlikely that it will have been chipped away sufficiently to expose much of the mantle. If such exposed mantles are rare, then so will be the achondrites derived from them.

Question 3.13

One reason is that the comet is in such a long-period orbit that it has not passed through the inner Solar System in recorded history. An alternative reason is that the comet has devolatilised, or become disrupted, so is no longer visible.

Question 4.1

In Figure 4.1 there is a clear gap in size between the pair of giant planets Jupiter and Saturn, and the other pair Uranus and Neptune. Moreover, because of their greater mass, the mean densities of Jupiter and Saturn have been more increased by compression than have those of Uranus and Neptune, with Jupiter more compressed than Saturn. Thus, in equal states of low compression, Jupiter and Saturn could be distinctly less dense than Uranus and Neptune. This points to different compositions. Therefore, on the basis of size and composition, Jupiter and Saturn could form one subgroup, and Uranus and Neptune another. {This is the case, as you will see later. Note that no reference was made to the *uncompressed* mean densities of the giant planets. This is because much of their mass would be gaseous in such a state, and an uncompressed gas will expand indefinitely.}

Question 4.2

At double the distance the $1/r^2$ term in equation (4.7) is reduced by a factor of 4, but the J_2 term, which is proportional to $1/r^4$, is reduced by a factor of 16. Therefore, the ratio of these terms is reduced by a factor of 4. Any further extra terms fall off no less rapidly, and so the field is more closely that of a spherically symmetrical body.

Question 4.3

For a hollow cylindrical shell, with a radius R, every mass element δM is a distance R from the longitudinal axis. Thus the sum of $\delta M R^2$, which is C, is $M R^2$ where M is the total mass of the shell. Therefore $C/M R^2 = 1$.

Question 4.4

A self-exciting dynamo generates a large magnetic dipole moment if there is an electrically conducting fluid in the interior of the planetary body, an energy source to sustain convection in the fluid, and rapid rotation of the planet to coordinate the motions. The Earth and the four giant planets rotate relatively rapidly (Table 1.1), and in Section 1.2.2 it was stated that the giants are fluid throughout. A reasonable hypothesis is that these five planets not only rotate sufficiently rapidly, but also contain electrically conducting fluids in convective motion.

For the other planets one or more of these conditions must not be met. Venus, which is the Earth's twin in size and mass (Section 1.2.1) differs in that it rotates slowly, so this might explain the absence of a large magnetic dipole moment. {Other reasonable hypotheses are that internal energy sources are insufficient to sustain convection, or that, in spite of the similarity in size and mass, there are no conducting fluids.} Mercury also rotates slowly, so that might be the explanation for its lack of a large magnetic dipole moment too. {The alternatives given for Venus apply here as well.} Mars rotates about as rapidly as the Earth, so reasonable hypotheses are that it either contains no electrically conducting fluids, or if it does, the internal energy sources are insufficient to sustain convection. Pluto {which is assumed to lack a large dipole moment} rotates slowly, and it is a small body a long way from the Sun so could well be solid throughout.

{Chapter 5 will explore these reasonable hypotheses in more detail.}

Question 4.5

{See Figure Q4.5. With an entirely fluid interior, only P waves exist. The P wave speed will change with depth. The case of an increase with depth is shown here. Half-way to the centre, the P wave speed will increase in accord with equation (4.8) by the factor $\sqrt{2/1.2}$, which is 1.29. The speed is then assumed to further increase to the centre.}

Question 4.6

We use equation (4.11) with ρ_m and R for Jupiter taken from Table 1.1, and G taken from Table 1.6. Thus

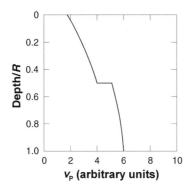

Fig. Q4.5

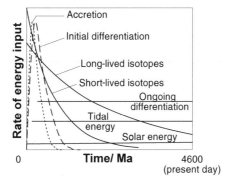

Fig. Q4.8

$$p_c \approx \frac{2\pi \times 6.672 \times 10^{-11} \text{N m}^2 \text{ kg}^{-2}}{3} (1330 \text{ kg m}^{-3})^2 (7.149 \times 10^7 \text{m})^2$$

$$\approx 1.3 \times 10^{12} \text{ N m}^{-2} = 1.3 \times 10^{12} \text{ Pa} = 1300 \text{ GPa}$$

{This is about four times the pressure at the Earth's centre. Models of Jupiter yield the more accurate value of about 5×10^{12} Pa, so equation (4.11) underestimates the pressure, as expected.}

At shallower depths the pressure must be less — pressure increases with depth.

Question 4.7

As the water is compressed at constant temperature the solid + gas phase boundary is encountered {just to the left of the triple point}. The water then solidifies. On further compression the solid + liquid phase boundary is encountered {this is because this boundary slopes to the left from the triple point}. The water then melts to became a liquid. Further compression increases the density of the liquid.

Question 4.8

{Much of the detail in Figure Q4.8 is arbitrary. The important features are:

- The rates of energy release from accretion and initial differentiation rise to an early maximum and then decline to zero, with differentiation later than accretion
- The rate of energy release from short-lived radioactive isotopes is now negligible, whereas that from long-lived isotopes, though declining, is still significant
- Ongoing differentiation, tidal energy (where it is significant), and solar energy are more sustained, with solar energy (in this case) making a relatively minor contribution.

The details will differ greatly from one planetary body to another.}

Question 4.9

(a) From equation (4.6), the difference Δg in the magnitude of the gravitational field across a body with a radius R, due to another body with a mass M a distance r away, is

$$\Delta g = GM\left(\frac{1}{(r-R)^2} - \frac{1}{(r+R)^2}\right)$$

This can be rearranged to give

$$\Delta g = GM\left(\frac{(r+R)^2 - (r-R)^2}{(r-R)^2(r+R)^2}\right)$$

Expanding the brackets on the top line and cancelling terms gives

$$\Delta g = GM\left(\frac{4rR}{(r-R)^2(r+R)^2}\right)$$

If r is much greater than R, then $(r-R)$ and $(r+R)$ are each equal to r to a good approximation. Thus

$$\Delta g \approx GM\left(\frac{4rR}{r^4}\right) = 4GM\left(\frac{R}{r^3}\right)$$

(b) For the Earth, the ratio of the solar tidal field to the lunar tidal field is $4GM_\odot(R_E/a_E^3)$ divided by $4GM_M (R_E/a_M^3)$, where R_E is the radius of the Earth, $M_\odot$ is the mass of the Sun, a_E is the semimajor axis of the Earth's orbit, M_M is the mass of the Moon, and a_M is the semimajor axis of the Moon's orbit. The ratio is thus $M_\odot a_M^3/M_M a_E^3$. Putting in the values from Tables 1.1 and 1.2, this ratio is 0.46, or 46%.

Question 4.10

(a) Radiative transfer is important only at high temperatures, so it is unlikely. Local regions of molten material are also unlikely, so advection is unlikely. With low-temperature gradients there is no convection below the surface layer, which itself is probably at too low a pressure for (solid-state) convection. Therefore, conduction is

the only likely means by which energy is being transferred to the surface layer.
(b) The thin surface layer must have a lower heat transfer coefficient than the rest of the body.

Question 4.11

(a) Accretional energy was endowed in a short period a long time ago, so the temperature will long since have become uniform throughout the interior.
(b) Long-lived radioactive isotopes will still be releasing energy in the core, so there will be a temperature increase towards the core.
(c) Because A/M is smaller in the larger body, it will have cooled less, so the temperature increase towards the centre will be greater.

Question 5.1

Were the Earth's outer core to solidify:

- There would be no electrically conducting fluid and so (according to the self-exciting dynamo theory) no magnetic dipole moment
- Seismic S waves would be able to travel throughout the interior, unless the latent heat released on solidification melted the lower mantle.

{The contraction of the core on solidification would also wreak havoc with the crust. Further crustal havoc might result from the latent heat released, which could make the solid-state convection in the mantle more vigorous.}

Question 5.2

Venus might have an asthenosphere as extensive as that of the Earth, i.e. extending throughout most of the mantle. Support for this view is that Venus probably has a similar composition to the Earth, with comparable pressures and temperatures at a given depth. In this case the 'plasticity profile' could be similar in the two planets. That the lithosphere is not thick is indicated by the recent geological activity.

If Mercury has an asthenosphere at all, it will extend over a small range of radii, and be located deep down, where pressures and temperatures are sufficiently high. Mercury is a small world, and in spite of its high mean density its internal pressures are not as high as in the Earth (Table 5.2). Its surface indicates that the lithosphere is thick.

Question 5.3

If Mars rotated very slowly it would be less flattened by rotation, and so the gravitational coefficient J_2 would be smaller and consequently more difficult to measure. Also, the more spherical form of Mars would result in a reduction in the torques that cause precession of the rotation axis, and a consequent increase in the precession period. At present the precession period is 0.1711 Ma, so would be more difficult to measure if it were even longer.

We would thus have a far less precise value of C/MR_e^2 and consequently a much poorer constraint on the variation of density with depth.

Question 5.4

Planetary body A will have a cooler interior than planetary body B for either one or both of the following reasons:

(1) Body A has had less copious energy sources
(2) Body A has lost energy at a greater rate.

Accretional energy per unit mass is roughly proportional to $\rho_m R^2$ (Section 4.5.1), and so the Earth would have received more accretional energy per unit mass than the considerably smaller Mars and Moon. The Earth's core is also a greater proportion of the mass of the Earth than the cores of Mars and the Moon in proportion to those bodies, so any energy of differentiation would have been greatest in the Earth. Insufficient information is given about radioactive sources and any other energy sources, to decide whether, overall, the Earth has had more copious energy sources than Mars and the Moon.

Equation (4.12) shows that, other things being equal, the energy loss rate per unit mass has been greater from Mars and the Moon than from the Earth because of their greater area-to-mass ratios. However, the generally higher temperatures in the Earth have promoted solid-state convection, and advection could also have been more effective (Table 4.5), thus increasing the Earth's heat transfer coefficient, and hence obscuring the issue.

{In fact, the smaller sizes of the Moon and Mars are indeed thought to be the main reason for their cooler interiors.}

Question 5.5

Pluto is an icy–rocky body because if formed well beyond the ice line in the solar nebula (Section 2.2.2). It is small because of the increasing difficulty of forming and combining planetary embryos with increasing heliocentric distance (Section 2.2.5).

Question 5.6

Adapting the answer to Question 4.9, the tidal field produced by the Earth across the Moon is given by $4GM_E(R_M/a_M^3)$, where M_E is the mass of the Earth, R_M is the radius of the Moon, and a_M is the semimajor axis of the Moon's orbit. In the case of tides produced *in the Earth*, the Sun is only 46% as effective as the Moon (Question 4.9(b)), and it will be even less effective for the Moon because the Earth/Sun mass ratio is much greater than the Moon/Sun mass ratio. So we can neglect the tide produced in the Moon by the Sun.

The tidal field produced across Io by Jupiter is given by $4GM_J(R_I/a_I^3)$, where M_J is the mass of Jupiter, R_I is the radius of Io, and a_I is the semimajor axis of Io's orbit. The tide produced in Io by the Sun and other bodies is negligible by comparison.

The ratio of the tidal field produced across the Moon by the Earth, to that produced across Io by Jupiter is thus $M_E R_M a_I^3 / M_J R_I a_M^3$. Putting in the values from

Tables 1.1 and 1.2, this ratio is 0.004. The Moon being much less geologically active than Io, it must be the case that the far greater tidal field across Io more than offsets the greater eccentricity of the lunar orbit (0.0549 versus 0.004).

Question 5.7

As Jupiter's interior cools the rate of settling of helium will increase, releasing energy of differentiation. It is triggered by cooling because the lower the temperature of the metallic hydrogen mantle, the lower the miscibility in it of helium. The heat from this differentiation will tend to sustain the internal temperatures, and so their decline will be retarded.

Question 5.8

The important radioactive isotopes are elements found in rocky materials (Table 4.4). Rocky materials are a minor ingredient of the giant planets, and are therefore a minor energy source.

Question 5.9

(a) If the Earth had no atmosphere then the atmospheric source of electrons and ions for the Earth's magnetosphere would be replaced by the Earth's surface. It is possible that the nature of the plasma belts would change {they would}. There would be no aurorae.
(b) If the solar wind had a lower speed v and a lower number density n of charged particles, then the factor $\mu^{1/3}/n^{1/6} \, v^{1/3}$ would be increased, and the magnetosphere would be larger. The number density of solar wind particles in the magnetosphere would be reduced.

Question 6.1

Because of its slow rotation, Titan will not be rotationally flattened, and in view of its large size it is likely to be approximately spherical. Therefore, a sphere centred on Titan's centre of mass, and lying within the range of radii at the surface, could serve as a zero of altitude. Alternatively, a specific value of mean atmospheric pressure could be used, somewhere near the mean surface pressure. {A body does not need to be rotationally flattened for atmospheric pressure to be used in this way.}

Question 6.2

If the atmosphere of Venus were transparent to radar, then radar could be used to establish surface morphology and composition. If circularly polarised radar were used then extra information on surface composition would be obtained. {The atmosphere of Venus is indeed transparent to radar, and much of what we know about the surface comes from radar.}

Question 6.3

(a) With a crater density on Chryse Planitia of 2.2×10^{-4} per km^2 for craters greater than 4 km diameter, there are $(10^5 \, \text{km}^2)(2.2 \times 10^{-4} \, \text{per km}^2)$ such craters on a typical area of $10^5 \, \text{km}^2$. This is 22 craters greater than 4 km diameter.

(b) If the Earth–Moon data in Figure 6.10 apply to Mars, then Chryse Planitia was resurfaced about 3500 Ma ago. However, the impact rate on Mars throughout much of Solar System history has been greater than in the Earth–Moon system. Therefore Chryse Planitia must have been resurfaced more recently than 3500 Ma ago. {Quite when is uncertain. One estimate is 3400 Ma.}

Question 6.4

Silica is not subject to partial melting or partial crystallisation because it is a single mineral, not a mixture of minerals.

Question 6.5

For the summit crater in Figure 6.13(b) to be of impact origin the impact would have had to be in the centre of a pre-existing mountain — rather unlikely. If the composition were typical of extrusive volcanism, then this would strongly indicate that the crater was a volcanic caldera. If around the crater the layers of rock were not turned over (as in Figure 6.5(c)) then this would be strong evidence against an impact origin.

Question 6.6

If a planet has an atmosphere but no water then the gradational processes that will occur are

- Mass wasting {water is not the only way of loosening material}
- Aeolian processes.

{It is, of course, hydrological processes that will be absent.}

Question 7.1

The anorthositic composition of the lunar highlands is thought do have been derived by differentiation from a magma ocean of peridotite composition. If the highlands had a peridotite composition this would imply that no such magma ocean formed.

Question 7.2

(a) Mare infill derived through gradation from the highlands would have the mineralogical composition of the highlands, i.e. anorthosite rather than basalt.

(b) Because of lithospheric thickening and interior cooling, lava has been unavailable since about 3200 Ma ago, and so the only infill of an impact basin formed today would be through gradation.

Question 7.3

{This question is similar to Question 5.4.}

A thicker lithosphere is a consequence of a cooler interior, so we have to look at energy gains and losses. Accretional energy per unit mass is roughly proportional to $\rho_m R^2$ (Section 4.5.1), and so Mercury would have received less accretional energy per unit mass than the considerably larger Earth and Venus. This might have been offset by energy of differentiation — Mercury's core is a greater proportion of the mass of Mercury than the cores of the Earth and Venus in proportion to those bodies (Figure 5.1, Section 5.1). We have no surface analyses of Mercury so can say little about radiometric heating, and so it is not possible to say overall, if Mercury has had less copious energy sources than the Earth and Venus.

Equation (4.12) (Section 4.5.4) shows that if the heat transfer coefficients have been comparable then the energy loss rate has been greater from Mercury than from the Earth and Venus, because Mercury has a higher area-to-mass ratio. However, the generally higher temperatures in the Earth and Venus have promoted solid-state convection, thus increasing these two planets' heat transfer coefficients, and hence obscuring the issue.

{In fact, the higher area to mass ratios of Mercury is thought to be the main reason for its cooler interior, and consequently for its greater lithospheric thickness.}

Question 7.4

The northerly hemisphere of Mars has many volcanic and tectonic features, and the martian meteorites indicate that magma was solidifying as recently as 180 Ma ago. Therefore, though there is no direct evidence for volcanic and tectonic activity today, it is possible that it is still ongoing, in which case the section on Mars belongs to Chapter 8 on active surfaces.

Question 7.5

Had the Martian crust in the southerly hemisphere been as thin as that in the northerly hemisphere then it is possible that the southerly hemisphere would resemble the northerly hemisphere. In this case there would be no heavily impact cratered surface, and the evidence of the Martian channels for a warmer, wetter epoch would be absent. Instead, there would be more volcanic and tectonic features.

Question 7.6

To cause cryovolcanism, the outer shell of each body needs to have its temperature raised so that (partial) melting could occur. If Pluto were move closer to the Sun then solar radiation could accomplish this. In the case of Ganymede, if it were moved closer to Jupiter, or if its orbital eccentricity were increased, then tidal heating could cause cryovolcanism.

Question 8.1

(a) Plate tectonics promotes rapid recycling of oceanic crust, and promotes volcanism that resurfaces continental crust. Therefore, most of the Earth's surface is young and impact craters consequently rare.

(b) The rift valley is explained as a new constructive margin. {Iceland is oceanic crust, so this is not a rift valley within continental crust.}

(c) Mountain ranges in the midst of continents are often sites where two slabs of continental crust have met at a subduction zone, and crumpled.

(d) At shallow depths near mid-oceanic ridges and ocean trenches there is a comparatively large amount of upward and sideways motion, and partial melting. This general activity is expected to generate stresses and fractures in the crust and in the solid mantle beneath, which will be observed as seismic activity.

(e) Figure 8.2 shows that the Aleutian Islands border a destructive margin, and so they could be an island arc. {This is the case.} Island arcs are caused by partial melting and volcanism at destructive margins where there are no continents at the margin.

Question 8.2

Few if any impact craters on Venus are older than about 300 Ma. Large impacts have been so rare since the end of the heavy bombardment 3900 Ma ago that it is very unlikely that Venus would have experienced one in the last 300 Ma.

Question 8.3

On Earth granites are intrusive rocks formed at destructive margins by the partial melting of continental crust. One feature of these margins is subduction. Subduction probably occurs on Venus only in a few locations, involving small quantities of crust, so perhaps granites are formed at such sites.

Question 8.4

If Io were devoid of sulphur then the magma would be devoid of what seems to be major volatile components, such as sulphur dioxide and sulphur itself. Therefore, the volcanic plumes would not reach as high, and perhaps the volcanism would be almost entirely effusive. The surface would be less colourful, because much of the present colour comes from a veneer of sulphur and its oxides.

Question 8.5

The decreasing tidal heating with increasing distance from Jupiter leads one to expect that the surface of Io will show signs of greater activity than Europa, with Callisto showing the least signs of activity. This is as observed. Io has extensive silicate volcanism, and a young surface, devoid of impact craters. Europa has cryovolcanism, liquid or slushy oceans of water, and perhaps some volcanic activity on the rocky floor of the oceans. Ganymede displays some evidence of having been episodically partially resurfaced by cryovolcanism, perhaps at times when tidal heating was greater than today. Callisto shows little sign of cryovolcanic resurfacing since soon after its formation — it is heavily cratered.

Question 8.6

It is only in the outer Solar System that temperatures can always have been low enough for the very volatile N_2 to have been retained as a solid {Section 5.2.2}. N_2 has only been detected on satellites beyond Saturn, and on Pluto itself. If there is subsequent modest internal heating, this can vaporise some of the N_2, giving nitrogen cryovolcanism, though sufficient internal heating would be restricted to the larger of this group of bodies. Triton is the largest of all, but still only has weak N_2 volcanism in spite of powerful tidal heating early in its history. Pluto is next down in size, and is thought to be inactive, partly because of its smaller size, and partly because it probably never experienced as much tidal heating as Triton {Section 7.4.1}.

Question 9.1

(a) If the number of carbon dioxide molecules in the Martian atmosphere were to decrease, the *area* of the spectral absorption line in Figure 9.1 would decrease.
(b) If the temperature of the Martian atmosphere were to increase, the Doppler width of the line would increase. If Doppler broadening was not initially determining the line shape it would increasingly do so as the temperature rose. {Such a rise could also reduce the line area, if a proportion of the CO_2 was thereby in the wrong state.}
(c) If the pressure of the Martian atmosphere were to increase the collisional broadening of the line would increase. If collisional broadening was not initially determining the line shape it would increasingly do so as the pressure rose.

Question 9.2

(a) If W_{int} is negligible, then, from equations (9.3) and (9.4)

$$L_{out} = F_{solar} A_p (1 - a)$$

Substituting this expression for L_{out} into equation (9.7) gives

$$T_{eff} = \left(\frac{F_{solar} A_p (1 - a)}{A \sigma} \right)^{1/4}$$

For a spherical body the projected area $A_p = \pi R^2$ and the total surface area $A = 4\pi R^2$. Thus

$$T_{eff} = \left(\frac{F_{solar}(1 - a)}{4\sigma} \right)^{1/4}$$

(b) The missing value is for the Earth. The Earth's average distance from the Sun is 1.496×10^{11} m (Table 1.1), and so, from equation (9.5) and Table 1.6

$$F_{solar} = \frac{3.85 \times 10^{26}\ W}{4\pi(1.496 \times 10^{11}\ m)^2} = 1370\ W\,m^{-2}$$

Thus, using the data in Tables 1.6 and 9.2

$$T_{eff} = \left(\frac{1370\ W\,m^{-2}(1 - 0.30)}{4 \times 5.67 \times 10^{-8}\ W\,m^{-2}K^{-4}}\right)^{1/4} = 255\ K$$

Question 9.3

(a) To rewrite equation (9.12) in terms of density the perfect gas equation of state (equation (9.2)) needs to be used in the rearranged form

$$\rho = p\left(\frac{m_{av}}{kT}\right)$$

Because m_{av} and T are assumed to be the same at all z in equation (9.12) they are the same at the surface as elsewhere, and so

$$\rho = \rho_s \times \exp(-z/h)$$

where ρ_s is the density at zero altitude.

{Thus, because density is proportional to pressure, the form of the equation is the same for density as for pressure.}

(b) A planetary atmosphere is approximately isothermal in the mesosphere {the Earth is an exception — Section 10.1.1}. If (as in Figure 9.8) the mesosphere is in the homosphere then (except for any condensable substances) it will also have uniform composition. It is thus in the mesosphere that the pressure is expected to vary with altitude in accord with equation (9.12).

Question 9.4

Though solar UV radiation is a small fraction of the total luminosity (Figure 9.5(b)), it is the main energy source for the mesosphere and thermosphere, and so these domains would be cooler, and the increase of temperature with increasing altitude in the thermosphere would be less steep. Also, without solar UV radiation there would be only a feeble ionosphere {from cosmic ray impacts}.

Question 9.5

At a partial pressure of 100 Pa, the 'Solid + gas' line in Figure 9.9 is at about 253 K. Therefore, if the partial pressure did not change with altitude the cloud base would be at an altitude of $(293\ K - 253\ K)\ 7\ K\,km^{-1}$, i.e. 5.7 km. This is a large change in altitude, and so the pressure will be less, and consequently the partial pressure of water too. A better estimate is to follow the shape of the path from point B in Figure 9.9, starting at 273 K and 100 Pa. This shifted path intercepts the 'Solid + gas' line at a temperature of about 247 K. In this case the altitude of the cloud base would be about 6.6 km.

In either case, the cloud will consist of water ice crystals, though if supercooling persists, then it will consist of liquid water droplets.

Question 9.6

(a) An atmosphere consisting of single atoms would be a weak absorber of mid-infrared radiation, and therefore the greenhouse effect would be weak.

(b) Water molecules consist of three atoms, and so water vapour is a strong absorber of mid-infrared radiation. Therefore, if a dry atmosphere were to have water vapour added to it the greenhouse effect would be larger, and T_s would rise. {However, if clouds became a lot more extensive, this rise could be offset by the increase in albedo and the consequent decrease in temperature.}

Question 9.7

(a) With a given exosphere radius, the escape speed is proportional to $\sqrt{M}$ (equation (9.17)) where M is the mass of a planetary body. Therefore, the thermal escape rate is dependent on M. Chemical escape and hydrodynamic escape must also depend on M, because the greater the escape speed, the less likely it is that molecules attain it, even if their speeds are non-thermal. {Impact erosion is a more difficult case. Greater planetary mass makes it less likely that atmosphere will be ejected, but greater mass also results it greater impact energy, and for a population of small potential impactors, greater planetary mass increase the collection rate.}

(b) The temperature at the base of the exosphere affects the thermal escape rate, the greater the temperature, the greater the rate.

(c) The greater the mass of a molecule, the less likely it is to have the escape speed. The mass thus affects the rates of thermal and chemical escape. The greater the mass of a molecule, the less likely it is to be entrained in outflow, and so the rate of hydrodynamic escape is reduced too.

(d) The solar UV flux is essential for at least some forms of chemical escape, and any reduction in this flux will reduce the rate of chemical escape. {A reduction in solar UV will also affect photochemical reactions deep in the atmosphere, and at the surface.}

Question 9.8

(a) Equation (9.20) gives the required speed. We need first to obtain the equatorial surface speed v_e. This is obtained from the Earth's equatorial radius and sidereal rotation period in Table 1.1. Thus

$$v_e = \frac{2\pi \times 6.378 \times 10^6 \text{ m}}{0.9973 \times 24 \times 3600 \text{ s}} = 465 \text{ m s}^{-1}$$

Thus from equation (9.20)

$$v_{rel} = \frac{465 \text{ m s}^{-1}}{\cos(30°)} - 465 \text{ m s}^{-1} \cos(30°) = 134 \text{ m s}^{-1}$$

{Quite a wind!}
In the northern hemisphere the parcel moves from west to east, and in the southern hemisphere from west to east also.
(b) If the Earth were to rotate much more slowly, then as well as lower east–west speeds, the Coriolis effect would be less disruptive on the Hadley circulation, and so the equatorial Hadley cells might extend to higher latitudes. Also, the cyclonic–anticyclonic activity at mid-latitudes might be less.

Question 10.1

Figure 10.1 shows that in the Earth's stratosphere the lapse rate is negative (temperature *increases* with increasing altitude), and so there will be no convection, and consequently no condensation.

Question 10.2

(a) The Earth's biosphere sustains both of the major present constituents of the atmosphere — O_2 and N_2. Without the biosphere the oxygen will largely be lost in oxidising surface substances and certain volcanic gases, and most of the nitrogen will end up as the nitrate ion in the oceans. The Earth's atmosphere will then consist mainly of water vapour, argon (Table 9.1), and CO_2. {The carbon cycle will become altered so that the atmospheric mass of CO_2 is increased, and it will become the main constituent.}
(b) Without knowing the new masses of the greenhouse gases H_2O and CO_2 in the atmosphere, and any changes in albedo {due to changes in cloud cover and in ice and snow at the surface} it is impossible to estimate the change in the Earth's mean surface temperature.
(c) Solar UV radiation at the surface would be greater because of the greatly reduced oxygen content of the atmosphere, and the consequent reduction in oxygen atoms in the thermosphere and ozone in the stratosphere.

Question 10.3

When the Earth's axial inclination is zero, the flux density on the Earth's surface at the poles is always extremely small because the Sun is always on the horizon. With its present axial inclination, though at the poles the Sun is below the horizon for about half the year, it is well above the horizon in midsummer. Consequently, the solar flux density at the poles, averaged over a year, is less when the Earth's axial inclination is zero than it is at present.

Question 10.4

(a) At midsummer in the southern hemisphere the Hadley cell on Mars extends from the subsolar latitude (in the southern hemisphere) to the northern (winter) hemisphere. After midsummer the subsolar latitude moves northwards, and with it the Hadley cell. Around the equinox the Hadley cell breaks down, but with the approach of midsummer in the northern hemisphere the Hadley cell becomes re-established, now extending from a northerly subsolar latitude to the southern hemisphere. Thereafter, the cell migrates southwards, disappears around the equinox, and becomes re-established as at the start of this answer.

(b) As the axial inclination of Mars varies, the midsummer subsolar latitude swings towards and away from the equator, being nearer the equator at low inclinations. For example, the location of the Hadley cell at midsummer in the southern hemisphere lies further north at low inclinations, and further south at high inclinations.

Question 10.5

For a positive lapse rate it is necessary that there is some heating from below. With no solar radiation penetrating the clouds, heating would be from above (ignoring the small outflow from the interior). Consequently the temperature would vary little with altitude, i.e. the lapse rate below the clouds of Venus would be close to zero. {This point was made in relation to planetary interiors, in Section 4.5.4.}

Question 10.6

From Figure 10.12, the global mass fraction of water of Venus today has a lower limit of about 2×10^{-9}, so 10^2–10^3 times this value is a lower limit of 2×10^{-7}–2×10^{-6}. On Earth the lower limit is much higher, about 2×10^{-4}, so before the loss Venus would probably still have been drier than the Earth.

Question 10.7

Figure 10.12 shows that the lower limits of the global mass fractions of CO_2 for Venus, the Earth, and Mars are, respectively, 1×10^{-4}, 6×10^{-5}, and 3×10^{-6}. The atmospheric quantity on Venus is only slightly less than the lower limit on its global value, whereas for the Earth it is less by a factor of about 10^5, and by a factor of about 100 for Mars.

The late stages of accretion would have brought CO_2 to these three planets. Even if all of this endowment was lost through hydrodynamic escape, impact erosion, T Tauri activity, and blow-off, a late veneer would have resupplied the planets. It is more likely that some of the earlier endowment was retained inside the planet, and that a proportion of this was outgassed, to add to the late veneer. Since the end of the heavy bombardment there has been little by way of loss to space, and so we can see why the quantities of CO_2 on Venus, the Earth, and Mars are substantial.

If the lower limits in Figure 10.12 are turned into ratios of one planet to another, then the broad differences in these ratios have their origin in plausible differences in the proportions of the different planetesimals that brought a late veneer to the three planets. The individual global mass fractions are not yet well explained.

Question 10.8

If the solar luminosity increased considerably, the Earth's atmosphere might well come to resemble that of Venus. The mass of water vapour in the atmosphere would increase, and this would increase the greenhouse effect, leading to positive feedback, until most or all of the oceans had evaporated to give a massive atmosphere initially dominated by water vapour. The mass of carbon dioxide would then increase as sedimentary carbonates decomposed in the high surface temperatures. The proportion of water would gradually fall through photodissociation of H_2O and the escape to space of the liberated hydrogen.

In the case of Mars, a significant proportion of the surface and near-surface reservoirs of water and CO_2 would be liberated, and the enhanced greenhouse effect would lead to higher surface temperatures, and there would be large open bodies of liquid water. For Mars to end up something like Venus, the luminosity of the Sun would have to rise more than for the Earth to become like Venus. This is because of the greater solar distance of Mars. In any case a Venus-like phase might be short-lived because of the lower escape speed of Mars.

Question 10.9

The differences between the atmospheres of Titan and Triton are believed to be due largely to their different distances from the Sun. Therefore, if they were to swap places they would also swap atmospheric characteristics. Much of the nitrogen and methane in Titan's atmosphere would condense onto its surface, and nitrogen and methane ices on the surface of Triton would be evaporated into the atmosphere.

Question 11.1

To obtain the Jovian mixing ratios, the number fractions in Table 9.1 have to be divided by the number fraction of H_2. This gives the following table, with the values from Table 11.1 added for comparison.

Species	Mixing ratio from Table 9.1	Mixing ratio in Table 11.1
H_2	1	1
He	0.156	0.156
CH_4	2.1×10^{-3}	2.1×10^{-3}

The values in Table 9.1 are for the whole atmosphere above the altitude at which the pressure is 10^5 Pa, whereas those in Table 11.1 apply at an altitude at which the pressure is a few times 10^5 Pa. The values are the same because the mixing ratios of He and CH_4 are independent of altitude until altitudes above which there is very little mass. This would not be the case for NH_3, because it condenses to form clouds at an altitude at which the pressure is about 10^5 Pa (Figure 11.1). Therefore, the mixing ratio will be much less above the NH_3 cloud than below it.

Question 11.2

(a) If the minor constituents of the atmospheres of Jupiter and Saturn were absent there would be no cloud formation. The absence of clouds would decrease the albedo, and this would lead to greater solar heating of the upper molecular envelope, and (presumably) to greater temperatures there, except perhaps where the latent heat of condensation of cloud particles is presently important. Without clouds it would be less easy to discern the circulation.

(b) The solar-driven component of the circulation would be changed because of the greater solar heating of the upper envelope, and the absence of condensation. {Plausible details are beyond the scope of this question.}

Question 11.3

The assumption that the proportion of N_2 is negligible in the atmospheres of Uranus and Neptune might fail for Neptune. The evidence is the detection of HCN in the upper troposphere of Neptune. HCN can be formed from N_2, and the mixing ratio of HCN for Neptune is large enough to require a source of N_2 from the interior. The mixing ratio of HCN for Uranus is certainly much less, and this difference between the two planets is consistent with other evidence for the lack of convection at some depths in Uranus. If indeed there is N_2 in Neptune's atmosphere, then the mixing ratio of helium must be reduced correspondingly. {We shall return to this important point in Section 11.3.}

Question 11.4

Neon, because it is volatile and unreactive, would have largely resided in the nebular gas, with only a small proportion trapped in the planetesimals. Because of its great abundance, most of the hydrogen was also in the gas. Similar proportions of the hydrogen and the neon in the nebula would have been captured, and therefore it is to be expected that the Ne/H ratio is the same in the giant planets as in the young Sun.

Because neon is soluble in helium, it will become depleted in the outer envelopes if helium settling has occurred. Therefore, the measured Ne/H ratios in the outer envelopes of Uranus and Neptune are expected to be about the same as in the young Sun, whereas the ratio for Jupiter should be lower, and that for Saturn lower still. {The Galileo probe found that in the outer Jovian envelope the Ne/H ratio is only about 10% of the value for the young Sun, presumably for the reason given.}

Glossary

Cross-references to other glossary entries are *italicised*.

accretion The acquisition by a larger body of smaller bodies. It is an essential stage in planet-building in *solar nebular* theories.

achondrite A type of *stony meteorite* that lacks chondrites (see *chondrite*).

adiabatic process A process during which no *heat* is transferred to or from a substance. If a vertical temperature gradient exceeds the adiabatic gradient, then *convection* occurs.

advection A process of energy transfer by the local bulk motion of relatively warm liquids, such as molten rock or water.

angular momentum The angular momentum of a body of mass m moving at a speed v at perpendicular distance r from an axis has a magnitude mvr. If no angular momentum is transferred to or from a system, its angular momentum is constant — this is the principle of conservation of angular momentum.

aphelion The point in the orbit of a body at which it is furthest from the Sun.

asteroids Small rocky bodies, largely confined to the asteroid belt. This belt is located between Mars and Jupiter. Among those outside the belt are the near-Earth asteroids.

asthenosphere A plastic region in the interior of a body, usually located in the *mantle*, and perhaps covering a great range of depths.

astronomical unit (AU) The *semimajor axis* of the Earth's orbit around the Sun.

aurora A dynamic display of light in the upper atmosphere of a body, mainly in the polar regions. Aurorae are caused by energetic charged particles that enter the upper atmosphere and collide with the atoms.

axial inclination The angle between the rotation axis of a body and the perpendicular to its orbital plane.

basaltic–gabbroic rocks *Extrusive* (basalt) and *intrusive* (gabbro) *rocks*, consisting largely of the *silicates* feldspar and pyroxene.

biosphere The assemblage of all living things and their remains. The Earth is the only planet known to have a biosphere.

black body See *ideal thermal source*.

blow-off The loss of nearly all of a planetary atmosphere through a large impact from space.

caldera   A crater created by *volcanism*, and not by impact.

carbonaceous chondrite   A type of *chondrite* (meteorite) that contains *carbonaceous material* and *hydrated minerals*. Carbonaceous chondrites have never been subjected to much heating or compression, and overall they have been little altered since they formed. Therefore they are primitive bodies. A particularly primitive subgroup is the C1 chondrites.

carbonaceous materials   Carbon and *organic compounds*.

carbonate   A chemical compound containing the chemical unit CO_3.

Centaurs   A group of *asteroids* with orbits that lie among the giant planets.

centre of mass   The point in a body (or in a system of bodies) that accelerates under the action of an external force as if all the mass were concentrated at that point.

chalcophiles   Chemical elements that tend to form compounds with sulphur, e.g. zinc.

chemical escape   The escape to space of a component of an atmosphere, promoted by chemical reactions that give the reaction products high speeds.

chondrite   A type of *stony meteorite* that contains small globules of silicates— chondrules—that have solidified from molten droplets. Most chondrites are classified as ordinary, but a small, important proportion are *carbonaceous chondrites*.

clathrate   An icy solid plus another substance enclosed within its crystal structure, such as carbon dioxide enclosed in water ice.

clay mineral   A silicate *mineral* that has been chemically modified by water.

climate   The medium-term (e.g. 30-year) average of temperature, precipitation, etc. plus the variability of these factors.

collisional broadening   The broadening of spectral lines due to collisions between atoms and molecules in the source region. Also called pressure broadening.

column mass   (1) The total mass in a column of unit cross-sectional area running perpendicular to the disc of the *solar nebula*. (2) The total mass of atmosphere in a column of unit cross-sectional area stretching vertically upwards from the surface.

comets   Small, *icy–rocky bodies* that develop huge fuzzy heads and tails when they are in the inner Solar System.

condensation flow   A component of atmospheric circulation whereby an atmospheric constituent flows to where it is condensing.

convection   The transport of energy by bulk flows arising from vertical temperature gradients.

Coriolis effect   As viewed in a rotating frame of reference, the tendency of a body to accelerate in a direction that is perpendicular both to the motion of the body and to the axis of rotation.

cosmic rays   Atomic particles that pervade interstellar space, moving at speeds close to that of light. They are primarily nuclei of the lighter elements, notably hydrogen.

crust   The relatively thin outermost solid layer of a body, chemically distinct from the underlying layer.

cryovolcanism   *Volcanism* involving *icy materials*.

day   By definition, exactly $24 \times 60 \times 60$ seconds in length. It is very nearly equal to the mean *solar day*, which varies slightly over a period of years.

delamination The melting and subsequent loss of lithosphere through any thickening that pushes it deeper, to where the temperatures are higher.

differentiation A process whereby denser substances settle downwards, and less dense substances float upwards.

Doppler effect The dependence of the observed wavelength (or frequency) of radiation on the speed of the source relative to the observer along the line of sight. The Doppler shift is the change in wavelength. The Doppler effect can lead to the broadening of a spectral line — Doppler broadening.

eccentricity A measure of the degree of departure of an *ellipse* from circular form. It is defined as $\sqrt{1 - b^2/a^2}$ where a is the *semimajor axis* and b is the semiminor axis.

eclipse The passage of a body into the shadow of another body. As seen from the Earth, in a total solar eclipse the Moon completely obscures the Sun, whereas in a total lunar eclipse the whole Moon enters the umbral shadow of the Earth.

ecliptic plane The plane of the Earth's orbit.

Edgeworth–Kuiper belt A belt of *comets* inferred to extend from just beyond the orbit of Neptune to the inner *Oort cloud*. Only its innermost, largest members have been observed.

effective temperature A temperature defined by $(L_{out}/A\sigma)^{1/4}$, where L_{out} is the radiant power that a body emits to space, A is its total area, and σ is Stefan's constant.

ellipse A closed curve that has the shape of a circle viewed obliquely. Half its long dimension is its *semimajor axis*, half of its short dimension is its semiminor axis.

embryos Bodies of the order of the mass of the Moon or Mars that appear at a fairly late stage in the growth of planets in solar nebular theories (*solar nebula*).

enthalpy See *latent heat*.

equation of state For a substance in equilibrium, the relationship between the pressure on the substance and its density and temperature.

equinox The point in an orbit where the rotation axis of a body is perpendicular to the line from the body to the Sun. At any latitude, day and night have equal length at this point.

escape speed The minimum speed a particle requires to escape from a body.

exosphere A layer in an atmosphere from where atoms and molecules can readily escape directly to space if they are travelling outwards at greater than the *escape speed*.

extrusive rocks *Rocks* that result from *volcanism*.

fault A crack or break in a *lithosphere*, along which relative motion can occur.

first point of Aries The direction from the Earth to the Sun when the Earth is at the *vernal equinox*.

flux density The power in the electromagnetic radiation that is incident on unit area of a receiving surface.

fractional crystallisation The crystalliation of a mineral from a melt that consists of a mixture of minerals.

Gaia hypothesis The hypothesis that active biospheric control tends to preserve optimum conditions for the Earth's biosphere.

Galactic tide   The *tidal force* exerted on bodies in the Solar System by the stars and interstellar matter in the Galaxy. It is thought to have been important in modifying the orbits of comets.

Galilean satellites   The four large satellites of Jupiter: Io, Europa, Ganymede, Callisto. They are named after their discoverer, Galileo Galilei.

general relativity   A theory by Albert Einstein that is superior to *Newton's laws*, though Newton's (simpler) laws can be used for most purposes in the Solar System.

geometrical albedo   A particular measure of the reflectance of a surface with respect to solar radiation.

giant planet   One of the four planets Jupiter, Saturn, Uranus, Neptune. These massive planets consist largely of hydrogen, helium, and *icy materials*.

global mass fraction   The total mass of a substance divided by the total mass of the body of which it is a component.

graben   A valley produced either by slumping between two parallel *faults*, or by collapse above a linear disturbance below the surface.

gradation   Any process by which material is eroded from a surface, and then transported and deposited elsewhere.

granitic–rhyolitic rocks   *Extrusive* (rhyolite) and *intrusive* (granite) *rocks* consisting largely of feldspar (a *silicate*) and quartz.

gravitational coefficient   A factor that defines the size of each extra term that has to be added to GM/r^2 to represent the *gravitational field* of a body.

gravitational field   The gravitational force per unit mass at any point in space; equivalently, the acceleration of any unrestrained mass placed at the point.

greenhouse effect   The phenomenon in which the radiant power emitted by the surface of a body exceeds that emitted by the body to space. This excess arise from the absorption and scattering of infrared radiation by atmospheric constituents. The greenhouse effect is often quantified as $T_s - T_{eff}$, where T_s is the global mean surface temperature, and T_{eff} is the *effective temperature*.

Hadley cell   A large convection cell in an atmosphere that arises from the decrease in solar heating with increasing latitude.

half-life   The time it takes half of the nuclei of a radioactive isotope to decay into another isotope.

heat   Energy transferred in which the transfer is random at a microscopic level.

heavy bombardment   The heavy bombardment of bodies in the Solar System that persisted until about 3900 Ma ago. In *solar nebular* theories it is due to the mopping up of remnant *planetesimals*. A possible peak in the bombardment before 3900 Ma ago is called the late heavy bombardment.

heavy elements   All chemical elements other than hydrogen and helium.

heterosphere   The layer in a planetary atmosphere where, aside from any substances that condense, the composition depends on altitude. It lies above the *homosphere*.

homosphere   The layer in a planetary atmosphere where, apart from any substances that condense, the composition is the same at all altitudes. It lies below the *heterosphere*.

hydrated mineral   A *mineral* that includes water, either as attached molecules of H_2O or as attached hydroxyl, OH.

hydrocarbon   A compound consisting of carbon and hydrogen. One of the simplest is methane, CH_4.

hydrodynamic escape   The escape of an atmospheric constituent through its entrainment in the high-volume flow rate of a rapidly escaping lighter constituent, notably hydrogen.

hydrostatic equilibrium   The equilibrium state of a body when it has responded to forces in the manner of a fluid, i.e. in the manner of material with zero shear strength.

hyperbolic orbit   An open orbit in which the two arms at infinity are diverging. A parabolic orbit is a marginal case, where the two arms are parallel at infinity. Elliptical and circular orbits are closed.

hypsometric distribution   A histogram showing the frequency distribution of different altitudes of the surface of a body.

ice age   A long period on Earth of cooler climate, particularly outside the tropics, where glacial conditions reach to mid-latitudes. An ice age is punctuated by warmer intervals called interglacial periods.

ice line   The distance from the protoSun beyond which water condenses in the *solar nebula*.

icy materials   A group of *volatile* substances, such as water, ammonia, methane, carbon dioxide, nitrogen, etc.

icy–rocky bodies   Bodies consisting of *icy* and *rocky materials*, with icy materials accounting for a significant proportion of the mass.

ideal gas   See *perfect gas*.

ideal thermal source   A body that absorbs all the electromagnetic radiation that falls on it. One consequence is that the spectrum of the radiation emitted by the body is uniquely determined by its temperature.

igneous rocks   *Rocks* produced from *magma*.

impact erosion   The loss of a proportion of a planetary atmosphere through the impacts of numerous small bodies from space.

inert gases   A group of chemically unreactive elements: helium, neon, argon, krypton, xenon, radon.

internal energy   The energy within a body, as opposed to the energy of the body as a whole. It consists of the random kinetic energy of its atoms and molecules, plus their potential energy of interaction.

intrusive rocks   *Rocks* that form from *magma* that solidifies beneath the surface of a body.

ionosphere   A layer in a planetary atmosphere sufficiently ionised for the layer to display special properties. Solar UV radiation is an important cause of ionisation.

iron meteorite   A *meteorite* consisting almost entirely of iron alloyed with a few per cent nickel.

isostatic equilibrium   A state in which, above some depth in a body, there are equal masses in equal-area vertical columns. The minimum depth in any region for which this condition holds is the depth of compensation.

J_2   The *gravitational coefficient* of a particular term additional to GM/r^2 in the *gravitational field* of a body. For *planetary bodies* the J_2 term happens to be the largest extra term.

Kepler's laws of planetary motion Three empirical rules that describe fairly accurately the orbital motions of the planets. They are explained by *Newton's laws*.

Kirkwood gaps Ranges of *semimajor axes* in the asteroid belt where there are few *asteroids*.

Lagrangian points Five points at locations fixed with respect to two bodies in orbit around each other. A third body with a small mass can be placed at each of these points, and remain there.

lapse rate The rate of decrease of temperature with increasing altitude in a planetary atmosphere. The *adiabatic* value is the adiabatic lapse rate.

latent heat *Heat* that produces no temperature change. Instead, the energy goes to increase the potential energy contribution to the *internal energy*.

lava Molten rock that reaches a surface due to *volcanism*, or icy liquids that reach a surface due to *cryovolcanism*.

lithophiles Chemical elements that tend to be found in silicates and oxides, e.g. magnesium and aluminium.

lithosphere The rigid outer region of a body, usually comprising the *crust* and upper *mantle*.

long-period comet A *comet* with an orbital period in excess of 200 years.

magma Partially or wholly molten *rocky* or *icy materials*.

magnetic field The force field sustained by electric current. A current loop generates a field that has the form of a dipole field far from the loop, and a strength proportional to the magnetic dipole moment of the loop.

magnetosphere The magnetic 'sphere of influence' of a body, bounded by the magnetopause. The Earth and the giant planets have extensive magnetospheres.

main belt A zone within the asteroid belt with a particularly high concentration of *asteroids*.

main sequence star A star in that phase of its lifetime when it is sustained by hydrogen *nuclear fusion*.

major planet One of the nine largest bodies in orbit around the Sun (excluding satellites in orbit around a planet). In order from the Sun they are: Mercury, Venus, Earth, Mars, Jupiter, Saturn, Uranus, Neptune, Pluto.

mantle The layer of a body that underlies any *crust*, and is chemically distinct from the crust.

mascon A region on the Moon with an excess of mass.

Maxwell distribution The distribution of molecular speeds in a gas in thermal equilibrium.

mesosphere A layer in a planetary atmosphere where the *lapse rate* is less than the adiabatic value, and so *convection* does not occur. It lies above any *troposphere* but below any *thermosphere*.

metallic hydrogen A high-pressure form of hydrogen in which the electrons detach from the nuclei. The substance then has the properties of a metal.

metamorphic rock An *igneous* or *sedimentary rock* that has been modified in any way short of complete melting.

meteor A small body that enters the Earth's atmosphere at high speed. They are commonly seen as bright, transient streaks of light across the night sky.

meteorite　A body, or its fragments, that has come from beyond the Earth and has survived passage to the Earth's surface. A micrometeorite is a meteorite less than a few millimetres across; some of these have recondensed after being vaporised or melted in the atmosphere.

meteoroid　*Asteroids* smaller than about a metre across.

mineral　A naturally occurring substance with a basic unit that has a particular chemical composition and structure. A *rock* is an assemblage of one or more minerals, in solid form.

minor planets　See *asteroids*.

mixing ratio　The number of molecules of a constituent in a local volume divided by the number of molecules of some other constituent in the same volume. In the giant planet atmospheres, mixing ratios are given with respect to H_2.

Newton's law of gravity　Enunciated by Isaac Newton. If two point masses M and m are separated by a distance r then there is a gravitational force of attraction between them with a magnitude GMm/r^2, where G is the universal gravitational constant.

Newton's laws of motion　Enunciated by Isaac Newton. The three laws are
(1) An object remains at rest or moves at constant speed in a straight line unless it is acted on by an unbalanced force.
(2) If an unbalanced force of magnitude F acts on a body of mass m, then the acceleration of the body has a magnitude F/m, and the direction of the acceleration is in the direction of the unbalanced force.
(3) If body A exerts a force of magnitude F on body B, then body B will exert a force of the same magnitude on body A but in the opposite direction.

nuclear fusion　A nuclear reaction in which a more massive nucleus is formed from less massive nuclei. The pp chains that power the Sun have the overall effect of converting four protons into a nucleus of the helium isotope ^4He.

number fraction　The number of molecules of an atmospheric constituent divided by the number of molecules in the whole atmosphere.

Oort cloud　A thick spherical shell of *comets* inferred to extend from about 10^3 to 10^5 AU from the Sun.

opposition　As viewed from the Earth, a body is in opposition when it is at that point in its orbit that places it nearest to being in the opposite direction to the Sun. At inferior and superior conjunctions, the body is nearest to being in the same direction as the Sun. These terms can be applied to vantage points other than the Earth.

orbital elements　Five quantities that specify the shape, size, and orientation of an orbit. For an elliptical orbit they are: the *semimajor axis*, the *eccentricity*, the *orbital inclination*, the longitude of the ascending node, and the longitude of perihelion.

orbital inclination　The angle between an orbital plane and a reference plane. The reference plane for planetary orbits is the *ecliptic plane*.

orbital resonance　Two bodies in orbit around a third body are in orbital resonance when their orbital periods form a simple ratio, such as 1:2, 2:3, etc.

organic compound　A compound of carbon and hydrogen, often with other elements. *Hydrocarbons* are a sub-class of organic compounds.

outgassing The volcanic emission of gases from the interior of a body.

oxidation A variety of chemical processes including those in which oxygen is incorporated into the reaction products. The opposite process — the extraction of oxygen — is one form of a chemical process called reduction.

partial melting The process in which only some of the constituents in a mixture become molten.

partial pressure The pressure exerted by the molecules of any one component in a mixture of gases.

perfect gas A gas that obeys the relatively simple *equation of state* $p = \rho k T / m_{av}$, where p, ρ, and T are respectively the gas pressure, density, and temperature, k is Boltzmann's constant, and m_{av} is mean mass of the molecules in the gas. Real gases approach perfect gas behaviour at low densities or high temperatures.

peridotite A *rock* consisting largely of the *silicates* olivine and pyroxene. The Earth's outer mantle consists largely of pyroxene. The *minerals* in the lower mantle have different crystal structures but the chemical composition is still that of peridotite.

perihelion The point in its orbit at which a body is nearest to the Sun.

phase diagram A diagram that shows the equilibrium phase of a substance at each pressure and temperature. Solid, liquid, and gas are three phases. The solid phase region is subdivided into different crystal structures.

photodissociation The disruption of molecules by electromagnetic radiation. UV radiation is particularly effective.

photoionisation The ionisation of atoms and molecules by electromagnetic radiation. UV radiation is particularly effective.

photometry The measurement of the radiation received from a source in a few, broad ranges of wavelengths. Compare *spectrometry*.

photon The particle of electromagnetic radiation in those circumstances where the radiation can be regarded as a stream of particles.

photosynthesis The process in which certain lifeforms, notably green plants and blue–green bacteria, synthesise carbohydrates from smaller molecules with the aid of solar radiation. Through photosynthesis, green plants and blue–green bacteria release molecular oxygen.

planetary albedo The fraction of the intercepted solar radiation that a body reflects back to space.

planetary body A body large enough for its own gravity to ensure that it is approximately spherical.

planetesimals 'Little planets' — small bodies 0.1–10 km across, formed from dust in the *solar nebula*. In solar nebular theories, many are incorporated into *embryos* on route to forming planets and the cores of *giant planets*.

plasma A highly ionised medium. In the Solar System, examples include the solar interior, the *solar wind*, and the particle content of *magnetospheres*.

plate tectonics The sculpturing of the Earth's surface through the relative motion of lithospheric plates. The plates are thought to be driven by convection in the mantle.

polar moment of inertia (C) The moment of inertia of a body with respect to its rotation axis.

power   The rate at which energy is transferred by any means.

precession   Generally, the periodic change in some orientation. The precession of *perihelion* is the motion of the perihelion direction around the plane of an orbit. The precession of the rotation axis is the coning of the axis at fixed *axial inclination*; this leads to the precession of the *equinoxes*.

prograde motion   The predominant direction of orbital motion and rotation in the Solar System — anticlockwise as viewed from above the Earth's North Pole.

radiation pressure   The pressure exerted by *photon* bombardment.

radiative transfer   The transfer of energy by means of electromagnetic radiation.

radiogenic heating   Heating resulting from the decay of radioactive isotopes.

radiometric dating   The dating of a variety of events from the relative abundances of radioactive isotopes and the isotopes they create when they decay. The *half-life* of the radioactive isotopes enables absolute ages to be obtained.

Rayleigh scattering   The scattering of radiation by objects that are much smaller than the wavelengths in the radiation.

refractory substance   A substance with low *volatility*.

regolith   The mixture of dust and small pieces of rock that is abundant on the surface of the Moon (and other bodies).

residence time   The average time that a molecule spends in a reservoir, e.g. the atmosphere.

retrograde motion   The opposite of *prograde motion*.

Roche limit   The distance between a smaller body and a larger body within which the smaller body is disrupted by the *tidal force* exerted by the larger body. It applies only if the smaller body is held together by gravitational forces.

rock   An assemblage of one or more *minerals*, in solid form.

rocky materials   A group of relatively *refractory substances*, including silicates, iron–nickel, etc.

saturation vapour pressure   The pressure on the saturation line of a substance, i.e. the line in the *phase diagram* on which liquid and gas, or solid and gas co-exist.

sedimentary rock   A *rock* produced from a sediment by chemical action, and in some cases by pressure too.

seismic wave   A mechanical wave in a *planetary* (or smaller) *body*. They also occur in the Sun. Two important types are P waves and S waves.

self-exciting dynamo   A mechanism thought to sustain the magnetic fields that originate deep in the Sun and in certain planets. It involves the conversion of kinetic to magnetic energy, which also happens in a dynamo.

semimajor axis   Half of the long dimension of an *ellipse*.

shield volcano   A volcano that in profile resembles a warrior's shield. Shield volcanoes result from a sequence of low-viscosity lava flows, erupted at modest rates.

shock wave   A wave with a very steep front, so that the material in front of the wave is undisturbed, whereas the material immediately behind the wave is highly disturbed. One way in which a shock wave is generated is when a projectile (or other disturbance) encounters a body at a speed greater than that of *seismic waves* in the body.

short-period comet A *comet* with an orbital period less than 200 years.

sidereal orbital period The time it takes a body to complete one orbit from a viewpoint fixed with respect to the distant stars. For the Earth this is the sidereal year.

sidereal rotation period The rotation period with respect to the distant stars. For the Earth it is called the mean rotation period.

siderophiles Chemical elements that tend to be present with iron, e.g. nickel.

silicate A chemical compound that has a basic unit consisting of atoms of one or more metallic elements and atoms of silicon and oxygen.

solar activity A collective term for those solar phenomena that vary (currently) with a period of about 11 years. The number of *sunspots*, flares, and prominences, and the luminosity of the Sun are among various aspects of this activity.

solar day The period of rotation of the Earth with respect to the Sun. The mean solar day is the mean length of the solar days averaged over a year. The mean solar day varies very slightly, whereas the *day* is exactly $24 \times 60 \times 60$ seconds in length.

solar nebula The disc of gas and dust that is presumed to have encircled the young Sun. In solar nebular theories, the Solar System forms from such a disc.

solar wind A thin, gusty stream of high-speed particles (mainly protons and electrons) that escapes from the Sun.

spectrometry The measurement of the radiation received from a source in numerous, narrow, contiguous wavelength ranges. Compare *photometry*.

spherical symmetry When the only variation in a quantity is with radius from some centre. In a body with a spherically symmetrical mass distribution, the density varies only with radius from the centre of the body.

steady state A system is in a steady state with respect to certain parameters (e.g. temperature), when the values of the parameters are not changing. A steady state is achieved when inputs that tend to increase a parameter are balanced by those that tend to decrease it.

stony meteorite A type of *meteorite* that consists mostly of silicates, often with small quantities of iron–nickel alloy.

stony–iron meteorite A type of *meteorite* that consists mostly of silicates and iron–nickel alloy.

stratosphere A layer peculiar to the Earth's atmosphere, immediately above the *troposphere*, where the temperature increases with increasing altitude. The *mesosphere* lies above the stratosphere, and in it the temperature decreases with increasing altitude. The temperature bulge is the result of the absorption of solar UV radiation by ozone.

sublimation A phase change in which a substance goes directly from solid to gas.

sunspot A small patch on the Sun's photosphere at a lower temperature than the rest of the photosphere. The number visible goes through a cycle with a mean period (currently) of 11 years. Sunspots are one aspect of *solar activity*.

synchronous orbit An orbit for which the period equals the rotation period of the body that is being orbited.

synchronous rotation When the rotation period of a body A with respect to a body B equals the orbital period of A around B.

synchrotron emission Electromagnetic radiation emitted by electrons travelling at very high speeds through a magnetic field.

synodic period For the Earth and another body, the time interval between similar spatial configurations, e.g. *oppositions* and conjunctions. The term can be applied to vantage points other than the Earth.

T Tauri phase The phase in a star's lifetime just before the *main sequence* phase. It is characterised by instability, with powerful stellar winds (the T Tauri wind) and high UV flux.

tectonic processes Processes that cause relative motion or distortion of the lithosphere.

terrestrial bodies The large bodies that are also predominantly rocky, i.e. the *terrestrial planets* plus the Moon, Io, and Europa.

terrestrial planet One of the four planets Mercury, Venus, the Earth, Mars. They consist largely of *rocky materials*.

thermal conduction The process by which *heat* is transferred through the direct contact of two regions that are at different temperatures.

thermal escape Escape of an atmospheric constituent arising from the *Maxwell distribution* of speeds extending to speeds in excess of the *escape speed*.

thermal tide A component of atmospheric circulation whereby the atmosphere flows to regions where the pressure has been lowered by cooling. This mass redistribution makes the atmosphere subject to an extra *torque* exerted by *tidal forces*.

thermosphere An upper layer in a planetary atmosphere where temperature increases with altitude, to reach high values.

tidal force A differential gravitational force. Any distortion produced by a tidal force is called a tide. Tidal energy is an important source of *internal energy* in some bodies.

Titius–Bode rule An empirical rule that describes the increasing spacing of the planetary orbits with increasing distance from the Sun.

torque A system of forces that tends to cause twisting or rotation. *Tidal forces* can exert torques. One outcome is *precession*.

troposphere The lowest layer in a planetary atmosphere where temperature decreases as altitude increases and where *convection* is a common occurrence.

uncompressed mean density The mean density that a body would have were its interior not compressed by the general increase of pressure with depth.

Van Allen radiation belts The belts of *plasma* that surround the Earth. They lie within the Earth's *magnetosphere*.

vernal equinox The *equinox* that occurs on Earth in March. Also, the position of the Earth in its orbit at this equinox. See also the *first point of Aries*.

volatility A property of a substance that can be measured by the maximum temperatures at which it is in equilibrium in a condensed phase. The lower these temperatures, the greater the volatility.

volcanism The processes by which gases, liquids, or solids are expelled from the interior of a body.

year Used here to mean the tropical year. The tropical year is the time interval between the Earth's *vernal equinoxes*. Its duration is 365.2422 *days*. The sidereal year is the Earth's orbital period with respect to the distant stars, and its duration is 365.2564 days. These years differ because of the *precession* of the equinoxes.

Electronic Media

There is a huge and rapidly growing range of electronic media, and a high rate of updating and obsolescence. The list below is a small selection of what is currently available by way of multimedia and websites that cover the Solar System.

Multimedia (and other Astronomy Software)

An annotated list of current astronomy software is available at the Sky Publishing Corporation website http://www.skypub.com/ It is updated monthly. Printed copies can be obtained from John E. Mosley, 7303 Enfield Avenue, Reseda, CA 91335, USA.

The Planets, CD-ROM for PCs or Macs, Byron Preiss Multimedia, New York, USA. Covers most aspects of the Solar System.

Dance of the Planets, CD-ROM for PCs, ARC Science Simulation, Colorado, USA. An extensive planetary tour. http://www.zephyrs.com

Starry Night Deluxe 2, CD-ROM for Macs, Sienna Software Inc, Ontario, Canada. Plenty on the Solar System, and lots more besides.

RedShift 3, CD-ROM for PCs or Macs, DK Multimedia, London, UK. Covers the Solar System and lots more. http://www.dk.com

Views of the Solar System, CD-ROM for PCs or Macs, C. J. Hamilton, National Science Teachers Association, Virginia, USA. An illustrated tour of the Solar System.

Voyage to the Outer Planets, A two CD-ROM set for PCs, NRSpace Software Inc, Washington State, USA. A structured set of the Voyagers 1 and 2 images.

Mars Navigator, CD-ROM for PCs or Macs, Engineered Multimedia Inc., Georgia, USA. Concentrates on Mars, but also has material on the rest of the Solar System.

Mars VR, CD-ROM for PCs or Macs, RVR Productions, California, USA. An exploration of the Mars Pathfinder site.

A Few Websites

http://www.skypub.com Sky Publishing Corporation, publishers of *Sky and Telescope*, the leading popular astronomy monthly in the USA and elsewhere. http://www.astronomynow.com The website of *Astronomy Now*, the leading UK popular astronomy monthly.

http://www.nasa.gov/search/index.html The NASA web search engine. ftp://ftp.hq.nasa.gov/pub/pao/pressrel/ NASA press releases. http://www.jpl.nasa.gov/ Lots available here — news of missions, etc. http://photojournal.jpl.nasa.gov/ A good selection of NASA images.

http://www.stsci.edu/ The Space Telescope Science Institute website. http://www.esrin.esa.it/ The European Space Agency website.

http://www.obspm.fr/planets Information and news on exoplanets.

Further Reading

*Books more advanced than this one are asterisked.

Books on the Solar System in General (in order of publication date)

Encyclopaedia of the Solar System, P. Weissman, L. McFadden and T. Johnson (eds), Academic Press, 1998
Encyclopaedia of Planetary Sciences, J. H. Shirley and R. W. Fairbridge, Chapman and Hall, 1997
The NASA Atlas of the Solar System, R Greeley and R Batson, Cambridge University Press, 1997

The New Solar System, J. K. Beatty, C. C. Petersen and A. L. Chaikin (eds), Sky Publishing, 1998
Physics and Chemistry of the Solar System, J. S. Lewis, Academic Press, 1997
The Planetary System, D. Morrison and T. Owen, Addison–Wesley, 1996
Mission to the Planets, P. Moore, Cassell, 1996
Planets: A Smithsonian Guide, T. R. Watters, Macmillan, 1995
The Solar System, T. Encrenaz, J.-P. Bibring and M. Blanc, Springer-Verlag, 1995
Planetary Landscapes, R. Greeley, Chapman and Hall, 1994
Moons and Planets, W. K. Hartmann, Wadsworth, 1993
The Grand Tour, R. Miller and W. K. Hartmann, Workman Publishing, 1993
The Planets: A Guided Tour, N. Henbest, Viking Penguin, 1992
Solar System Evolution: A New Perspective, S. R. Taylor, Cambridge University Press, 1992

Books on Particular Bodies or Topics

Discovering the Secrets of the Sun, R. Kippenhahn, John Wiley, 1994
Guide to the Sun, K. J. H. Phillips, Cambridge University Press, 1992

Venus Revealed, D. H. Grinspoon, Helix Books, 1997
Atlas of Venus, P. Cattermole and P. Moore, Cambridge University Press, 1997

Venus II, S. W. Bougher, D. M. Hunten and R. J. Phillips (eds), University of Arizona Press, 1997

Venus: The geological story, P. Cattermole, The Johns Hopkins University Press, 1994

Planet Earth, C. Emiliani, Cambridge University Press, 1992

Meteorite Craters and Impact Structures of the Earth, P. Hodge, Cambridge University Press, 1994

The Once and Future Moon, P. D. Spudis, Smithsonian Institution Press, 1996

Lunar Sourcebook, G. Heiken *et al.* (eds), Cambridge University Press, 1991

Eclipse! P. S. Harrington, John Wiley, 1997

Totality — Eclipses of the Sun, M. Littman and K. Willcox, University of Hawaii Press, 1991

Mars and the Development of Life, A. Hansson, John Wiley, 1997

The Planet Mars, W. Sheehan, University of Arizona Press, 1996

Water on Mars, M. H. Carr, Oxford University Press, 1996

Mars, H. H. Kieffer *et al.* (eds), University of Arizona Press, 1992

Atlas of Volcanic Landforms on Mars, C. A. Hodges and H. J. Moore, US Geological Survey, 1994

Jupiter: the Giant Planet, R. Beebe, Smithsonian Institution Press, 1996

The Giant Planet Jupiter, J. H. Rogers, Cambridge University Press, 1995

Uranus, J. T. Bergstralh, E. D. Miner and M. S. Matthews (eds), University of Arizona Press, 1991

Uranus: the planet, rings, and satellites, E. D. Miner, Ellis Horwood, 1990

The Planet Neptune, P. Moore, John Wiley, 1996

Neptune and Triton, D. P. Cruikshank (ed.), University of Arizona Press, 1995

Atlas of Neptune, G. E. Hunt and P. Moore, Cambridge University Press, 1994

Far Encounter: the Neptune System, E. Burgess, Columbia University Press, 1991

Pluto and Charon, A. Stern and J. Mitton, John Wiley, 1997

Pluto and Charon, S. A. Stern and D. J. Tholen (eds), University of Arizona Press, 1997

Satellites of the Outer Planets, D. A. Rothery, Oxford University Press, 1999

Meteorites: Messengers from Space, F. Heide and F. Wlotzka, Springer-Verlag, 1995

Rocks from Space, O. R. Norton, Mountain Press, 1994

Meteors, N. Bone, Sky Publishing, 1993

Asteroids: Their Nature and Utilisation, C. T. Kowal, John Wiley, 1997

Rogue Asteroids and Doomsday Comets, D. Steel, John Wiley, 1997

Hazards due to Comets and Asteroids, T. Gehrels *et al.* (eds), University of Arizona Press, 1995

Asteroids II, R. P. Binzel, T. Gehrels and M. S. Matthews (eds), University of Arizona Press, 1989

The Quest for Comets, D. H. Levy, Avon Books, 1995

Rendezvous in Space: the Science of Comets, J. C. Brandt and R. D. Chapman, W. H. Freeman, 1992

Physics & Chemistry of Comets, W. F. Huebner (ed.), Springer-Verlag, 1990

The Origin of Comets, M. E. Bailey, S. V. M. Clube and W. M. Napier, Pergamon Press, 1990

Protostars and Planets III, E. H. Levy and J. I. Lunine (eds), University of Arizona Press, 1993

Orbital Motion, A. E. Roy, Hilger Revised/IoP Publishing, 1988

Fundamentals of Atmospheric Physics, M. L. Salby, Academic Press, 1996
Physics of Atmospheres, J. T. Houghton, Cambridge University Press, 1989

Introduction to Planetary Volcanism, G. Mursky, Prentice Hall, 1996
Planetary Volcanism, P. Cattermole, John Wiley, 1996
Volcanoes of the Solar System, C. Frankel, Cambridge University Press, 1996

Exploration of Terrestrial Planets from Spacecraft, Y. Surkov, John Wiley, 1997

Elementary Problems and Answers in Solar System Astronomy, J. A. Van Allen, Duckworth/University of Iowa Press, 1993

Periodicals

Sky and Telescope (monthly), *Astronomy* (monthly), and *Astronomy Now* (monthly) are all aimed at a wide readership and contain articles on the Solar System, plus other areas of astronomy. *Scientific American* (monthly) and *New Scientist* (weekly), also aimed at a wide readership, contain occasional articles and short items on the Solar System. There is a large number of technical periodicals, available at university libraries, and aimed at the research community. One of more general interest is the *Annual Review of Earth and Planetary Sciences*.

Index

Note that in this index, general features and principles are separated from specific bodies. For example, under 'atmospheres' there are no entries for specific atmospheres, such as that of the Earth. To find entries for the Earth's atmosphere you would need to look under 'Earth'. You should also check under groups of bodies, e.g. 'terrestrial planets'.

1992 QB$_1$ 104
accretion 64
    heterogeneous 144
    homogeneous 143
accretional energy 142–143
Adams, John Couch 13
adiabatic gradient, *see* convection
adiabatic process 148
advection 150
albedo
    geometrical 86
    planetary 280
altimetry 188
angular momentum 55
    principle of conservation 298
anorthosite 216
aphelion, *see* orbit
asteroids 10
    Amor group 85, 102
    Apollo group 85, 102
    asteroid belt 81
    Aten group 85
    C class 92, 111
    Centaurs 85, 104
    differentiation 93, 111
    geometrical albedos 91
    Hilda group 82
    Hirayama families 84
    interior heating 93
    Kirkwood gaps 81
    M class 92, 111
    main belt 21, 81
    mean densities 90
    near-Earth 85, 110
    number 80
    orbits 21, 81–86

    origin 67
    rotation 90
    S class 92, 111–112
    shapes 88–90
    sizes 86–87
    surface composition 92
    Tholen classes vs semimajor axis of orbit
        92–93
    Tholen classification 91
    total mass 80–81, 369
    Trojans 84
    V class 91–92, 111–112
asthenosphere 160
astrometric technique 50
astronomical unit 10, 16, 18
atmospheres
    adiabatic lapse rate 285
    blow-off 296
    chemical escape 296
    cloud formation 288–291
    clouds 288
    column mass 274
    condensation flow 316
    Coriolis effect 298–299
    data, *see* Tables 9.1, 9.2
    energy gains 280–281, 285–288
    energy losses 281–282, 285–288
    exosphere 288
    gains and losses 292–297
    gains from volatile-rich bodies 296
    greenhouse effect 291–292
    Hadley cell 297
    Hadley circulation 297–298
    heterosphere 288
    homosphere 288
    hydrodynamic escape 296

atmospheres (*continued*)
  impact erosion 296, 383
  ionosphere 288
  isothermal scale height 284
  lapse rate 285
  mesosphere 288
  methods of studying 276–280
  number fraction 274
  outgassing 293
  precipitation 290
  pressure and density vs altitude 283–284
  reservoirs 292–293
  residence time 292
  stationary eddies 300
  temperature vs altitude 285–288
  thermal escape 293–295
  thermal tide 316
  thermosphere 288
  troposphere 287
  waves 300
aurora 182
aurora australis and borealis, *see* Earth,
  magnetosphere
axial inclination 29

basalt 204
basaltic-gabbroic rocks 204
Beta Pictoris 54
biosphere, *see* Earth
black body, *see* ideal thermal source
Bode's law, *see* Titius–Bode rule
Bode, Johann Elert 23
bolide 105
Boltzmann's constant 276, 294
Boznemcova 112
breccia 216
brown dwarf 51

Callisto
  central pressure 169
  incomplete differentiation 172
  interior model 169, 172
  observational data on the interior 172,
    Table 4.2
  surface 243–244
carbonaceous materials 92
carbonates 209
carbonatite lavas 260
centre of mass 25
Ceres
  discovery 80
  mean density 90, 120
chalcophiles 137
chaos 48

Charon
  interior 173
  origin 71, 104
  surface 242
chemical affinities 136–137
chemical elements, relative abundances, *see*
  Solar System
Chiron 85–86, 104
chondrule 107
circular polarisation, *see* radar
circumstellar discs 54
clathrates 100
clay minerals 208
climate 300–301
  changes 301
  determinants 301
clouds, *see* atmospheres
coagulation 63
column mass 61, 274
comets 94
  collisions with the Sun and planets 102
  coma 94, 97
  devolatilisation 101
  dust tail 94, 98–99
  Edgeworth–Kuiper (E–K) belt 14, 99,
    103–104
  Halley-type 97
  hydrogen cloud 94, 98
  ionised gas tail 94, 98–99
  Jupiter family 97
  long-period 95–96
  nucleus 94–95, 98–101
  Oort cloud (Öpik–Oort cloud) 14, 99,
    102–103
  orbits 95–97
  origin of Edgeworth–Kuiper belt 71
  origin of Oort cloud 71
  remnants 101–102
  short-period 96–97, 102
  size of nucleus 99–101
  source of Halley-type 104–105
  source of the hydrogen cloud 93
  source of the long-period comets 102
  source of the short-period comets 103
  tails 94, 98–99
  transient brightening 101
conduction, *see* thermal conduction
convection 148–150
  adiabatic gradient 148–149
  solid state 150
Coriolis effect 298–299
Coriolis, Gaspard Gustave de 299
coronagraph 3
cosmic ray exposure age 110
cosmic rays 110, 182
craters, *see* impact craters

critical point, *see* phase diagram
crust 155
    formation 201–202

day 31
    mean rotation period (Earth) 30
    mean solar 31
    solar 30
Deimos 90
delamination 252
dense clouds 52
    compression, contraction, fragmentation 53
densities
    mean 119
    uncompressed mean 120, 139
    variation with depth (or with radius) 124–125
    of some important substances, *see* Table 4.3
depth of compensation 126
differentiation 93, 144, 200–201
    heat generation 143–144
    partial 201
Doppler effect 50, 188
dust 81, 105

Earth
    asthenosphere 160, 248–249
        convection 160
    atmosphere
        carbon dioxide 306–309
        circulation 310
        evolution 332–334, 359–360
        nitrogen 309
        origin and early evolution, *see* terrestrial planets
        oxygen 307, 309, 333–334
        ozone 304–305
        properties, *see* Tables 9.1, 9.2
        reservoirs, gains and losses 306–310, 322, 332–334, 359–360
        vertical structure, heating and cooling 303–306
    biosphere 306, 333
    carbon cycle 306–309
    climate changes 310–313, 332–333
    clouds 305
    conservative margin 249–250
    constructive margin 250
    continental crust
        creation and destruction 251–252
        thickening 251
    core composition 158–159

crust 247–249, 251–252
crust and surface composition 247, 249
destructive margin 250
evolution of interior and surface 255
first point of Aries 19, 33
Gaia hypothesis 333
Great Rift Valley 204–205
greenhouse effect 305–306, 332–333
heat transfer to surface 150, 254
hypsometric distribution 248–249
ice ages 310–313
impact craters 191–192, 195, 253
interior model 156, 158–160
internal energy sources 160
internal temperatures, pressures, densities 157
island arcs 250, 252–254
isostatic equilibrium 249
lithosphere 160, 247–249
magnetic field generation 159
magnetic field reversals 129
magnetosphere 182–183
mantle composition 159, 247
mantle plumes 254
observational data on the interior 156–161, Table 4.2
orbit 16, 18, 19
origin of life 332–333
photosynthesis 307–308, 333
plate tectonics 249–254
    causes of plate motion 254
    evidence for it 252–254
precession of the equinoxes 33
precession of the rotation axis 33
respiration and decomposition (biosphere) 307–308
rotation 30
    increase in period 32
rotation axis 29
seasonal changes 32
seismic low speed layer 160
seismic wave speeds 133–134
solid state convection 160, 254
spectrum (of radiation emitted to space) 283
stratosphere 304–305
Van Allen radiation belts 182–183
vernal equinox 19, 30, 33
volatiles (*see also* terrestrial planets)
    inventories 306–309, 321–322, 324–325
    inventory changes 306–310, 322, 324–330, 332–334, 359–360
eclipses
    annular solar 37
    forthcoming total solar 37
    lunar 37

eclipses (*continued*)
   partial solar 37
   solar 37
   total lunar 37
   total solar 37
ecliptic plane 19
Edgeworth, Kenneth Essex 104
Edgeworth–Kuiper (E–K) belt, *see* comets
effective temperature 281–282, 291–292
Einstein, Albert 28
elemental mass ratio 356
ellipse 16
embryo 64
Enceladus
   interior 269
   surface 269–270
   tidal heating 270
energy sources 7, 140–144
equation of state 137
   perfect gas 276
equinoxes 30
Eros 94
escape speed 293
Europa
   asthenosphere 268–269
   atmosphere 267, 340
   central pressure and density 169
   crust 267–268
   interior model 169, 171
   internal heating 171, 268
   life 269
   observational data on the interior 171,
     Table 4.2
   oceans of liquid water 171, 268–269
   surface 267–269
exoplanets 50
   formation 77–78
   list, *see* Table 2.2
extreme ultraviolet 303

faults, *see* tectonic processes
feldspars 201
first point of Aries, *see* Earth
flux density 86
fractional crystallization 201

gabbro 204
Galactic tide 71
Galilean satellites 12
   atmospheres 267, 340
   interiors 169–173
   origin 172–173
   surfaces 243–245, 265–269
Galileo Galilei 12

Galle, Johann Gottfried 13
Ganymede
   central pressure and density 169
   interior model 169, 172
   magnetic field generation 172
   observational data on the interior 172,
     Table 4.2
   surface 243–245
Gaspra
   asteroid class 92
   surface 88
general relativity 28
Giacobini–Zinner 17
giant molecular clouds 52
giant planets 9, 12, 120
   atmospheres vs interiors 342
   atmospheric properties, *see* Tables 9.1, 9.2,
     11.1
   fine structure of rings 77
   formation 69–71, 356–359
   formation and evolution of rings 73–77
   formation of satellites 72–73
   interior models 173–175
   kernels 69
   observational data on the interiors, *see*
     Table 4.2
   overall elemental mass ratios 356–359,
     Table 11.2
   overall He/H ratios 353, 357
   ring particles 75–76
   rings 74
global mass fraction 321
graben (rift valley) 204
gradation 206
   aeolian processes 207
   chemical alteration by water 208
   hydrological processes 207–208
   mass wasting 206–207
   U-shaped valleys 208
   weathering 208
granite 204
granitic–rhyolitic rocks 204
gravitational coefficients 123–124
gravitational field 123
greenhouse effect, *see* atmospheres

Hadley, George 297
Hale–Bopp 96, 99
   orbit 96
half-life 109
Halley's comet 97
   composition of nucleus 100
   geometrical albedo 100
   mean density of nucleus 100
   orbit 96
   shape and size of nucleus 99–100

Halley, Edmond 97
heat 142
heat sources, *see* energy sources
heat transfer coefficient 150
heavy bombardment 66, 200
heavy (chemical) elements 15
Hebe 112
helioseismology 133
Herschel, William 13
Hidalgo 85, 101
Hirayama, Kiyotsugu 82
Huyghens, Christiaan 13
hydrated minerals 61, 326
hydrocarbons 76, 217
hydrostatic equilibrium 125
hydroxyl 208
hypsometric distribution 227

icy materials 61
icy-rocky bodies 120, 167–169, 173
  atmospheres 337–339
  atmospheric origins 339–340
  atmospheric properties, *see* Tables 9.1,
    9.2
  surfaces 241–245
Ida
  asteroid class 92
  mean density 90
  surface 89
ideal gas, *see* perfect gas
ideal thermal source 2, 146–147, 281
impact craters 191, 235
  ejecta blankets 195–196
  formation of central peaks or peak rings
    194–195
  formation of multiring basins 195
  formation of terraces 194
  heavy bombardment 200
  late heavy bombardment 200
  mechanisms of formation 192–196
  morphologies 194–196
  rays 195–196
  saturation of a surface 198
  secondary craters 195–196, 198
  use in determining surface ages 196–200
inert gases 15, 324
inferior conjunction 34
infrared excess 151
interiors
  central pressures 139
  energy (heat) transfer to surface 147–151
    specific planetary bodies, *see* Table 4.5
  energy losses 146–147
  energy sources 142–146
  modelling principles 117

modelling constraints
  from available materials 135–141
  from gravitational field data 119–127
  from magnetic field data 127–129
  from seismic wave data 129–135
  on temperatures 151–152
observational data, *see* Table 4.2
pressures 137–139
size effect on temperatures 93–94, 143,
  153
temperatures 152–153
internal energy 142
interstellar medium 52
Io
  blotchy plains 266
  central pressure and density 169
  evolution of interior and surface 267
  heat transfer to surface 150, 267
  interior model 169–171
  lithosphere 170
  mountains (non-volcanic) 266
  observational data on the interior
    169–170, Table 4.2
  solid state convection 170
  surface composition 267
  tectonic features 267
  tidal heating 170–171, 265–266
  volcanic activity 170, 265–266
isostatic equilibrium 126

J_2 123–124
Jupiter
  atmospheric circulation 348–351
  atmospheric composition 176, 342–343,
    345–347, 387
  atmospheric vertical structure 343–344
  belts and zones 348
  cloud coloration 351–353
  clouds 342–344, 347–348, 352
  core 175, 177
  Galilean satellites formation 72
  Great Red Spot 351
  heat transfer to surface 178, 343–344, 350
  interior circulation 349–350
  interior model 174–178
  internal energy sources 177
  internal temperatures 177–178
  internal temperatures, pressures, densities
    175
  magnetic field generation 177
  magnetosphere 183–184
  metallic hydrogen mantle 176–177
  molecular hydrogen envelope 176
  observational data on the interior
    174–177, Table 4.2

Jupiter (*continued*)
  ovals 350–351
  overall composition 175
  overall He/H ratio 179
  Shoemaker–Levy 9 impact 344
  spots 351

Kepler's laws 15–18, 22–24, 26
Kepler, Johannes 16
kinetic energy 192
Kirkwood, Daniel 81
Kuiper belt, *see* comets, Edgeworth–Kuiper belt
Kuiper, Gerard Peter 104

Lagrange, Joseph Louis 84
Lagrangian points 84
late heavy bombardment 200
latent heat 142, 286
Le Verrier, Urbain Jean Joseph 13
light year 51
lithophiles 137
lithosphere 160
Lovelock, James Ephraim 333

magma 200
magma ocean 200–201
magnetic dipole moment 127
magnetic field 8, 127
  axis 127
  Cowling theorem 128
  dipole field 127–128
  equatorial plane 129
  generation in planetary body interiors
    128–129
  near to and far from a planetary body
    128–129
  poles 129
  remanent magnetism 129
  self-exciting dynamo mechanism 128
magnetosphere 180–182
  bow shock 181
  magnetopause 181
  magnetosheath 182
  reconnected lines 181
  size 181
  sources of plasma 182
main sequence star 53
major planets 12
mantle 155
Mars
  aeolian processes 226–227, 234, 239
  albedo features 226–227
  atmosphere
    circulation 316
    dust 314

  evolution 334–335, 360
  origin and early evolution, *see* terrestrial planets
  properties, *see* Table 9.1, 9.2
  reservoirs, gains and losses 315, 322–323, 334–335, 360
  vertical structure, heating and cooling 313–315
central temperature, pressure, density 158
channels 235–237
chaotic terrain 235
climate changes 234, 237, 316–317, 334–336
clouds 314
contrasting hemispheres 227–233
core composition 163, 240
core formation 240
crust and surface composition 163, 227, 229, 238–239
crustal thicknesses 227
differentiation 163
dust storms 227
dykes 229
evolution of interior and surface 240
fretted terrain 233
greenhouse effect 314–315, 334–335
hypsometric distribution 227–228
impact basins 232
impact craters 232
interior model 156, 163–164
internal temperatures 164
layered sediments 234, 237
life 238, 335–336
lithosphere 164
magnetic field 163–164
mantle composition 163, 240
meteorites from Mars 239–240
observational data on the interior 163–164, Table 4.2
observations at the surface 156–157, 238–239
Olympus Mons 231
origin of satellites 68
polar regions 233–234
regional domes 229, 231
seasonal effects 226
shorelines 237
surface ages 227, 229
surface temperatures 226, 233
tectonic features 229–231
Tharsis region 229
Valles Marineris 229–230
volatiles (*see also* terrestrial planets)
  inventories 315, 317, 321–322, 324–325
  inventory changes 315, 322–330, 334–335, 360

volcanic activity   229, 231
volcanic features   231
water-related features   233–239
weathering   232–233
mass measurement   25, 119
Mathilde
asteroid class   92
mean density   90
surface   89
Matthew effect   64
maximum eastern elongation   35
maximum western elongation   34
Maxwell distribution   293–294
Maxwell, James Clerk   294
Mercury
atmosphere   336
Caloris basin   221–222
central temperature, pressure, density   158
core and mantle composition   162
crust and surface composition   162, 220
evolution of interior and surface   225–226
gradation   221
heavily cratered terrain   221, 223–224
impact basins   221–222
impact craters   220–221
intercrater plains   224
interior model   156, 162–163
interior thermal history   162
lava flows   221, 225
lithosphere   162
magnetic field generation   162
observational data on the interior
   156–157, 162, Table 4.2
precession of the perihelion of the orbit
   28
rotational resonance   220
smooth plains   221
surface temperatures   220
tectonic features   225
volatiles   336
   polar   225
volcanic features   225
metallic hydrogen   176–177
meteorites   105
achondrites   92, 107
C1 chondrites   325–326
calcium–aluminium inclusions   108
carbonaceous chondrites   92, 108
chondrites   107
classes   106–108
cosmic ray exposure ages   110
falls   105
finds   105
irons   106–107
isotope ratios   62, 108
lunar   113

Martian   113, 239–240, 336
orbits of the parent meteoroid   105, 110
ordinary chondrites   108
solidification ages   109–110
sources   110–113
stones   107
stony-irons   107
Widmanstätten pattern   107
meteoroids   12, 81
meteors   105
showers   113–114
sources   114
micrometeorites   106
composition   114
sources   113–114
micrometeoroids   81, 105
minerals   61
list of important ones, *see* Table 6.1
minor planets, *see* asteroids
mixing ratio   342
moment of inertia   124
polar   124–125
Moon
ages of rocks   218
atmosphere   336
central temperature, pressure, density
   158
core composition   164
crust and surface composition   216–218
crustal thicknesses   215
evolution of interior and surface   219–220
far side   212–213, 215
far side–near side contrasts   215
fines   216
formation   68, 166
gradation   215
highland rocks formation   216–217
impact basins   212
   ages   218
impact crater densities versus surface age
   199
impact cratering rate versus time   199, 219
impact craters   212, 218–219
interior model   164–167
internal temperatures   165–166
lava flows   213–214
librations   212
linear rilles   213, 215
lithosphere   166
magnetic field   166
mantle composition   166, 217–218
mantle convection   215
Mare Imbrium   213–214
mare infill and rock formation   213–214,
   217
Mare Orientale   213, 215

Moon (*continued*)
   maria 211–215, 220
     infill ages 218–219
   mascons 212
   observational data on the interior
     164–166, Table 4.2
   orbit 36
   overall composition 166
   polar ice 215
   regolith 216
   seismic activity 135, 165
   seismic wave speeds 134–135, 165–166,
     216–217
   sinuous rilles 213
   South Pole-Aitken basin 212, 215
   surface fracturing 165–166, 216
   surface temperatures 212
   surface, absolute ages 199, 218–220
   tectonic features 213, 215
   transient lunar phenomena 219
   volatiles 215, 336
   volcanism 214–215, 217, 219–220
Murchison meteorite 106

Neptune
   atmospheric circulation 354–355
   atmospheric composition 342–343,
     353–354, 356
   atmospheric vertical structure 352–353
   bands 354
   clouds 352–354
   discovery 13
   heat transfer to surface 353, 355
   interior model 174, 179–180
   internal energy sources 180
   internal temperatures 179–180
   internal temperatures, pressures, densities
     175
   magnetic field generation 180
   magnetosphere 184
   observational data on the interior
     174–175, 179–180, Table 4.2
   overall composition 179
   spots 354–355
Newton's law of gravity 22
Newton's laws of motion 22
Newton, Isaac 22
non-gravitational forces 26, 76
nuclear fusion 7
number fraction 274

oblate spheroid 123
observational selection effect 51
olivine 61
Oort cloud, *see* comets
Oort, Jan Hendrick 102

Öpik, Ernst Julius 102
opposition 35
orbit
   aphelion 16
   argument of perihelion 20
   changes in orbital elements 28
   eccentricity 16
   elements 19
   elliptical 16
   hyperbolic 24
   inclination 20
   longitude of perihelion 20
   longitude of the ascending node 20
   node 20
   parabolic 24
   perihelion 16
   period 17
   semimajor axis 16
   unbound 24
orbital resonance 81–82
organic compounds 92
outgassing 293
oxidation 297

Pallas
   discovery 80
   mean density 90, 120
partial melting 160, 201–203
partial pressure 289
perfect gas 276
peridotite 159
perihelion, *see* orbit
period of orbit, *see* orbit
phase (of a substance) 140
phase boundary, *see* phase diagram
phase diagram 139–141, 288–290
   ammonia–water mixture 202
   methane 338
   molecular hydrogen 176
   phase boundary 140
   saturation vapour pressure 290
   triple point 140
   water 289
Phobos 90
   surface 89
Pholus 86
photochemical reactions 296
photodissociation 97
photoionisation 97
photometry 91
photon 7
Piazzi, Giuseppe 80
Planck's constant 8
planetary bodies 120
planetary nebula 360
planetary rings, *see* giant planets

planetary satellites   10
  formation of giant planet satellites   72–73
  icy-rocky satellites   104, 167–169, 173
  protosatellite discs   72, 167
  small satellites   173
  surfaces of small icy-rocky satellites, *see*
    Table 7.2
planetesimals   64
  icy-rocky   64, 69
  rocky   64
plasma   6
plate tectonics, *see* Earth
Pluto
  atmosphere   339–340
  central pressure, temperature, density   169
  discovery   14
  interior model   168–169
  observational data on the interior   168,
    Table 4.2
  orbit   17
  origin   71, 104
  surface   242, 339
power   142
Poynting–Robertson effect   76
pp chains   7
precession of the perihelion   28
precession of the rotation axis   33, 124–125
prograde direction   10
protoplanets   70
protostars   53
  bipolar outflows   58
  circumstellar discs   54
  nuclear fusion in the interior   53
protoSun, *see* Sun
pulsar   50
  planets   50
pyroxene   159

radar   188, 190–191
radial velocity technique   50
radiation pressure   98
radiative transfer   8, 147
radiogenic heating   143–144
radiometric dating   109
Rayleigh scattering   351
red giant   360
refractoriness   60, 141
retrograde direction   14, 73
rhyolite   204
rift valley (graben)   204
Roche limit   73–75
rocks   61
  extrusive   204
  igneous   204
  intrusive   204
  list of important ones, *see* Table 6.1

metamorphic   209
  sedimentary   209
rocky materials   61
rotational flattening   123

satellites, *see* planetary satellites
saturation vapour pressure, *see* phase
    diagram
Saturn
  atmospheric circulation   348–351
  atmospheric composition   342–343,
    345–347
  atmospheric depletion of helium   178–179,
    353, 357
  atmospheric vertical structure   343–345
  belts and zones   348
  cloud coloration   351–352
  clouds   342–345, 348, 352
  heat transfer to surface   343–344, 350
  interior circulation   349–350
  interior model   174–175, 178–179
  internal energy sources   178
  internal temperatures   178
  internal temperatures, pressures, densities
    175
  magnetic field generation   178
  magnetosphere   184
  metallic hydrogen mantle   178
  molecular hydrogen envelope   178
  observational data on the interior
    174–175, 178–179, Table 4.2
  ovals   350–351
  overall composition   175
  overall He/H ratio   179
  spots   351
sediments   207–208
seismic waves   129, 131
  generation   131
  P wave   131
  P wave speed   132
  paths in a planetary body interior   132
  S wave   131
  S wave speed   133
  use in investigating interiors   133
shock wave   193
Shoemaker–Levy 9   100, 102, 344
shooting star   105
sidereal orbital period   17
sidereal rotation period   30
siderophiles   137
silicates   61
solar nebula   55
  column mass   61, 69
  condensation of dust   60–62
  dust sheet formation   63
  embryo formation   64

solar nebula (*continued*)
evaporation of dust 59–60
giant planet formation 69–71
ice line 61, 69
icy materials 167
initial mass 55, 59
planetesimal formation 64
terrestrial planet formation 64–67
solar nebular theories 55
successes and shortcomings 77–78
Solar System
angular momentum 57–58
future evolution 359–360
general properties, *see* Tables 1.1–1.5, 2.1
orbits 11
origin, *see* solar nebula, and solar nebular
theories
sizes and densities of bodies 122
sizes of bodies 10
relative abundances of the chemical
elements 15, 55, 135–136
solidification age 109
solstices 29
spacecraft, *see* Table 4.1
Apollo missions 215
Cassini 243
Clementine 215
Galileo Orbiter 169, 243, 343
Galileo probe 343, 347
Giotto 99
Huyghens probe 243
IRAS 101
Luna missions 215
Lunar Prospector 215
Lunik III 212
Magellan Orbiter 255
Mariner 10 220
Mars Global Surveyor 229
Mars Pathfinder 238
NEAR 90
Sojourner (rover) 239
Viking Landers 238, 335–336
Voyagers 169, 242–243
spectral lines
collisional broadening (pressure
broadening) 278
Doppler broadening 278
information from spectral lines 276–279
spectrometry 91
speed of light 8
spherical symmetry 22
spiral density wave 53
star clusters 53
star formation 52–54
steady state 286–287
Stefan's constant 281

stellar winds 54
sublimation 59
Sun
age 6
chromosphere 3
composition of the interior 6
initial composition 357
composition of the photosphere 2
convection in the interior 8
corona 4
faculae 6
flares 3
future evolution 359–360
granulations 8
interior 6–9
luminosity 1, 9
luminosity increase 330–331, 359–360
magnetic field 8
neutrinos from the interior 9
nuclear fusion in the interior 7, 360
photosphere 1–2
prominences 3
protoSun 55
rotation 58
radiative transfer in the interior 8
rotation 3
solar activity 5
solar wind 5, 54, 181–182, 295–296, 336
spectrum (electromagnetic) 2, 283
sunspot cycle 3
sunspots 2
T Tauri phase 59, 66, 69, 295, 325
superior conjunction 35
surfaces
active and inactive 211
energy gains 280–281, 285–288
energy losses 281–282, 285–287
investigations by photometry and
reflectance spectrometry 190–191
investigations by radar 188–191
mapping in two and three dimensions
187–190
zero altitude definitions 189–190
synchronous orbit 76
synchronous rotation 32
synchrotron emission 182
synodic period 35
synthetic aperture radar 188–189

T Tauri phase (*see also* Sun) 54, 78
T Tauri wind (*see also* Sun) 54
tectonic processes 204–206
fold mountains 205
graben (rift valley) 204
normal fault 204

strike-slip fault 205
thrust fault, or reverse fault 204
tektites 113
temperature 142
size effect, *see* interiors
terrestrial bodies 120
terrestrial planets 10–11
atmospheric evolution in the distant
future 359–360
atmospheric origins and early atmospheric
evolution 324–330
early massive losses 325, 327–328
inert gas evidence 324–326
atmospheric properties, *see* Tables 9.1, 9.2
formation 64–67
interior properties, *see* Tables 5.1, 5.2
late veneers of volatiles 328–329
observational data on the interiors, *see*
Table 4.2
outgassing 329–330
overall chemical composition 157
possible consumption by the Sun 360
volatile acquisition during planet
formation 326–327
volatile inventories 321–326
thermal conduction 147
tidal energy (heating) 144–146
tidal force 27
Titan
atmosphere 337–338
atmospheric origin and evolution
339–340
central pressure 169
interior model 168–169
observational data on the interior 168,
Table 4.2
surface 242–243, 338–339
Titius, Johann Daniel 21
Titius–Bode rule 21, 80
Tombaugh, Clyde William 14
torque 28
Toutatis 90
surface 89
triple point, *see* phase diagram
Triton
atmosphere 271, 338–340
cantaloupe terrain 271
central pressure 169
cryovolcanism 272
interior model 168–169
internal energy sources 272
observational data on the interior 168,
Table 4.2
origin 73, 104
polar cap 271–272
surface 270–272, 338

Tunguska impact 102
turbulence 58

Uranus
atmospheric circulation 354–356
atmospheric composition 342–343,
353–354, 356
atmospheric vertical structure 352–353
bands 354
cause of large axial inclination 70
clouds 352–354
discovery 13
heat transfer to surface 180, 353, 355
infrared excess 180
interior model 174, 179–180
internal temperatures 180
internal temperatures, pressures, densities
175
magnetic field generation 180
magnetosphere 184
observational data on the interior
174–175, 179–180, Table 4.2
overall composition 179

Van Allen radiation belts, *see* Earth
Van Allen, James Alfred 182
Venus
atmosphere
circulation 319–320
evolution 330–332
origin and early evolution, *see* terrestrial
planets
properties, *see* Table 9.1, 9.2
reservoirs, gains and losses 319,
322–324, 330–332
vertical structure, heating and cooling
317–319
central temperature, pressure, density 158
core composition 160–161
coronae 262–263
crust 256, 262–265
crust and surface composition 161, 256,
261, 264
evolution of interior and surface 265
global resurfacing 258, 264–265
gradation 256, 258
greenhouse effect 318–319, 330–332
heat transfer to surface 150, 264–265
highlands 256–258, 264
hypsometric distribution 256–257
impact craters 258–259
interior model 156, 161–162
life 332
lithosphere 263–265

Venus (*continued*)
  lowlands or plains 256–257, 264
  magnetic field 161
  mantle composition 162
  mantle convection 263–265
  mesolands 256, 264
  observational data on the interior
    156–157, 161–162, Table 4.2
  surface texture 256, 258, 261
  surface images from landers 261
  tectonic features 261–263
  tectonic processes 263–264
  tesserae 261–262
  volatiles (*see also* terrestrial planets)
    inventories 319, 321–322, 324–325
    inventory changes 319, 322–332
  volcanic features 258, 260
  volcanic processes 263–264
vernal equinox, *see* Earth
Vesta
  and V class asteroids 112
  discovery 80
  mean density 90
  surface 89, 91
  surface composition 92, 112

volatility 60, 141
volcanism and magmatic processes 203–204
  calderas 204
  cryovolcanism 203
  dyke 204
  effusive volcanism 203
  explosive volcanism 203
  extrusive rocks 204
  igneous rocks 204
  intrusive rocks 204
  lava 203
  shield volcanoes 203
  volcanic cones 203
  volcanic pits 203
  volcanic plains 203
  volcanism 203

weathering 208
white dwarf 360
Widmanstätten, Alois von 107

year 34
  sidereal 18, 31
  tropical 34
Yucatan asteroid impact 85